United States Navy
Foundry Manual

Fredonia Books
Amsterdam, The Netherlands

Foundry Manual

by
United States Navy

ISBN: 1-4101-0900-3

Copyright © 2006 by Fredonia Books

Reprinted from the 1957 edition

Fredonia Books
Amsterdam, The Netherlands
http://www.fredoniabooks.com

All rights reserved, including the right to reproduce
this book, or portions thereof, in any form.

NAVY DEPARTMENT,
Bureau of Ships,
15 April 1957

The Foundry Manual of 1944 has been revised to reflect the advancement in foundry technology and to indicate current foundry practice. The revised manual contains information for persons who operate or are employed in a foundry.

J. B. Duval, Jr.
Captain, USN
Assistant to the Assistant Chief of
Bureau for Shipbuilding and Fleet
Maintenance

PREFACE

This Manual is intended primarily for use by foundry personnel aboard repair ships and tenders. The recommended practices are based on procedures proved workable under Navy conditions and are supplemented by information from industrial sources.

The Manual is divided into two general sections. The first section, chapters 1 through 13, contains information of a general nature, such as "How Metals Solidify," "Designing a Casting," "Sands for Molds and Cores," "Gates, Risers, and Chills," and "Description and Operation of Melting Furnaces." Subjects covered in these chapters are generally applicable to all of the metals that may be cast aboard ship.

The second section, chapters 14 through 21, contains information on specific types of alloys, such as "Copper-Base Alloys," "Aluminum-Base Alloys," "Cast Iron," and "Steel." Specific melting practices, suggestions for sand mixes, molding practices, gating, and risering are covered in these chapters.

This manual has been written with the "how-to-do-it" idea as the principal aim. Discussions as to the "why" of certain procedures have been kept to a minimum. This manual contains information that should result in the production of consistently better castings by repair ship personnel.

FOUNDRY MANUAL

TABLE OF CONTENTS

	Page
Chapter I. How Metals Solidify	1
The Start of Solidification	1
Contraction	1
Freezing Temperature of Metals	3
Crystallization	3
Heat Transfer	4
Gases in Metals	5
Summary	5
Chapter II. Designing a Casting	15
Strength Requirements	15
Stress Concentrations	15
Section Thickness	16
Directional Solidification	16
Wall Junctions	17
Good Casting Design	17
Summary	18
Chapter III. Patternmaking	25
Functions of the Pattern	25
Types of Patterns	25
Pattern Materials	25
Making the Pattern	28
Finishing and Color Coding	28
Maintenance, Care and Repair	28
Calculation of Casting Weight	29
Summary	29
Chapter IV. Sands for Molds and Cores	39
Molding Sands	39
Sand Properties	40
All-Purpose Sand	43
Properties of a 63 AFS Fineness Number Sand	43
Molding Sand Mixtures	44
Cores	45
Core Sand Mixes	48
Methods for Testing Sand	49
Summary	51
Chapter V. Making Molds	61
Molding Tools and Accessories	61
Types of Molds	64
Molding Loose-Piece Patterns	64
Molding Mounted Patterns	66
False-Cope Molding and the Use of Broken Parts as Patterns	66
Setting Cores, Chills, and Chaplets	67
Closing Molds	68
Summary	68

	Page
Chapter VI. Making Cores	83
Coremaking Tools and Accessories	83
Types of Cores	83
Internal Support	83
Facing, Ramming, Relief, and Venting of Cores	83
Turning Out and Spraying	84
Baking	84
Cleaning Assembly	85
Storage of Cores	85
Making a Pump-Housing Core	85
Summary	86
Chapter VII. Gates, Risers, and Chills	95
General Purpose	95
Gating System	95
Risers	98
Chills	104
Summary	105
Chapter VIII. Description and Operation of Melting Furnaces	121
Oil-Fired Crucible Furnace	121
Electric Indirect-Arc Furnace	123
Electric Resistor Furnace	127
Electric Induction Furnace	128
Sintering the Monolithic Lining and Making the First Steel Heat	131
Summary	131
Chapter IX. Pouring Castings	139
Types of Ladles	139
Ladle Linings	139
Pouring the Mold	140
Speed of Pouring	141
Pouring Temperature	141
Summary	142
Chapter X. Cleaning Castings	147
Removing Gates and Risers	147
Grinding and Finishing	147
Welding	148
Summary	148
Chapter XI. Causes and Cures for Common Casting Defects	149
Names of Defects	149
Design	149
Pattern Equipment	150
Flask Equipment and Rigging	151
Gating and Risering	151
Sand	152
Cores	154
Molding Practice	154
Pouring Practice	155
Miscellaneous	155
Summary	156

	Page
Chapter XII. Heat Treatment of Castings	179
Iron and Steel Castings	179
Nonferrous Castings	179
Brass and Bronze Treatment	180
Stress-Relief Anneal	180
Reasons for Heat Treatment	179
Aluminum	181
Iron and Steel	181
Monel	183
Summary	183
Chapter XIII. Composition of Castings	185
Specifications	185
Selection of Metal Mixtures	185
Raw Materials and Calculation of Charges	189
Summary	194
Chapter XIV. Copper-Base Alloys	213
Selection of Alloy	213
How Copper-Base Alloys Solidify	213
Patterns	214
Molding and Coremaking	214
Melting	217
Pouring	220
Cleaning	221
Causes and Cures for Common Casting Defects in Copper-Base Casting	221
Welding and Brazing	222
Summary	222
Chapter XV. Aluminum-Base Alloys	227
Aluminum Silicon	227
Aluminum Copper	227
Aluminum Zinc	227
Aluminum Magnesium	227
How Aluminum Solidifies	227
Patterns	227
Molding and Coremaking	228
Melting	230
Pouring	231
Cleaning	231
Causes and Cures for Common Defects In Aluminum Castings	231
Welding and Brazing	232
Summary	232
Chapter XVI. Nickel-Base Alloys	235
Monel	235
Modified S-Monel	235
How Nickel-Base Alloys Solidify	235
Patterns	235
Molding and Coremaking	235
Melting	236
Pouring	236
Cleaning	237
Causes and Cures for Common Casting Defects in Nickel-Base Alloy Castings	237
Welding and Brazing	237
Summary	237

	Page
Chapter XVII. Cast Iron	239
Selection of Alloys	239
How Gray Cast Irons Solidify	241
Patterns	241
Molding and Coremaking	242
Melting	244
Pouring	247
Cleaning	247
Causes and Cures for Common Defects in Iron Castings	247
Welding and Brazing	248
Summary	248
Chapter XVIII. Steel	251
Selection of Alloys	251
How Steels Solidify	252
Patterns	252
Molding and Coremaking	252
Melting	254
Pouring	256
Cleaning	256
Welding and Brazing	257
Summary	257
Chapter XIX. Copper	259
Selection of Metal	259
How Copper Solidifies	259
Patterns	259
Molding and Coremaking	259
Melting	260
Pouring	260
Cleaning	261
Causes and Cures for Common Casting Defects in Copper	261
Summary	261
Chapter XX. Babbitting With Tin-Base Bearing Metal	263
Selection of Alloys	263
Preparation of Bearings	263
Melting	264
Pouring	264
Finishing of Bearing	265
Bearing Failures	265
Summary	266
Chapter XXI. Process Control	269
Sand	269
Molding	269
Melting	269
Inspection and Test	270
Summary	270

LIST OF ILLUSTRATIONS
CHAPTER I

Figure	Title	Page
1	Schematic Illustration of the Solidification of Metal in a Mold	6
2	Volume Change During the Cooling of a 0.35 Percent Carbon Steel	7
3	Types of Shrinkage	7
	(a) piping	7
	(b) gross shrinkage	7
	(c) centerline	7
	(d) microshrinkage	7
4	Cooling Curves of a Pure Metal, a Solid Solution Alloy, and an Eutectic Alloy	8
5	Melting Points of Metals and Alloys	9
6	Effect of Section Size on Size of Crystals	10
7	Schematic Representation of Crystal Growth	11
8	Dendrite Growth	11
9	High Magnification of Shrink Area in an Aluminum Casting, Showing Dendrites	12
10	Crystal Growth in a Gun Metal Casting Dumped Before Solidification was Complete	12
11	Preferred Orientation in Chill Zone Crystals	12
12	Dendritic Solidification and Dendritic-Equiaxed Solidification	13
13	Solubility of Hydrogen in Iron and Nickel at One Atmosphere Pressure	13
14	Mechanism of Pinhole Formation in Steel	13

CHAPTER II

Figure	Title	Page
15	Effect of Section Size on Physical Properties	19
16	Use of Fillets	19
17	Blending of Thin and Heavy Sections	20
18	Wheel Design	20
19	Recommended Wheel Designs	21
20	Transitions in Section Size	21
21	Simple Directional Solidification	21
22	Taper as an Aid to Directional Solidification	21
23	Hot Spot Location by the Method of Inscribed Circles	22
24	Reduction of Cross Section in L and V Junctions	22
25	Reduction of Cross Section in an X Junction	22
26	Various Treatments for a T Junction	22
27	Coring to Reduce Section in a Rib Junction	22
28	Removal of Heavy Section by Redesign	22
29	Hub Cross Section - Heavy Section	23
30	Hub Cross Section - Improved Design	23
31	Bracket Casting	23
32	Aluminum Yoke Casting	23

CHAPTER III

Figure	Title	Page
33	One Piece Pattern	35
34	Split Pattern	35
35	Core Print Construction	36
36	Chaplet Location With Pads	36
37	Mold Broken Due to a Lack of Taper	36
38	Clean Pattern Draw With Correct Taper	36
39	Pattern Draft	36
40	Distortion Allowance in a Simple Yoke Pattern	37
41	Plaster Patterns and Core Boxes	37
42	Making a Simple Plaster Pattern	37
43	Calculating Casting Weight	37
44	Calculating Casting Weight	37

CHAPTER IV

Figure	Title	Page
45	Permeability as Affected by the Grain Size of Sand	53
46	Permeability as Affected by Sand Fineness and Moisture	53
47	The Effect of Sand Grain Shape on Permeability	53
48	Permeability as Affected by the Amount of Binder	53
49	The Effect of Bentonite and Fireclay on Permeability	53
50	Green Strength as Affected by the Fineness of Sand	53
51	Green Strengths of Sands With Varying Fineness Numbers	54
52	Green Strength as Affected by the Shape of Sand Grains	54
53	Green Strength as Affected by Moisture and Varying Bentonite Contents	54
54	The Effect of Bentonite and Fireclay on Green Strength of Foundry Sand	54
55	The Effect of Bentonite on Sands With Various Moisture Contents	54
56	The Effect of Western and Southern Bentonite on Green Strength and Dry Strength	54
57	Green Strength as Affected by Mulling Time	55
58	Relationship Between Moisture Content, Bentonite Content, Green Compressive Strength, and Permeability for an All-Purpose Sand of 63 AFS Fineness Number.	55
59	Relationship Between Moisture Content, Bentonite Content, Green Compressive Strength, and Dry Strength for an All-Purpose Sand of 63 AFS Fineness Number	56
60	General Green Compressive Strengths for Sands of Different Grain Class Numbers	56
61	Strength of Baked Cores as Affected by Baking Time and Baking Temperatures	56
62	Core Gas Generated by Two Different Core Binders	57
63	The Effect of Single Binders and Combined Binders on the Baked Strength of Cores	57
64	Rammer Used for Test Specimen Preparation	57
65	Permeability Test Equipment	58
66	Strength Testing Equipment	58
67	Equipment for Drying Sand Specimens for Moisture Determination	58
68	Jar and Stirrer for Washing Sand	58
69	Sand Washing Equipment Assembled	59
70	The Difference in Sand Grain Distribution for Two Foundry Sands Having the Same Grain-Fineness Number	59

CHAPTER V

Figure	Title	Page
71	Molder's Hand Tools	69
72	Additional Molder's Tools	70
73	Double-Headed Chaplets	71
74	Stem Chaplets	71
75	Perforated Chaplets	71
76	Recommended Chaplet Design for Emergency Use	71
77	Anchoring Cores With Chaplets	71
78	Pattern Set in Drag With Gating System Parts	71
79	Hand Packing Riddled Sand Around the Pattern	72
80	Ramming a Deep Pocket	72
81	Striking Off the Drag	73
82	Drag Ready for the Cope	73
83	Cope With Pattern and Gating Pieces Set	74
84	Ramming the Partially Filled Cope	74
85	Venting the Cope	75
86	Start of the Pattern Draw	75
87	Pattern Completely Drawn	76
88	Setting the Core	76
89	Cope and Drag Ready for Closing	77
90	Clamped Mold With Weights and Pouring Basin	78
91	Pouring the Mold	78
92	Finished Pump Housing Casting	79
93	Propeller Set in the Drag	79
94	Propeller in the Drag With Parting Line Cut	80
95	Drawn Cope	80
96	Mold Ready for Closing	81
97	As-Cast Propeller	82

CHAPTER VI

Figure	Title	Page
98	Arbor for a Medium-Size Core	87
99	View of Inside of Core Showing Hollowing to Make the Core More Collapsible When Metal is Poured Around It	87
100	Section of Mold Showing Use of Lifting Eye for Supporting Heavy Core	87
101	Typical Lifting Hooks for Lifting Cores	87
102	Core Boxes for Pump Housing Core	88
103	Ramming Up the Core	88
104	Striking Off the Core	89
105	Placing the Reinforcing Rods	89
106	Cutting Vents	90
107	Drag Core Turned Out	91
108	Cope Core Turned Out	92
109	Applying Core Paste	93
110	Assembling the Two Core Halves	93

CHAPTER VII

Figure	Title	Page
111	Parts of a Simple Gating System	106
112	Illustration of Gating Ratio	106
113	Gating Nomenclature	107
114	Unfavorable Temperature Gradients in Bottom Gated Casting	108
115	Defect Due to Bottom Gating	108
116	Bottom Gate	108
117	Reverse Horn Gate	108
118	Reverse Horn Gate	108
119	Bottom Gating Through Side Risers	109
120	Bottom Gating Through Riser with Horn Gate	109
121	Sprue With Well at Base	109
122	Simple Top Gating	109
123	Pencil Gate	109
124	Typical Parting Gate	109
125	Parting Gate Through the Riser	110
126	Whirl Gate	110
127	Simple Step Gate. (Not Recommended.)	111
128	Thirty-Degree Mold Manipulation	111
129	Complete Mold Reversal	111
130	Pouring Cups	111
131	Pouring Basin	111
132	Solidification Time vs A/V Ratio	111
133	Effectiveness of Square and Round Risers	112
134	Proper and Improper Riser Height	112
135	Poor Riser Size and Shape	112
136	Proper Riser Size and Shape	112
137	Riser Location at Heavy Sections	112
138	Cold Metal Riser. (Not Recommended)	112
139	Hot Metal Riser	113
140	Feeding Through a Thin Section	113
141	Flanged Casting with Open Riser	113
142	Flanged Casting with Blind Riser	113
143	Inscribed Circle Method for Riser Contact	113
144	Effect of Keeping Top Risers Open	113
145	Effect of Keeping Blind Risers Open	114
146	Casting Defects Attributable to Shrinkage Voids and Atmospheric Pressure	114
147	Blind Riser Principle	114
148	Individual Zone Feeding for Multiple Risers	114
149	Padding to Avoid the Use of Chills or Risers	115
150	Padding to Prevent Centerline Shrinkage	115

Figure	Title	Page
151	Typical Padding of Sections	115
152	Shrinkage on the Thermal Centerlines of Unpadded Sections	115
153	Use of a Core to Make a Padded Section	115
154	Effect of Insulated Risers	115
155	Reduction in Riser Size Due to Insulation	115
156	Comparison of Ordinary Riser, Insulated Riser, and Exothermic Riser	116
157	Typical Internal Chills	116
158	Typical External Chills with Wires Welded-On or Cast-In to Hold Chill in Place	116
159	Use of External Chills in a Mold for an Aluminum Casting	117
160	Use of External Chills on a Bronze Casting	117
161	As-Cast Aluminum Casting Showing Location of External Chills	118
162	Gear Blank Mold Showing Location of External Chills	118
163	Principle of Tapering Edges of External Chill	118
164	Effect of Chill Mass and Area of Contact	119
165	Typical Application of External Chills to Unfed L,T,V,X, and W Junctions	119
166	Preferred Method of Applying External Chills by Staggering	119

CHAPTER VIII

167	Pit-Type Crucible Furnace	133
168	Crucible for Tilting Crucible Furnace	133
169	Tilting Crucible Furnace	133
170	Cross Section of a Stationary Crucible Furnace	133
171	Undercutting a Refractory Patch	133
172	Proper Burner Location	134
173	Proper Fit for Crucible Tongs	134
174	Electric Indirect-Arc Furnace	134
175	General Assembly View of Electric Indirect-Arc Furnace	134
176	Accessory Equipment for Electric Indirect-Arc Furnace	135
177	Properly Charged Electric Indirect-Arc Furnace	135
178	Electric Resistor Furnace	135
179	Electrode-Bracket Assembly for Electric Resistor Furnace	136
180	Cross Section of Electric Induction Furnace	136
181	Flow Lines in an Induction Furnace Melt	136
182	Essential Parts of an Induction Furnace	137
183	Typical Electric Induction Furnace	137
184	Induction-Furnace Control Panel	137
185	Method of Lining Induction Furnace Using a Steel Form	137

CHAPTER IX

186	Lip-Pouring Ladle	143
187	Teapot Ladle	143
188	Lip-Pouring Crucibles	143
189	Teapot Crucibles	143
190	Lining a Teapot Pouring Ladle	143
191	Proper Pouring Technique	144
192	Poor Pouring Technique	144
193	Use of Pouring Basin and Plug	144
194	Skim Core in Down Gate	144
195	Skim Core in Pouring Basin	144
196	Pyrometer Field When at Correct Temperature, Too High a Setting, and Too Low a Setting	144
197	Effect of Pouring Temperature on Grain Size	145

CHAPTER XI

198	Sticker	173
199	Gross Shrink	173
200	Surface Shrink	173

Figure	Title	Page
201	Surface Shrink	173
202	Internal Shrink	174
203	Gating and Risering that Corrected Internal Shrink in Figure 202	174
204	Gross Shrink	174
205	Dross Inclusions	174
206	Blow	174
207	Expansion Scab	174
208	Erosion Scab and Inclusions	175
209	Metal Penetration and Veining	175
210	Hot Tear	175
211	Pin Holes	175
212	Rattails	176
213	Rattails	176
214	Buckle	176
215	Cracked Casting	176
216	Misrun	176
217	Blow and Expansion Scab	177
218	Sticker	177
219	Blows	177
220	Blow	178
221	Drop	178

CHAPTER XIII

Figure	Title	Page
222	Example of Charge Calculation for Ounce Metal	210
223	Example of Charge Calculation for Gray Iron	211

CHAPTER XIV

Figure	Title	Page
224	Horizontal Molding of a Bushing	223
225	Vertical Molding of a Bushing	223
226	Gating a Manganese Bronze Casting	223
227	Gating a Number of Small Castings in Manganese Bronze or Red Brass	223
228	Gating for a Thin Nickel-Silver Casting	223
229	Poor Gating System for a Cupro-Nickel Check Valve	223
230	Improved Gating That Produced a Pressure-Tight Casting	224
231	Globe Valve - Poor Risering Practice	224
232	Globe Valve - Improved Risering Practice	224
233	High Pressure Elbow - Poor Risering Practice	224
234	High Pressure Elbow - Improved Risering Practice	225
235	Risers for a Cupro-Nickel Valve Body	225
236	Tapered Chills on a Flat G Metal Casting	225
237	Tapered Chills on a G Metal Bushing	225
238	Examples of Gassy and Gas-Free Metal	225

CHAPTER XV

Figure	Title	Page
239	Enlargement-Type Sprue Base	233
240	Well-Type Sprue Base	233
241	Coarse-Grained Structure (Caused by Iron Contamination)	233
242	Porosity. (Caused by Excessive Moisture in the Sand)	233

CHAPTER XVI

Figure	Title	Page
243	Poor Gating and Risering Practice for a Nickel-Base Alloy Casting	238
244	Improved Gating and Risering for Nickel-Base Alloy Casting	238

CHAPTER XVII

Figure	Title	Page
245	Knife Gate	249
246	Lap Gate	249
247	Riser for a Gray Iron Casting Molded in the Cope and Drag	249
248	Riser for a Gray Iron Casting Molded in the Drag	249
249	Riser for a Gray Iron Casting Molded in the Cope	250
250	Plan View of Runner, Riser, and Ingate	250
251	Operating Log for Cast Iron Heats	250

CHAPTER XVIII

Figure	Title	Page
252	Iron-Carbon Diagram	258
253	Tapered Chill	258
254	Steel Rods Used for Determining the Pouring Temperature of Steel	258

CHAPTER XIX

Figure	Title	Page
255	The Effect of Various Elements on the Electrical Conductivity of Copper	262
256	Properly Deoxidized Copper Sample	262
257	Partially Deoxidized Copper Sample	262
258	Gassy Copper Sample	262

CHAPTER XX

Figure	Title	Page
259	Jig for Babbitting Bearings	267

Chapter I
HOW METALS SOLIDIFY

Making a casting involves three basic steps: (1) heating metal until it melts, (2) pouring the liquid metal into a mold cavity, and (3) allowing the metal to cool and solidify in the shape of the mold cavity. Much of the art and science of making castings is concerned with control of the things that happen to metal as it solidifies. An understanding of how metals solidify, therefore, is necessary to the work of the foundryman. The control of the solidification of metal to produce better castings is described in later chapters on casting design, gating, risering, and pouring.

The change from hot molten metal to cool solid casting takes place in three main steps. The first step is the cooling of the metal from the pouring temperature to the solidification temperature. The difference between the pouring temperature and the solidification temperature is called the amount of superheat. The amount of superheat determines the amount of time the foundryman has available to work with the molten metal before it starts to solidify.

The second step is the cooling of the metal through the range of temperature at which it solidifies. During this step, the quality of the final casting is established. Shrink holes, blow holes, hot cracks, and many other defects form in a casting while it solidifies.

The third step is the cooling of the solid metal to room temperature. It is during this stage of cooling that warpage and casting stresses occur.

THE START OF SOLIDIFICATION

Solidification of a casting is brought about by the cooling effect of the mold. Within a few seconds after pouring, a thin layer of metal next to the mold wall is cool enough for solidification to begin. At this time, a thin skin or shell of solid metal forms. The shell gradually thickens as more and more metal is cooled, until all the metal has solidified. Solidification always starts at the surface and finishes in the center of a section. In other words, solidification follows the direction that the metal is cooled.

The way in which metal solidifies from mold walls is illustrated by the series of steel castings shown in figure 1. The metal that was still molten after various intervals of time was dumped out to show the progress of solidification. All metals behave in a similar manner. However, the time required to reach a given thickness of skin varies among the different metals.

The speed of solidification depends on how fast the necessary heat can be removed by the mold. The rate of heat removal depends on the relation between the volume and the surface area of the metal. Other things being equal, the thin sections will solidify before the thick ones. Outside corners of a casting solidify faster than other sections because more mold surface is available to conduct heat away from the casting. Inside corners are the slowest sections of the casting to solidify. The sand, in this case, is exposed to metal on two sides and becomes heated to high temperatures. Therefore, it cannot carry heat away so fast.

Changes in design to control solidification rate sometimes can be made by the designer. If, however, a change in solidification rate is required for the production of a good casting, the foundryman is usually limited to methods that result in little or no change in the shape of the casting. The rate of solidification can be influenced in three other ways: (1) by changing the rate of heat removal from some parts of the mold with chills; (2) by proper gating and risering, mold manipulation, and control of pouring speed, and (3) by padding the section with extra metal that can be machined off later.

CONTRACTION

Metals, like most other materials, expand when they are heated. When cooled, they must contract or shrink. During the cooling of molten metal from its pouring temperature to room temperature, contraction occurs in three definite steps corresponding to the three steps of cooling. The first step, known as liquid contraction, takes place while the molten metal is cooling from its pouring temperature to its freezing temperature. The second, called solidification contraction, takes place when the metal solidifies. The third contraction takes place when the solidified casting cools from its freezing temperature to room temperature. This is called solid contraction. Of the three steps in contraction, the first liquid contraction causes least trouble to the foundryman because it is so small in amount.

Figure 2, which shows the change in volume of a steel alloy as it cools from the pouring temperature to room temperature, illustrates these contractions. In a similar way, most of the metals considered in this manual contract in volume when cooling and when solidifying. The amount of shrinkage in several metals and alloys is given in table 1. Notice that some compositions of gray cast iron expand slightly

TABLE 1. THE AMOUNT OF SHRINKAGE FROM POURING TEMPERATURE
TO ROOM TEMPERATURE FOR SEVERAL METALS AND ALLOYS

Name	Composition	Decrease in Volume During Solidification, percent	Total Decrease in Volume, percent
Copper	Deoxidized	3.8	10.7
Red brass	85 Cu, 5 Zn, 5 Pb, 5 Sn	6.3	10.6
Yellow brass	70 Cu, 27 Zn, 2 Pb, 1 Sn	6.4	12.4
Bearing bronze	80 Cu, 10 Sn, 10 Pb	7.3	11.2
Manganese bronze	56-3/4 Cu, 40 Zn, 1-1/4 Fe 1/2 Sn, 1 Al, 1/2 Mn	4.6	11.5
Aluminum bronze	90 Cu, 10 Al	4.1	11.2
Aluminum	Commercial	6.5	12.2
Nickel	98 Ni, 1-1/2 Si, 0.1 C	6.1	14.2
Monel	67 Ni, 32 Cu	6.3	13.9
Nickel silver	20 Ni, 15 Zn, 65 Cu	5.5	12.1
Carbon steel	0.25 C, 0.2 Si, 0.6 Mn	3.8	11.4
Nickel cast iron	13 Ni, 7 Cu, 2 Cr, 3 C	1.6	7.8
Gray cast iron	2.18 C, 1.24 Si, 0.35 Mn 3.08 C, 1.68 Si, 0.44 Mn 3.69 C, 2.87 Si, 0.59 Mn	4.85 1.94 -1.65 (expands)	

during solidification. This results from the formation of graphite, which is less dense than iron. The formation of graphite compensates for a part of the shrinkage of the iron.

Reservoirs of molten metal, known as risers, are required to make up for the contraction that occurs during solidification. If risers are not provided at selected spots on the casting, shrinkage voids will occur in the casting. These voids can occur in different ways, depending on the shape of the casting and on the type of the metal. Piping, the type of shrinkage illustrated in figure 3a, occurs in pure metals and in alloys having narrow ranges of solidification temperature. Piping in a riser is usually a good indication that it is functioning properly. Gross shrinkage, illustrated in figure 3b, occurs at a heavy section of a casting which has been improperly fed. Centerline shrinkage, illustrated in figure 3c, occurs in the center of a section where the gradually thickening walls of solidified metal from two surfaces meet. Centerline shrinkage occurs most frequently in alloys having a short solidification range and low thermal conductivity. Microshrinkage, which is also known as microporosity, occurs as tiny voids scattered through an area of metal. It is caused by inability to feed metal into the spaces between the arms of the individual crystals or grains of metal. This type of shrinkage, which is illustrated in figure 3d, is most often found in metals having a long solidification temperature range. Microporosity may also be caused by gas being trapped between the arms of the crystals.

After solidification, cast metal becomes more rigid as it cools to normal room temperature. This cooling is accompanied by contraction, which is allowed for by the patternmaker in making the pattern for the casting. Contraction in cast metals after solidification is resisted by the mold. Often, different cooling rates of thin and heavy sections result in uneven contraction. This uneven contraction can severely stress the partially solidified, and still weak, heavier sections. Resistance to contraction of the casting results in severe "contraction stresses" which may tear the casting or which may remain in the casting until removed by suitable heat treatment. Sharp internal corners are natural points for these stresses.

Some metals, such as steel, undergo other dimensional changes as they pass through certain temperature ranges in the solid state. In the case of castings with extreme variations in section thickness, it is possible for contraction to take place in some parts at the same time that expansion occurs in others. If the design of the junctions of these parts is not carefully considered, serious difficulties will occur in the foundry and in service.

FREEZING TEMPERATURE OF METALS

Molten metal has the ability to dissolve many substances, just as water dissolves salt. The most important elements that are soluble in molten iron are other metals and five nonmetals--sulfur, phosphorus, carbon, nitrogen, and hydrogen. When substances are dissolved in a metal, they change many of its properties. For example, pure iron is relatively soft. A small amount of carbon dissolved in the iron makes it tough and hard. Iron containing a small amount of carbon is called steel. More carbon dissolved in the iron makes further changes in its properties. When enough carbon is dissolved in the molten iron, the excess carbon will form flakes of graphite during solidification. This metal is known as cast iron. The graphite flakes lower the effective cross section of the metal, lower the apparent hardness, and have a notch effect. These factors cause cast irons to have lower strengths and lower toughness than steels.

One of the most important changes in a metal as it dissolves other substances is a change in the freezing characteristics.

Pure metals and certain specific mixtures of metals, called eutectic mixtures, solidify without a change in temperature. It is necessary, however, to extract heat for solidification to occur. The solidification of pure metals and eutectic mixtures is very similar to the freezing of water. Water does not begin to freeze until the temperature is lowered to 32°F. The temperature of the ice and water does not change from 32°F. until all of the water is converted to ice. After this, the ice can be cooled to the temperature of its surroundings, whether they are zero or many degrees below zero. This type of temperature change during cooling, shown in figure 4a, is typical of pure metals, eutectic mixtures, and water. Actual solidification temperatures are different for each material.

Most of the metals used by foundrymen are impure and are not eutectic mixtures. These metals solidify over a range of temperature known as the solidification range. Mixtures of metals have many of the solidification characteristics of mixtures of salt and water. Just as the addition of salt to water changes the temperature at which water starts to freeze, so does the addition of one metal to another change the freezing point of the second metal. An example of such a mixture of metals is the copper-nickel system shown in figure 4b (right). A given mixture of copper and nickel will be liquid until it reaches the temperature that crosses the line marking the upper boundary of Area A + L. In the Area A + L, the mixture will be partly liquid, and in the Area A, it will be entirely solid. It will be noted that the addition of copper to nickel lowers the freezing temperature. On the other hand, the addition of nickel to copper raises the freezing temperature. A metal system which has the same general shape as the copper-nickel system is said to have complete solid solubility. Like the mixture of water and salt, metal mixtures of this type must be cooled well below the temperature at which freezing begins before they are completely solidified. In its simplest form, the cooling curve looks like that in figure 4b (left). The range of temperature between the upper and lower line is the solidification range.

Most of the metal mixtures used in the foundry do not have cooling curves as simple as those shown in figures 4a and 4b. As an example, the addition of tin to lead lowers the freezing temperature of the mixture (see figure 4c, right). The addition of lead to tin also lowers the freezing temperature of the mixture. However, there is one specific mixture which has a lower freezing temperature than either lead, tin, or any other mixture of the two. The mixture that has the lowest freezing temperature is the eutectic mixture. A typical set of alloys that has an eutectic mixture is that of the lead-tin system shown in figure 4c (right). A cooling curve for one lead-tin alloy is also shown in figure 4c (left). In such mixtures, the mechanism of solidification is quite complicated.

The melting temperatures of important metals are shown in figure 5. The melting temperatures of many metals are so high that they create real problems in selecting materials for handling the molten metal and for making the mold.

CRYSTALLIZATION

A casting is made up of many closely packed and joined grains or crystals of metal. Within any particular crystal, the atoms are arranged in regular orderly layers, like building blocks. On the other hand, there is no orderly arrangement of atoms in molten metal. Solidification, therefore, is the formation and growth of crystals, layer by layer, from the melt. The size of the crystals is controlled by the time required for the metal to solidify and by its cooling rate in the mold. Obviously, the heavy sections take more time to freeze than the light sections. As a result, the crystalline structure of a heavy section is usually coarser than that of the lighter members. This may be seen in figure 6.

Although the physical properties of coarse-grained metals differ from those of fine-grained metals of the same chemical composition, this difference will not be considered in detail. As one example, coarse grains lower the strength of steel.

Metal crystals start to grow at the surface of the casting because this is where the molten metal first cools to its freezing temperature. Once a crystal starts to form, it grows progressively larger until its growth is stopped by other crystals around it or until there is no more molten metal to feed it. The growth of metal crystals is similar to the growth of frost crystals on a pane of glass.

A schematic drawing of the start and growth of metal crystals is shown in figure 7. The black square represents the original crystal center or nucleus which grows into a crystal or grain by the addition of layers of atoms from the melt. A three-dimensional sketch of crystal growth is shown in figure 8. Part (a) shows the crystal shortly after it has formed and has started to grow. In part (b), the crystal has become elongated and growth has started in two other directions. Still further growth is shown by part (c). The original body of the crystal has grown still longer and has become thicker in cross section. Two other sets of arms have started growing near the ends of the longest arms of the crystal. A still further stage of growth is shown in part (d). Crystals grow in this manner with continued branching and thickening of the arms. Because of its branching nature, the type of crystal shown in figure 8 is called a dendrite. When the metal is completely solidified, the arms will have grown and thickened until they have formed a continuous solid mass. A photograph of dendrites in a shrink area of an aluminum casting is shown in figure 9. The branching of the dendrite arms at right angles can be seen in this photograph. Close examination will also show where the growth of crystals was stopped by the growth of neighboring dendrites.

The first metal that solidifies at the mold surface will be composed of grains that are not arranged in any particular pattern and that grow about the same length in each direction. Such grains are called randomly oriented, equiaxed grains. The crystals of zinc on the surface of galvanized steel are a familiar example. Another example of crystal structure is shown in figure 10. The faces of the individual crystals can be seen easily and growth would have continued if it had not been dumped to reveal the crystals.

For a while after solidification begins at the surface of the casting, there will be a solid skin against the mold and the metal in the center will still be liquid. The growth of the metal crystals in the skin will take place by the building up of metal on some of the crystals of the surface layer which are favorably positioned for further growth. Figure 11 shows the small grains at the mold surface, with some of them positioned for further growth. The position for favorable growth is perpendicular to the mold wall and parallel to the direction of heat transfer from the casting. Properly oriented crystals will grow in toward the center because side growth will stop as soon as adjacent crystals meet. This type of crystal growth toward the center of the casting is known as columnar grain growth. Depending on the pouring temperature and the type of metal, growth of elongated grains may extend to the center of the casting. If the characteristics of the metal are such that it is impossible to feed properly the last parts of the dendrites, the casting defect known as centerline shrinkage is formed. This is shown in figure 12a. A point may be reached during solidification when the solidification temperature is reached by the entire remaining liquid metal. Nucleation and growth of crystals will then start throughout the melt and result in an equiaxed crystal structure in that part of the casting. Solidification which started as dendritic growth and finished as an equiaxed structure is shown in figure 12b.

HEAT TRANSFER

The solidification of molten metal in the mold is a result of the extraction of heat from the metal by the sand that surrounds it. This process of heat extraction is called heat transfer.

The transfer of heat from the molten metal to the sand and its transfer away from the casting is most rapid at the time the mold cavity is first filled. As the casting cools and solidifies, the transfer of heat is carried on at a reduced rate. The rapid heat transfer in the early period of solidification is due to the ability of the sand to store a large amount of heat. As the maximum capacity of the sand to store heat is reached, the sand becomes saturated with heat, and further transfer of heat from the casting to the mold is controlled by the ability of the sand to conduct the heat away. Because this is a much slower process than the absorption of heat by the sand, the transfer of heat away from the casting takes place at a lower rate. Many times, the rate of transfer is further slowed by an air gap which is formed when the solidified casting starts to contract and draw away from the mold. The presence of this air gap causes a further decrease in the rate of heat transfer. Chills produce an increased rate of solidification because of their increased heat-storage capacity, as compared to an equal volume of sand, and their ability to conduct heat at a rate much more rapid than that at which sand can conduct it.

GASES IN METALS

Many defects in castings are caused by gases which dissolve in the metal and then are given off during solidification. These defects may range in size and form from microscopic porosity to large blow holes. Because of the large volume that a small weight of gas occupies, very little gas by weight can cause the foundryman a lot of trouble. As an example, at room temperature and atmospheric pressure, 0.001 percent by weight of hydrogen in a metal occupies a volume equal to that of the metal, and at 2,000°F., the same amount of hydrogen would occupy a volume equal to four times that of metal.

Gases may be absorbed by the metal during smelting, refining, melting, and casting. Here, we are primarily concerned with the gas absorption during melting. The gases in any melting process often come from water vapor in the air, or from water which is introduced into the melt by careless foundry practice.

A gas frequently absorbed by metals is the hydrogen produced from water vapor. The solubility of hydrogen in nickel and steel at various temperatures is shown in figure 13. Notice that it is possible to dissolve more hydrogen in molten metal than in solid metal. Therefore, gas that is absorbed during melting may escape when the molten metal cools and solidifies. If the gas cannot escape from the metal freely, bubbles are trapped in the casting causing defects. The treatment of metals to reduce their gas content before they are poured into the mold is discussed in later chapters dealing with the specific metals.

Gas defects in castings are not always caused by gas that is dissolved in the molten metal. In some cases, these defects are caused by gases driven into the metal from the mold. The gases are trapped as the metal solidifies. In some cases, gas is generated by chemical reactions within the metal, such as may sometimes occur between carbon and oxygen in steel to form carbon monoxide.

A good example of the formation of a casting defect due to gas in pinhole formation in steel. This takes place as shown in figure 14. When the molten steel comes in contact with moist sand in the mold, a thin skin of steel is formed almost immediately. At the same time, the water in the sand is changed to steam with an increase in volume of approximately 5,000 times. The steam is highly oxidizing to the steel and reacts with it. As a result, iron oxide and hydrogen are formed. The iron oxide produces the scale which is seen on steel castings when they are shaken out of the mold.

The hydrogen which is formed in this reaction passes through the thin layer of solid steel and enters the still molten steel. The hydrogen in the molten steel can then react with iron oxide, which is also dissolved in the steel. This reaction produces water vapor. As the steel cools, it must reject some of this water vapor and hydrogen, just as an ice cube must reject gas as it freezes. A bubble is formed and gradually grows as more steel solidifies. The bubbles become trapped between the rapidly growing crystals of steel and cause the familiar pinhole defect.

SUMMARY

An understanding of the solidification or freezing of metals is important to the foundryman who wants to know how to make good castings.

Solidification of a casting starts by the formation of solid grains next to the surface of the mold. These grains grow inwardly from the surface until they meet other grains growing from other surfaces. When these growing surfaces meet, the casting is solid.

Improper foundry practice will cause many defects which can be explained and avoided if proper attention is given to the way in which the metal solidifies. Casting defects which can occur if the freezing characteristics of metals are not taken into account are as follows: (1) microshrinkage, (2) centerline shrinkage, (3) shrink holes, (4) certain types of gas holes, (5) piping, and (6) hot tears. These defects can be minimized if proper attention is given to the practices described in later chapters, particularly Chapter 2, "Design," and Chapter 7, "Gates, Risers, and Chills."

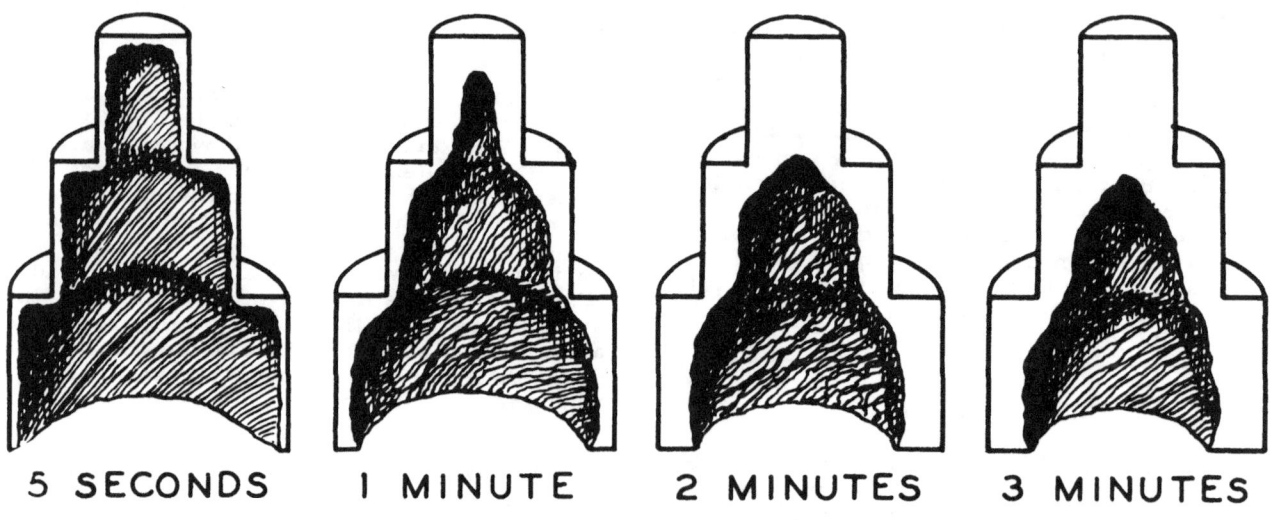
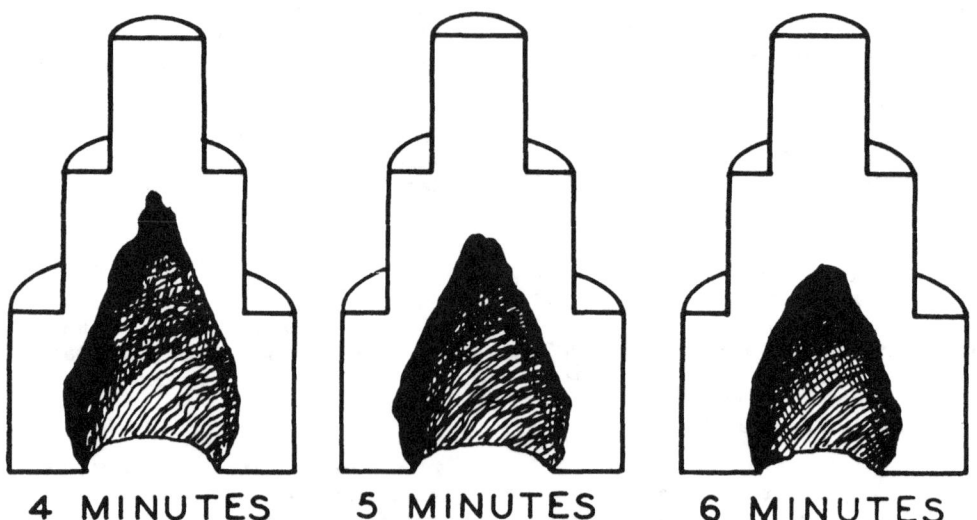

Figure 1. Schematic illustration of the solidification of metal in a mold.

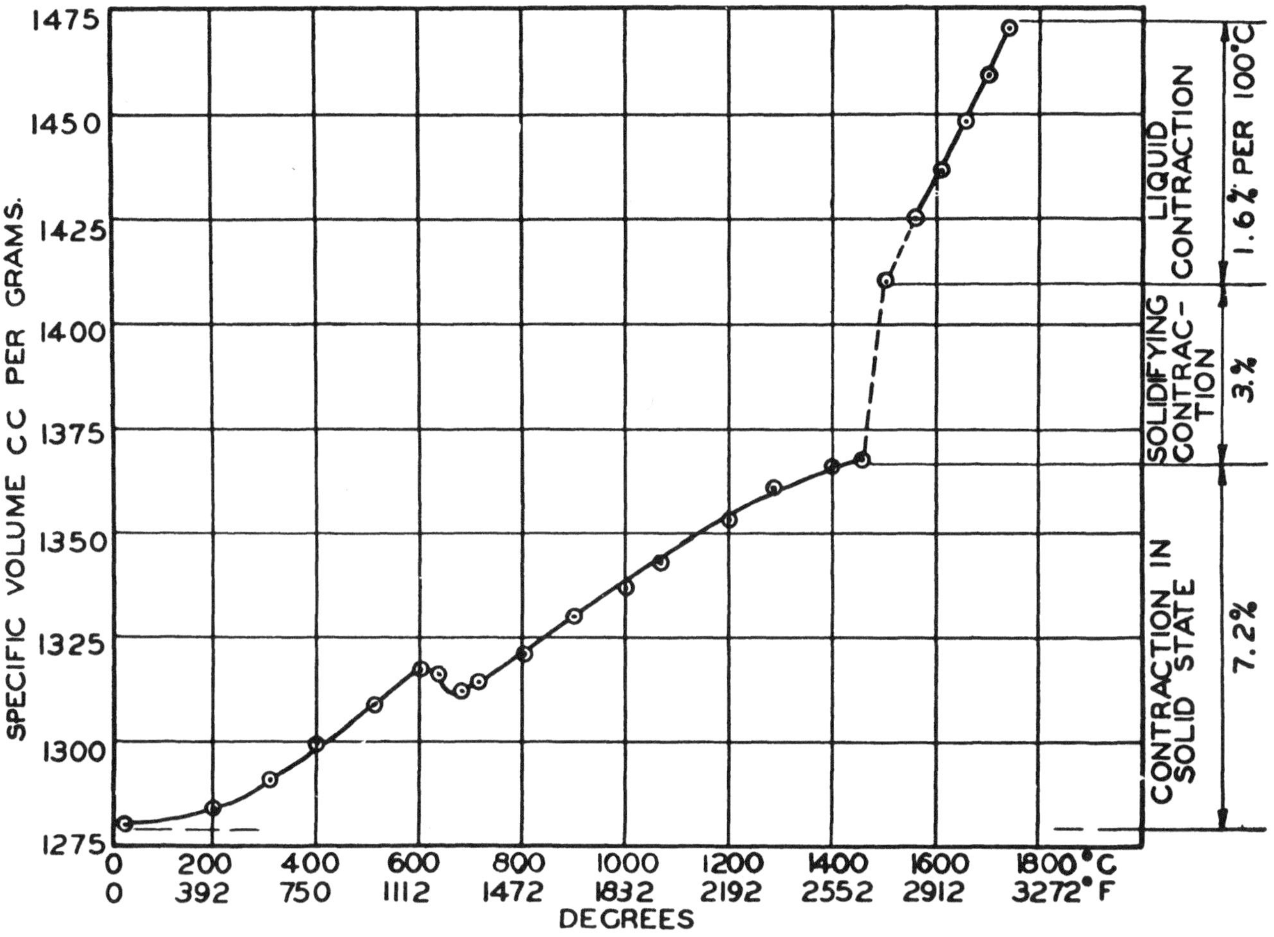

Figure 2. Volume change during the cooling of a 0.35 percent carbon steel.

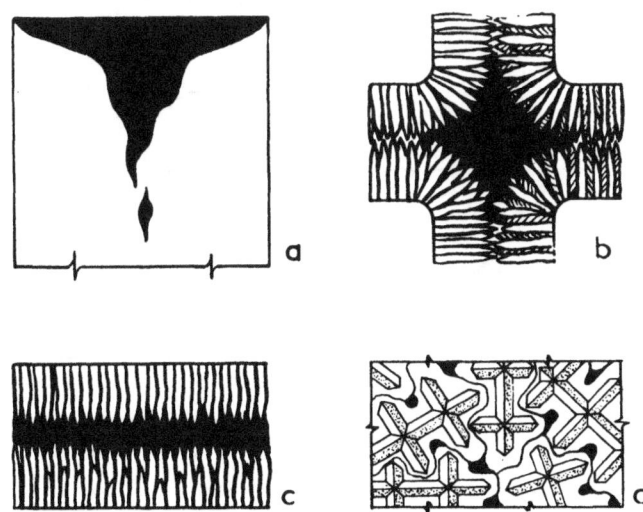

Figure 3. Types of shrinkage
    (a) piping   (b) gross shrinkage
    (c) centerline   (d) microshrinkage

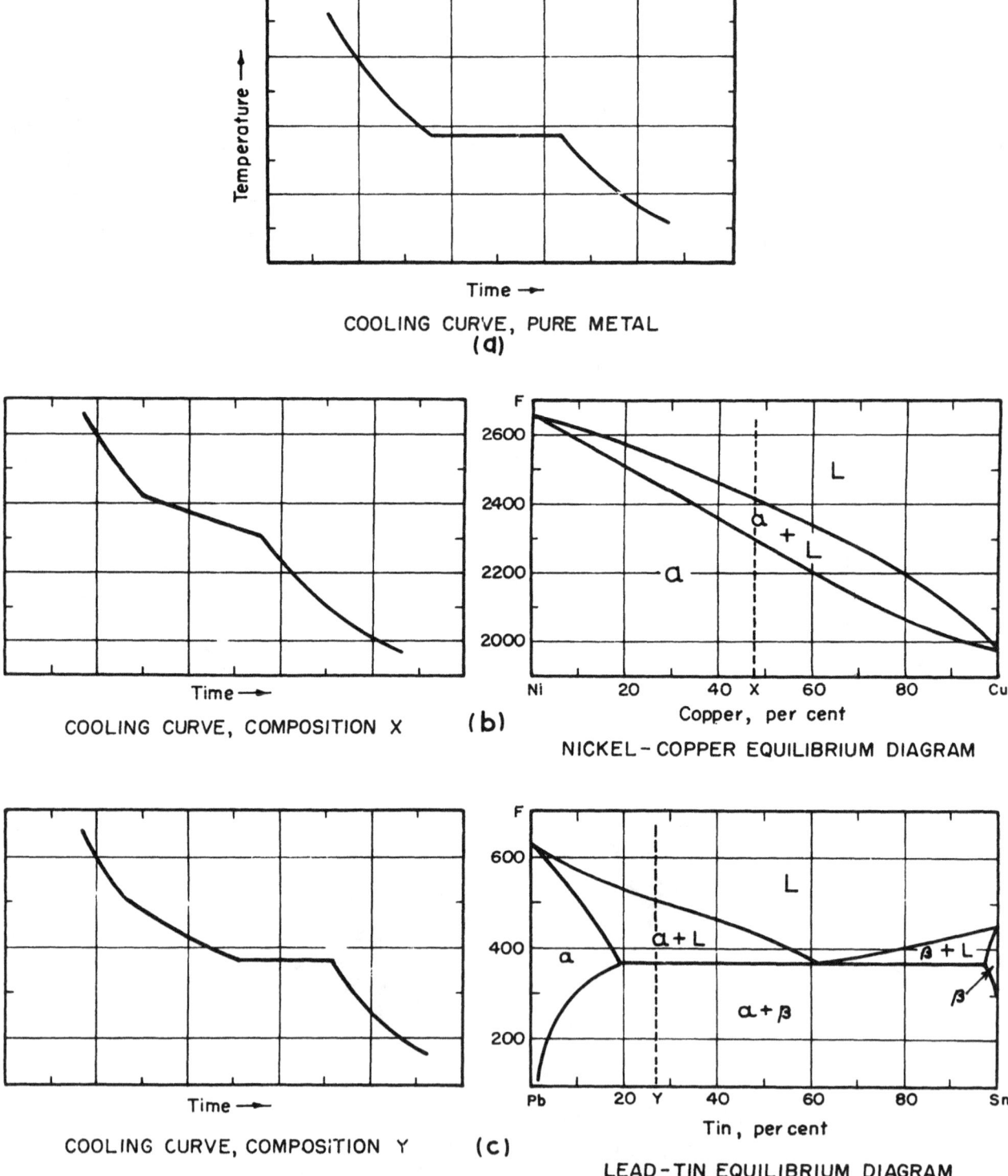

Figure 4. Cooling curves of a pure metal, a solid solution alloy, and an eutectic alloy.

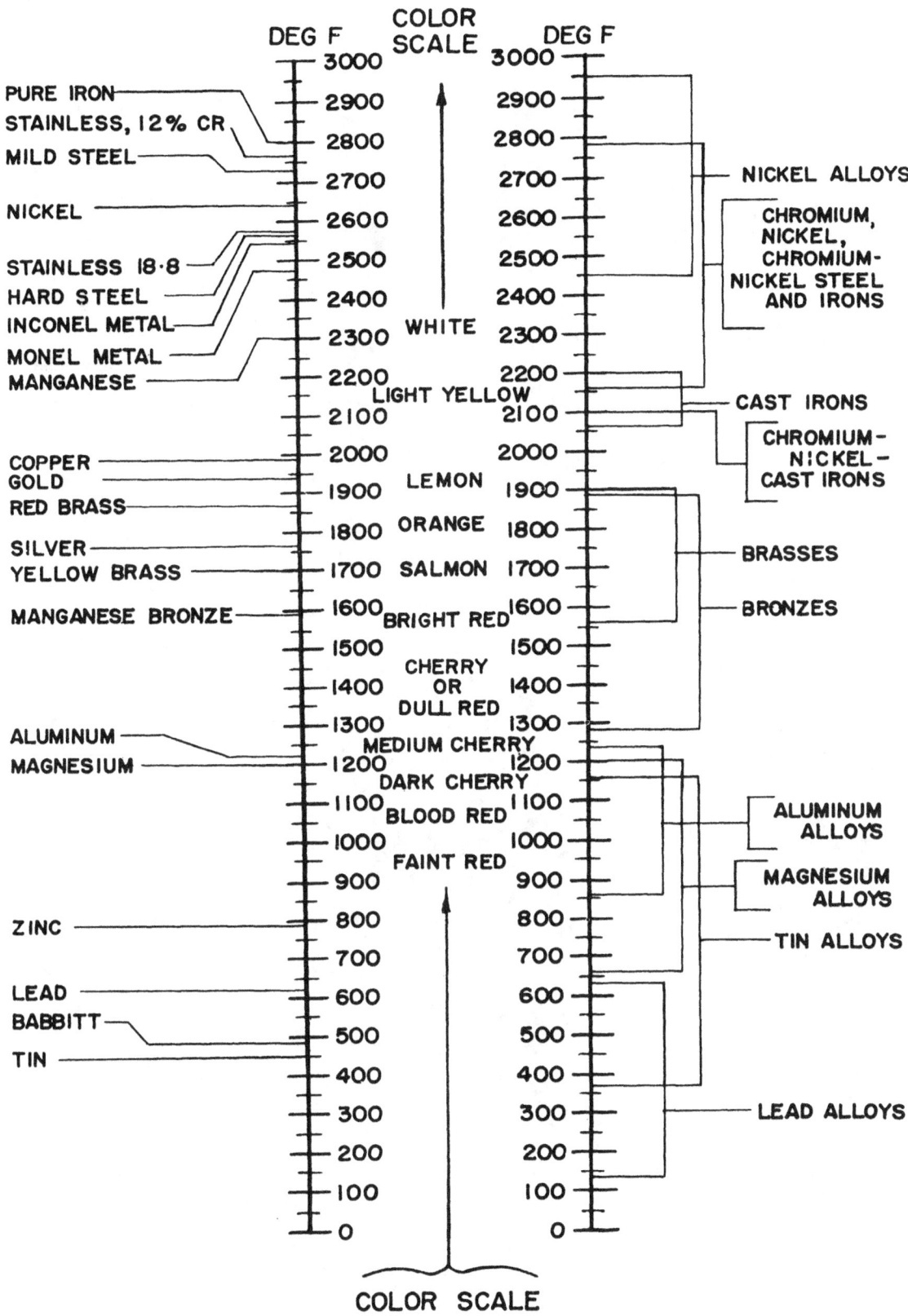

Figure 5. Melting points of metals and alloys.

Figure 6. Effect of section size on size of crystals

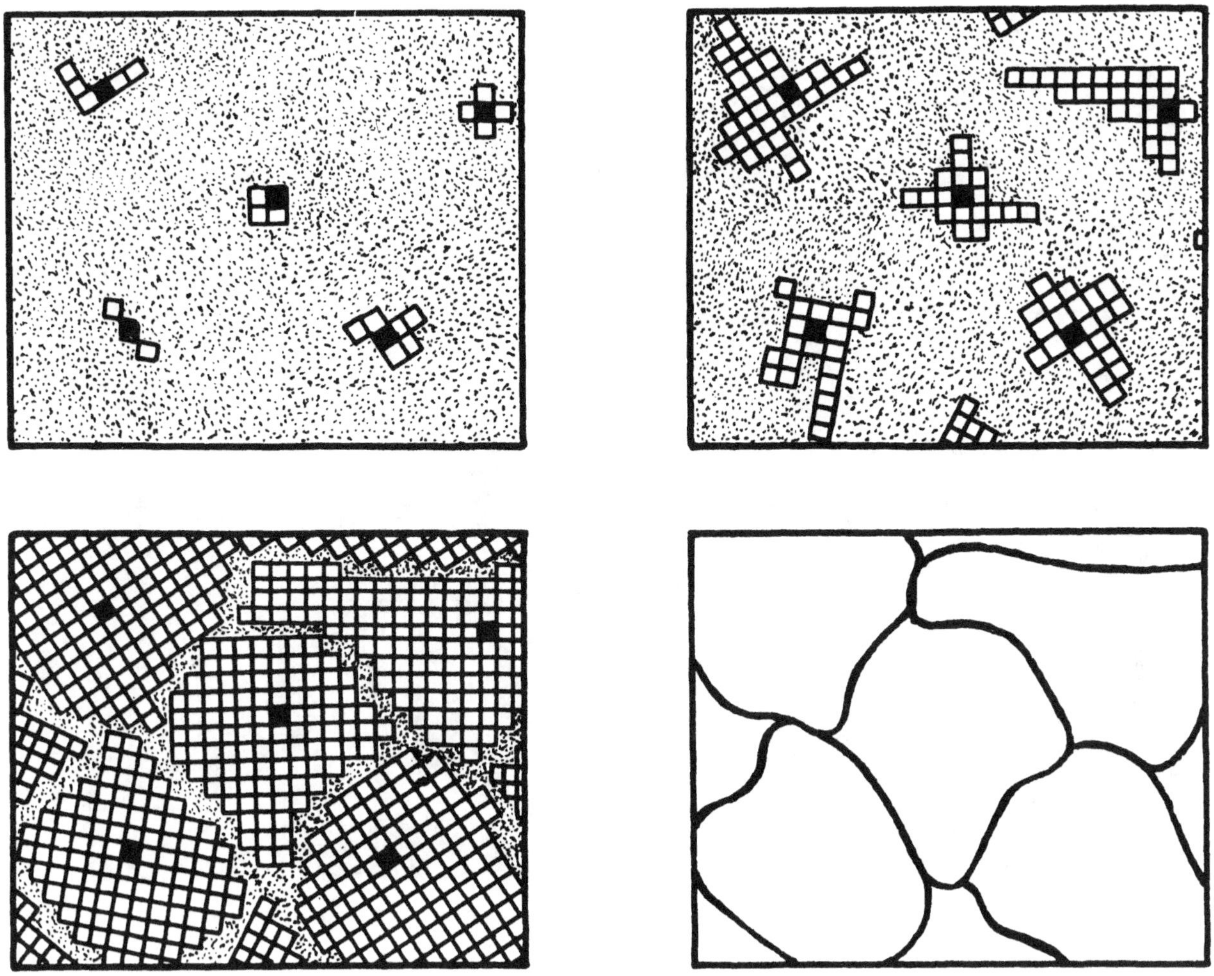

Figure 7. Schematic representation of crystal growth.

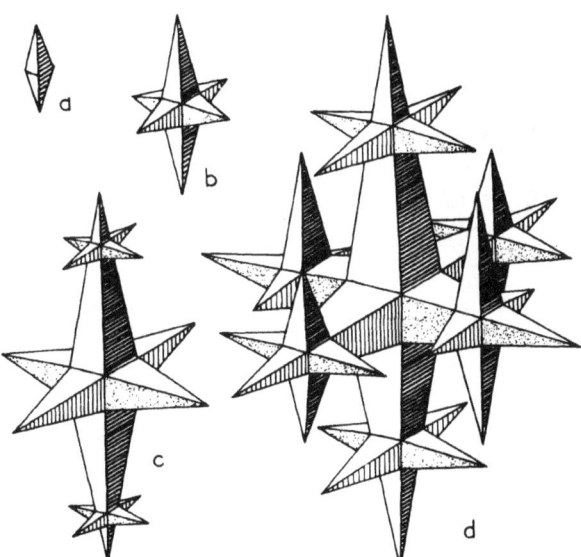

Figure 8. Dendrite growth.

- 11 -

Figure 9. High magnification of shrink area in an aluminum casting showing dendrites.

Figure 10. Crystal growth in gun metal casting dumped before solidification was complete.

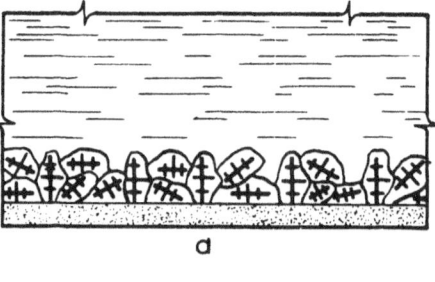
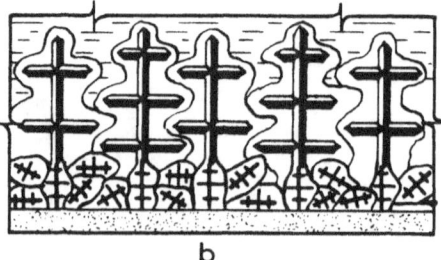

Figure 11. Preferred orientation in chill zome crystals.

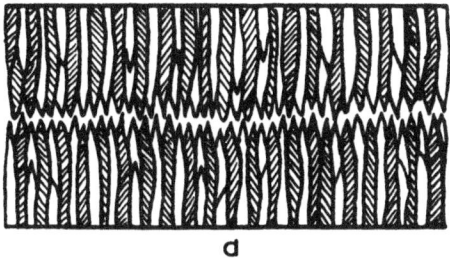

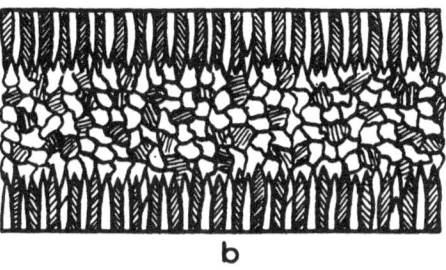

Figure 12. Dendritic solidification and dendritic-equiaxed solidification.

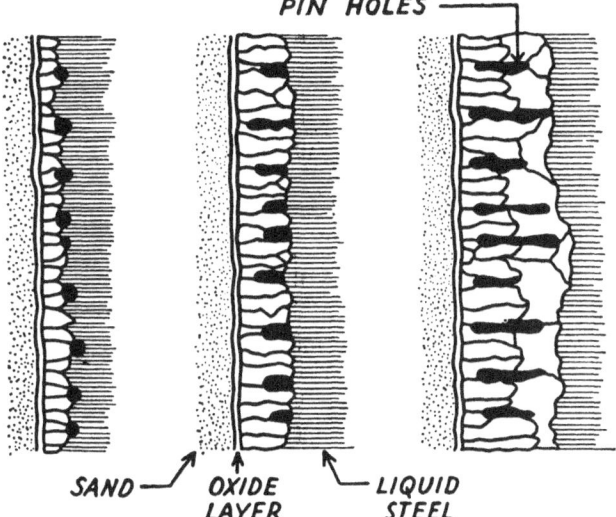

Figure 14. Mechanism of pinhole formation in steel

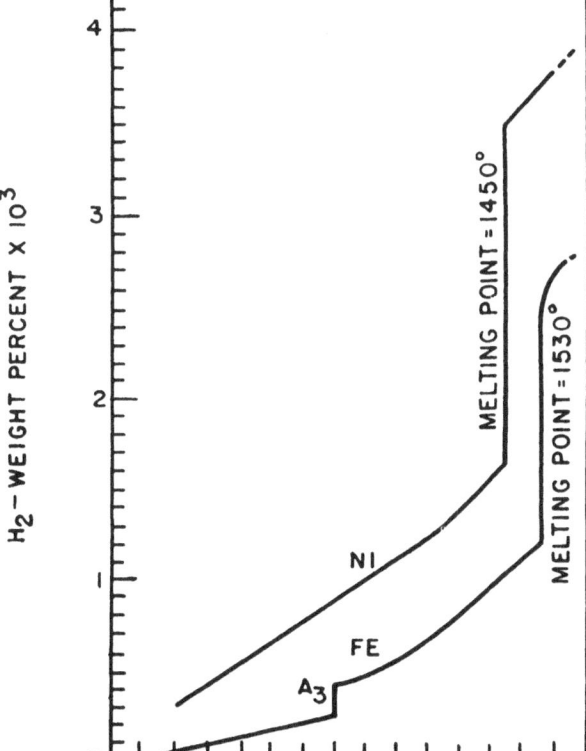

Figure 13. Solubility of hydrogen in iron and nickel at one atmosphere pressure.

Chapter II
DESIGNING A CASTING

The design of a casting might seem to be something far removed from the field of interest of a Navy molder. He is usually called upon to make a casting from a loose pattern or from the broken parts of an existing casting. Very rarely is he consulted as to what is good casting design from the foundryman's point of view. Nevertheless, an understanding of what constitutes good casting design will help the molder to make a consistently better product.

Design influences the soundness, freedom from dirt, shrinkage, porosity, hot tears, and cracks found in a casting, and thus affects its serviceability. A capable foundryman may produce satisfactory castings that violate some of the principles of good design, but he will never produce them with any degree of consistency. Superior craftsmanship of the foundryman should not be relied upon to overcome poor design.

Good casting design is based on two general considerations. The first thing to consider is the intended use of the casting, and the second is which alloy should be used. The intended use of the casting (that is, whether it is a supporting structure, moving part, pressure casting, or bearing) will be the major factor in determining the general shape of the casting. The amount of corrosion resistance, wear resistance, machinability, and strength that are needed will determine which alloy should be used. More often than not, a casting must meet a combination of requirements.

Many times, the same features of design which give trouble to the foundryman will also adversely affect the service life of the part. Therefore, the first step in the production of a casting should be a careful study of its design in the light of the information given in this chapter. This applies equally to a new design and to the replacement of a casting of an old design. In the replacement of a casting, the defective part should be thoroughly studied to determine if failure was in any way due to design faults; whether faulty design contributed to casting unsoundness, or whether it adversely affected the service strength of the solid part.

STRENGTH REQUIREMENTS

The amount of strength that is needed for a casting will be determined primarily by the part it plays in the structure or machine in which it is used. A casting should be designed so that the strength requirements are met with the proper safety factor. Care should be taken not to overdesign a casting. Many times when a casting fails, certain regions in the vicinity of the failure will be made larger with the idea that additional strength will be gained with an increase in thickness. In reality, this overdesign frequently produces casting defects which offset the desired increase in strength.

Sections that are heavier than necessary do not make use of all the strength that is available in the metal. As a general rule, a metal has lower strength per square inch of cross section when cast in thick sections than it does in thin sections. The effect of increasing section size on the strength and elongation of four different copper-base alloys is shown in figure 15. It is evident that the tin bronze and red brass are very sensitive to section thickness, while aluminum bronze and manganese bronze are less affected by section size. From this, it can be seen that the effect of section size on the properties of a casting must be considered if the casting is to make the best use of the metal poured into it.

STRESS CONCENTRATIONS

One of the major factors that cause the untimely failure of castings is the concentration of stresses that results from improper design. Stresses, of course, are the forces and loads that cause a casting to crack, tear, or break.

Sharp corners and notches should be avoided in castings because they are points of high stress. The liberal use of fillets and rounded corners of proper size is the easiest way to reduce the concentration of stresses in corners. A sharp corner will also produce a plane of weakness in a casting where crystal growth from two sides meet. This is shown in figure 16a. The combination of high stresses and the plane of weakness result in early failure of the casting. The partial removal of this plane of weakness by rounding the corners is shown in figure 16b, and its complete elimination, in figure 16c.

The junction of thin and heavy sections is another point of stress concentration. The stresses in this case result from the rapid solidification and contraction of the thin section. This contraction will set up very high stresses at the junction with the hotter, weaker, heavy section and may produce hot tearing. Where sections of different thicknesses are necessary, they should be blended together to reduce the stresses as much as possible. Recommended practices for the blending of junctions are shown in figure 17. Although shown for aluminum, the same practices should be followed for all metals.

There are some castings in which the design must allow for the absorption of casting stresses in order to produce a good casting. A spoked wheel is an example. Correct and incorrect designs for wheels are shown in figure 18. The original design (with straight spokes) caused hot cracks at the junction of the spokes with the rim and hub. The modified design (with a curved spoke) produced a casting without hot tears. The modified design permits the spokes to stretch and distort slightly without tearing under the stresses set up by contraction. Two other patterns made to prevent tearing in a wheel casting are shown in figure 19.

Contraction stresses often cause warping of the casting. When distortion cannot be solved directly by design, as with the wheel casting, it must be allowed for by the patternmaker after consultation with the molder. Correction of this type of distortion is covered in Chapter 3, "Patternmaking."

SECTION THICKNESS

The minimum thickness that can be cast is determined by the ability of the metal to flow and fill thin sections without the use of an excessive pouring temperature. The normal minimum sections that can be cast from several metals are listed in table 2.

TABLE 2. NORMAL MINIMUM SECTIONS FOR CAST METALS

Material	Normal Minimum Section Thickness, in.
Gray cast iron	1/8
White cast iron	1/8
Steel	3/32
Brass and bronze	3/32
Aluminum	1/8

These minimum dimensions for thin sections may vary slightly with composition of the alloy, pouring temperature, and size or design of the casting. The use of adequate but not excessive section thickness in a casting cannot be stressed too strongly, because it is a major factor in good design.

An effort should be made at all times to increase gradually the section size toward the reservoir of liquid metal in the riser. This will promote directional solidification described in the next paragraph. A sudden change in section thickness should be avoided wherever possible. Where a change in section thickness must be made, it should be gradual. A blending, or gradual change in section thickness reduces stresses at the junctions. Figure 20 shows various methods for changing from one section thickness to another.

DIRECTIONAL SOLIDIFICATION

Directional solidification means that solidification will start in one part of the mold and gradually move in a desired direction; it means that solidification will not start in some area where molten metal is needed to feed the casting. An effort is always made by the foundryman to get solidification to progress toward the riser from the point furthermost from the riser. Casting design is a determining factor in the control of the direction of solidification, and every effort should be made to apply the principles of good design to reach this objective.

A slab casting of uniform dimension, shown in figure 21, demonstrates directional solidification. The metal is poured through the riser, and as it flows over the mold surface, it gives up some of its heat to the mold. Such a condition will mean that when the mold is filled, the metal at the right end will not be as hot as the metal near the riser. The first metal to solidify will then be the metal at the right, as shown in figure 21a. The mold to the left of the casting will also have been heated by the molten metal flowing over it and its ability to conduct heat away from the casting will be reduced so that the cooling of the casting in that area will be retarded. Figure 21b shows the casting with solidification in a more advanced stage. Because of controlled solidification, this will probably be a sound casting. However, the reduction of area at the corner is undesirable from the structural design standpoint.

In actual practice, conditions are usually such that directional solidification cannot be obtained as simply as described above because of the properties of the metal or the design of the casting. In such cases, the desired directional solidification of the casting must be obtained by other methods. In designing a casting to control directional solidification, tapering sections can be used. The sections are tapered with the larger dimensions toward the direction of feeding. When a flat casting is poured, solidification will begin at about the same rate from both sides and centerline shrinkage will be found because of the lack of directional solidification. Solidification of this type is known as progressive solidification and is shown in figure 22a. If this casting did not have to have parallel sides, then a taper could be used to good advantage. Figure 22b shows the taper employed to obtain directional solidification. It will be noted that although solidification has taken place at the same rate from the opposing walls, the taper permits molten metal to feed the casting properly.

If it is impossible to design a casting to make full use of directional solidification, then other aids must be used. The most effective and most easily used is the chill. Chills are used to start or speed up solidification in a desired section of a casting. Their application and use are covered in Chapter 7, "Gates, Risers, and Chills." Another method of obtaining directional solidification in a casting is to taper the section intentionally and then remove the excess metal by machining. This method, called "padding," is also described in chapter 7.

WALL JUNCTIONS

Junctions such as "L" and "T" sections must be given special consideration when designing a casting. Because a junction is normally heavier than any of the sections which it joins, it usually cools more slowly than adjacent sections. The method of inscribed circles, illustrated in figure 23, can be used to predict the location of hot spots, which are locations of final solidification and possible shrinkage. In the L section, the largest circle which can be drawn in the junction is larger than the largest circles that can be drawn in the walls. The same is true of the T section, where the circle at the junction is even larger than the one for the L section. The larger circles in both of the junctions predict the location of a hot spot, which will be unsound unless special precautions are taken. Figure 23b shows similar junctions with the progress of solidification indicated by the shaded areas. These sketches were made from actual laboratory studies of the solidification of the junctions. The location of the small white area in each case indicates the location of the hot spot. These small spots are within the large circles inscribed at the junctions as shown in figure 23a.

The joining of two walls may result in an L, V, X, or T-shaped junction. If small fillets and rounded corners are used in the L or V-type junction, a heavy section will be formed. Radii should be used so that the thickness in the junction will be the same as that in the adjoining walls. This is shown in figure 24. The area within the dashed line shows the amount of metal which should be eliminated to avoid hot spots. The wall thickness at the junction can be reduced further by using radii which will produce a junction thinner than the adjoining sections. Such junctions would be used only if they were

the part of the casting from which solidification was to start. The first method is most commonly used. Chills may also be used to produce a sound junction. They are described in Chapter 7, "Gates, Risers, and Chills."

An X section has a still greater tendency toward hot spots and unsoundness than do the L or V sections. The only way to reduce the wall section in this type of junction is to use a core, as shown in figure 25b, to produce a hole in the junction. A method which is preferred, especially when the junction is a result of ribbed construction, is that of staggering the sections so as to produce T junctions which can be more easily controlled with chills. Figure 25c shows the staggered design. Various treatments for a T section are shown in figure 26. A cored hole can be used, as in figure 26a; the section thickness can be used, as in figure 26b; the external chills can be used, as in figure 26c; or internal chills can be used, as in figure 26d. Internal chills should not be used without authorization from the foundry supervisor.

Many times, a large casting will require ribs to provide added strength at certain locations. The use of ribs produces a hot spot at the junction because it is thicker. The heavy section may also be reduced by using a core to make a hole at the junction of the rib with the casting section, as shown in figure 27.

GOOD CASTING DESIGN

Casting designs often cannot be ideal because the casting must be designed to do a certain job. Everything should be done, however, to give the casting a section having a gradual taper, so that the best possible conditions for solidification can be obtained. A detailed discussion of a good casting design cannot be given here, but a few examples are given of design features which can be of help to the molder and patternmaker in making a better casting.

A casting having a tubular section joining a flat base is shown in figure 28. As originally designed, the tubular section had a heavier wall than the plate. Redesigning eliminated the heavy section in the casting. A hub casting is shown in figure 29. The inscribed circle shows the heavy section which would be difficult to feed and would probably cause a shrinkage defect. A cross section of the same casting is shown in figure 30 as it was redesigned to eliminate the heavy section and make the casting more adaptable to directional solidification.

Many times, a casting can be designed to permit easier molding as well as to improve the feeding. The bracket shown in figure 31 is such a casting. The original design did not have the shaded areas shown. This not only made the making of the mold difficult, but also resulted in heavy sections in the casting with the possibilities of shrinkage defects. By padding the area as shown by the shaded portions, the pattern was easier to draw and feeding of the lugs was simplified.

Another example of good casting design is shown in figure 32. Note that the thin sections are connected to the heavy sections which are located so that they may be easily fed.

SUMMARY

A few general rules can be made to assist the foundryman in producing a better casting. It must be remembered that in many cases, these rules cannot be followed to the letter. There also may be a conflict between rules. In such a case, a compromise must be made which will best suit the casting desired.

1. The casting thickness, weight, and size should be kept as small as possible, consistent with proper casting performance. (See "Strength Requirements," page 19, and figure 15.)

2. All sections should be tapered so that they are thickest near the risers. Sections should never be tapered so that thick sections are far from the risers. If proper tapering is impossible, the section should have uniform thickness. (See "Section Thickness," page 20, and figures 17 and 22.)

3. Abrupt changes in adjoining sections should never be allowed. (See figures 17 and 20.)

4. Heavy sections should not be located so that feeding must take place through thin sections.

5. Use ribs to avoid warpage or to add stiffness. Ribbed construction can often be used to replace a heavier section.

6. Where junctions produce thick sections of metal (hot spots), use cores or other methods to eliminate the heavy section (figures 23, 24, 25, 26, and 27, "Wall Junctions," page 22).

7. A casting should be made as simple as possible. The use of cores should be kept to a minimum. If a casting is complicated, consider the use of several simpler castings which can be welded together.

8. Avoid junctions of several walls or sections at one point.

9. Bosses, lugs, and pads should not be used unless absolutely necessary.

10. Allow for shrinkage and machine finish in dimensional tolerances. (Chapter 3, "Patternmaking," Table 4, page 27, Table 5, page 28.)

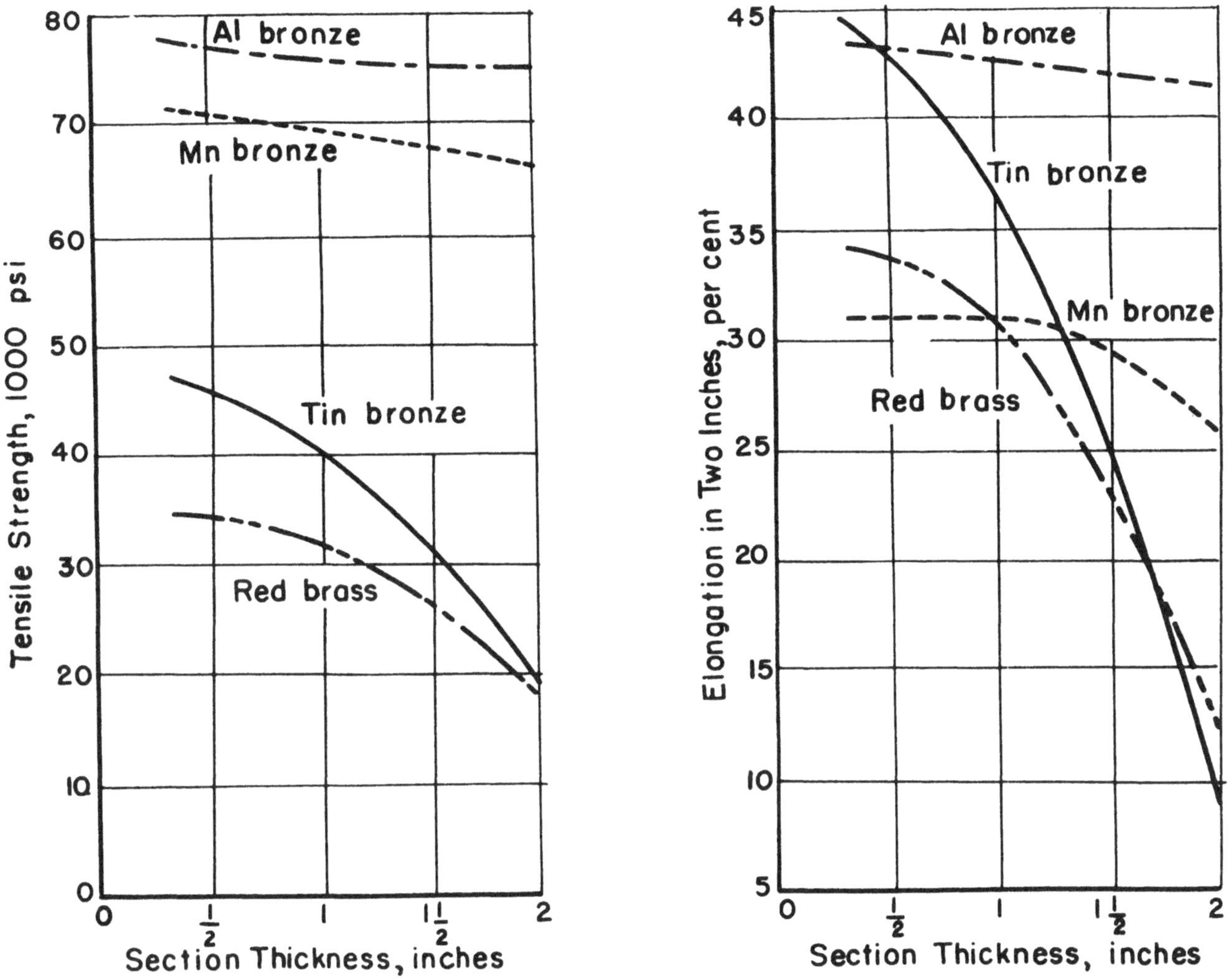

Figure 15. Effect of section size on physical properties.

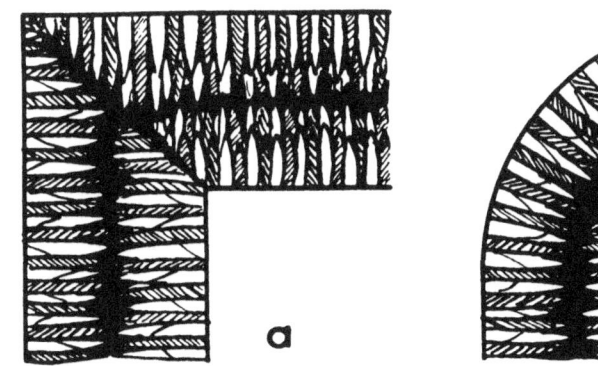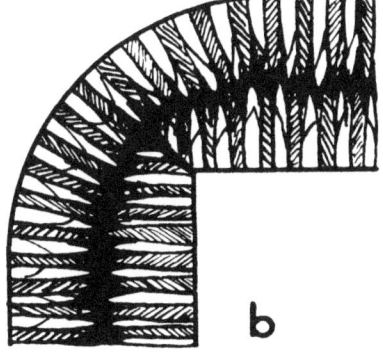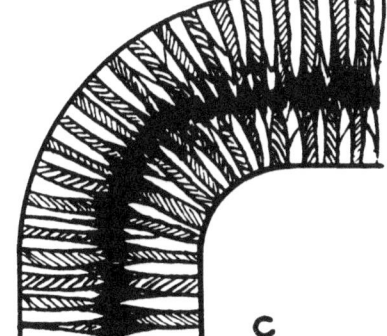

Figure 16. Use of fillets.

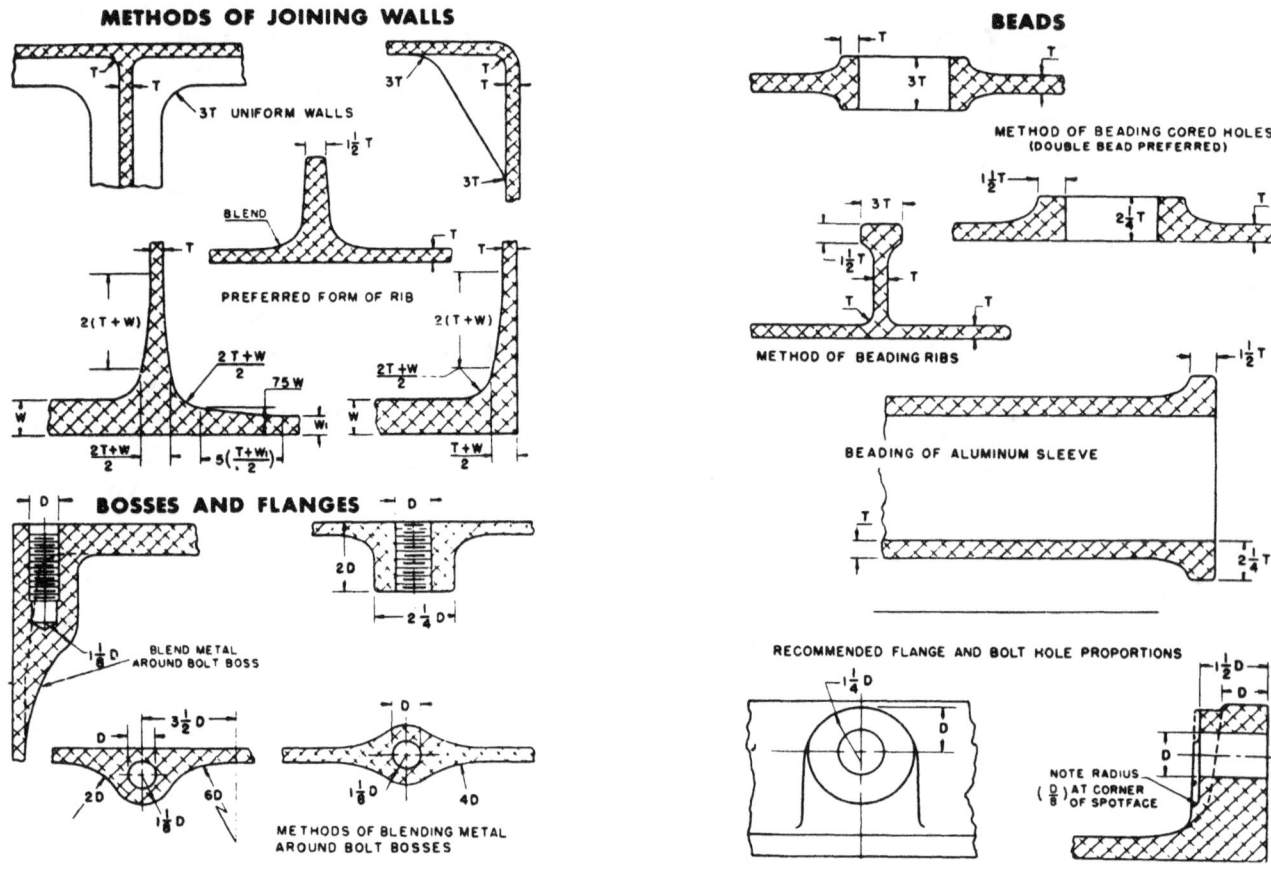

Suggested design details for aluminum alloy castings.

Figure 17. Blending of thin and heavy sections.

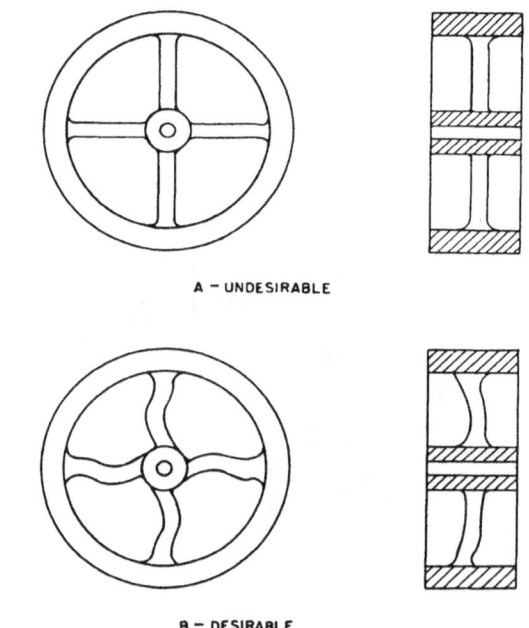

Figure 18. Wheel design.

- 20 -

Figure 19. Recommended wheel designs.

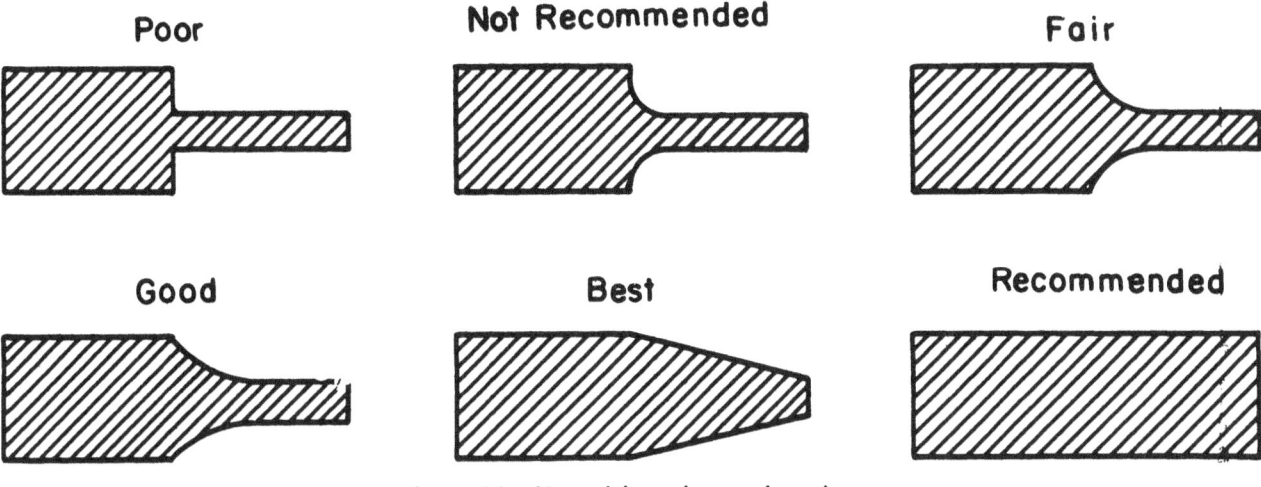

Figure 20. Transitions in section size.

Figure 21. Simple Directional solidification.

Figure 22. Taper as an aid to directional solidification.

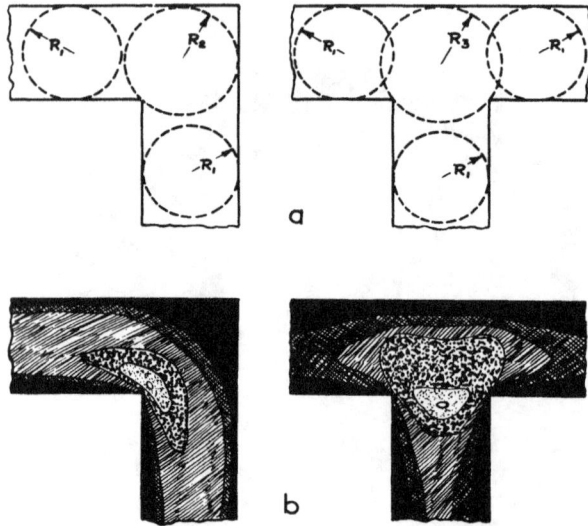

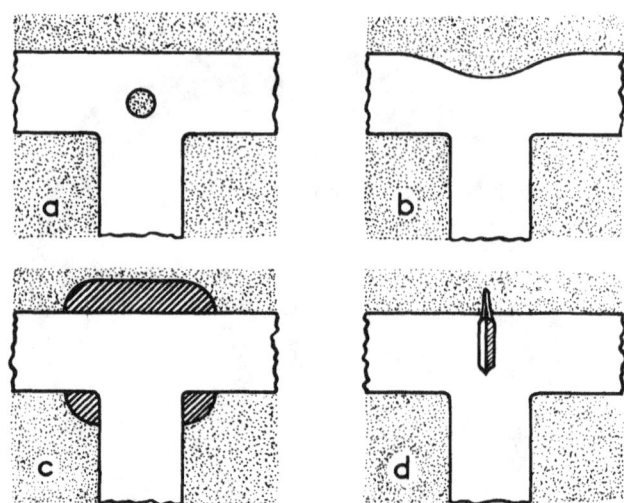

Figure 26. Various treatments for a T junction.

Figure 23. Hot spot location by the method of inscribed circles.

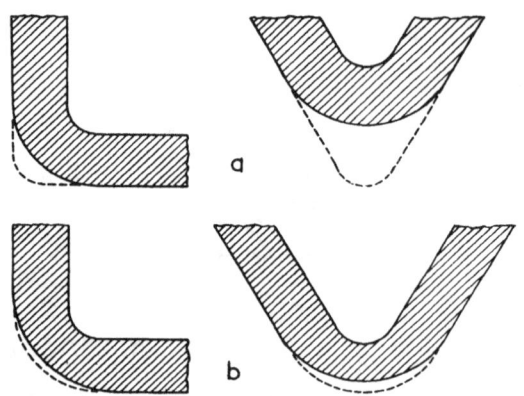

Figure 24. Reduction of cross section in L and V junctions.

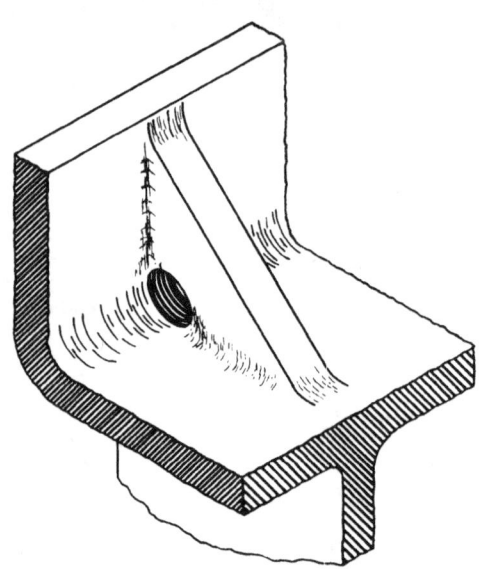

Figure 27. Coring to reduce section in a rib junction.

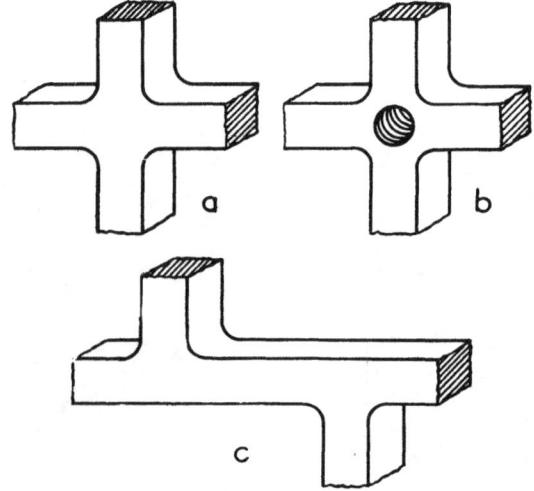

Figure 25. Reduction of cross section in an X junction.

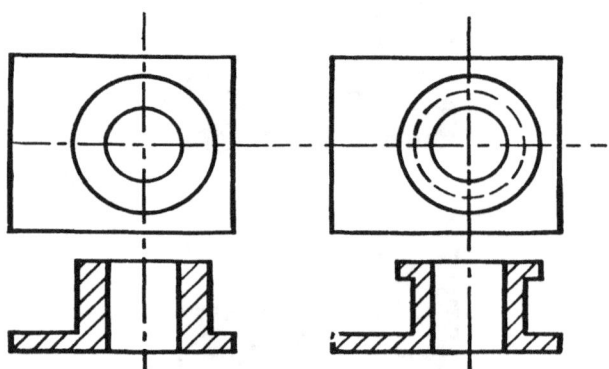

Figure 28. Removal of heavy section by redesign.

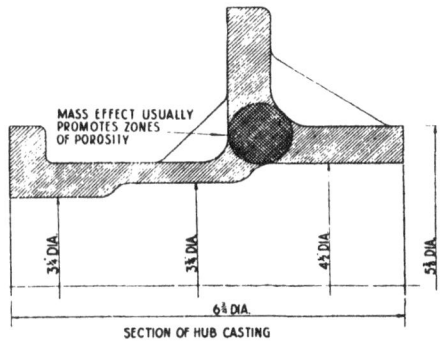

Figure 29. Hub cross section - heavy section.

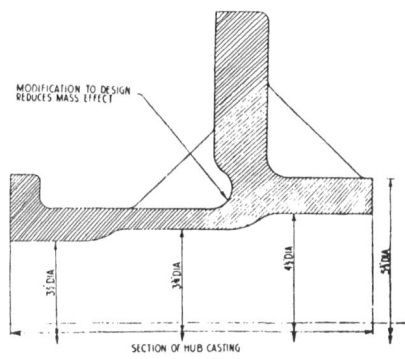

Figure 30. Hub cross section - improved design.

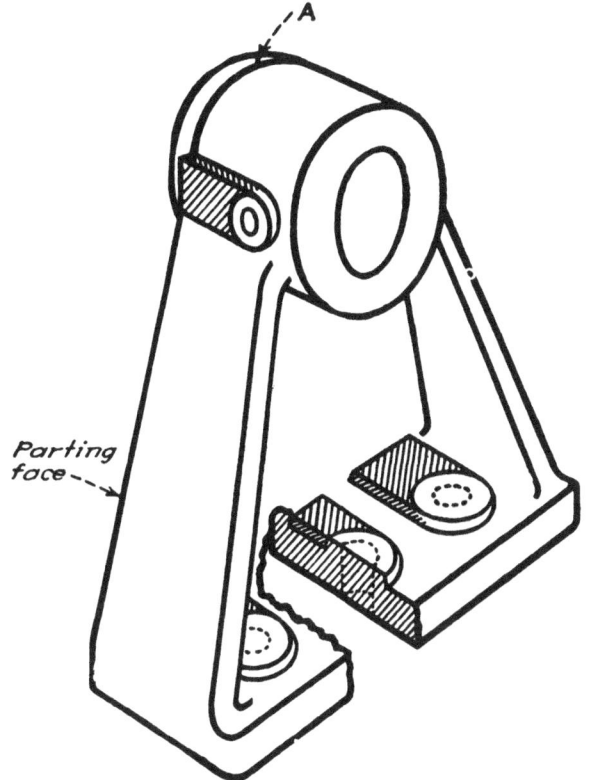

Figure 31. Bracket casting.

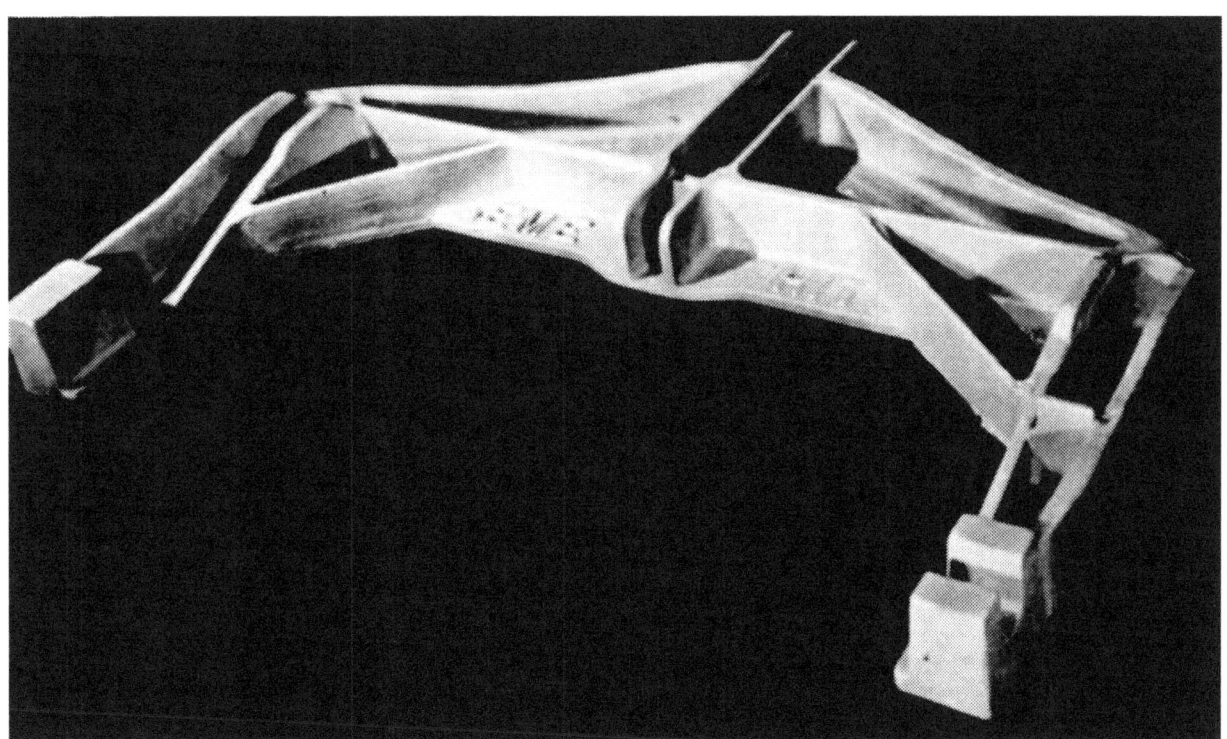

Figure 32. Aluminum yoke casting.

Chapter III
PATTERNMAKING

FUNCTIONS OF THE PATTERN

A pattern is used to form the mold cavity into which molten metal is poured to produce a casting. As such, it is a tool in the hands of the foundryman. A great deal of success in producing a good casting depends on the quality and design of the pattern. For example, a pattern that does not have the proper draft is difficult to draw from the sand without breaking the mold.

The design of the casting itself, as well as that of the pattern, must be taken into consideration to make molding less difficult. The casting design should be as simple as possible, since it will determine the ease with which a pattern can be drawn from the mold, the number of loose pieces required in the pattern, and the number of cores needed.

TYPES OF PATTERNS

There are three main types of patterns: loose patterns, mounted patterns, and core boxes.

Loose Patterns. The majority of molds made aboard repair ships are made with loose patterns, since castings required are usually few in number and not too often repeated. A loose pattern is the wood counterpart of the casting, with the proper allowance in dimensions for contraction and machining. A typical loose pattern is shown in figure 33. A loose pattern may be made in one piece or it may be split into the cope and drag pieces to make molding easier. A split pattern is shown in figure 34.

A loose pattern has the disadvantage of requiring a follow board or a false cope to make the parting line, or hand cutting the parting line. The different steps used to make molds from loose patterns are described in Chapter 5, "Making Molds."

The original casting or the broken parts of a casting which have been put together may be used in an emergency as a loose pattern. In such a case, the part to be used as a pattern must be built up to allow for the contraction of the cast metal and prevent the new casting from being too small. A material known as "Celastic" (see allowance list), supplied in sheets, can be applied to the metal part. When Celastic dries, it will adhere firmly and form a hard surface which may be sandpapered or sawed like wood. For directions on the use of Celastic, see the section on "Maintenance, Care, and Repair" in this chapter.

Mounted Patterns. Patterns fastened permanently to a flat board, called a match plate, are known as mounted patterns.

The main advantage of the mounted pattern over the loose pattern is that it is easier to use and store. For these reasons, a mounted pattern is generally warranted when several of the castings (say, five or more) are to be made during one "run" or when the casting is made at frequent intervals.

Another advantage of the mounted pattern is that a pattern of the gating system also can be mounted on the match plate. This practice of molding the gating system eliminates the loose sand that often results when gates are hand cut. As a result, the castings produced usually are better than those produced with the loose patterns.

Core Boxes. Core boxes are actually negative patterns. When looking at a pattern, one sees the casting in its actual shape. A core box on the other hand shows the cavity which will be created by the core. Core boxes are used not only to make cores for holes in castings but also to make parts of a mold. In some cases, a pattern cannot be made so that it can be drawn. In such a case, the part of the casting which would hinder drawing is made as a core that can be placed in the mold after the pattern proper has been withdrawn. The making and proper use of cores is described in Chapter 6, "Making Cores."

PATTERN MATERIALS

The most commonly used material for patterns is wood, because it is easy to work with and is readily available. Mahogany, white pine, and sugar pine are acceptable materials. Select kiln-dried white or sugar pine is most widely used because it is easily worked and is generally free of warping and cracking.

For pattern work, it is essential that the wood has a low moisture content, 5 to 6 percent if possible, in order to avoid warping and shrinking of the finished pattern.

Metal patterns are usually used as mounted patterns, with the gating included in the pattern. Their use is warranted only when a large number of castings must be made. Mounted metal patterns are difficult to make and require special skills. The one distinct advantage of a metal pattern is that it does not warp on storage and when removed from storage, no preparation other than cleaning is necessary before use.

A material which may be used for an emergency pattern, when only a small number of castings are required and there is not sufficient time to make a wooden pattern, is plaster or gypsum cement. Gypsum cement is made from gypsum rock, finely ground and heated to high temperatures. When mixed with water, it forms a plastic mass which can be molded, shaped, or cast. Plaster patterns have the disadvantage of being very fragile and require careful handling, therefore, it is recommended for use only in an emergency.

PATTERN LAYOUT

The process of actually laying out a pattern comes under the work of the patternmaker. The various parts of proper pattern layout are discussed briefly here to provide the molder with information which may prove useful in determining any nonconformity between the casting and the original drawing.

Parting Line. The parting line divides the pattern into the parts that form the cavity in the cope (top) and drag (bottom) of the mold. Whenever possible, a casting is designed so a straight parting line can be used; that is, a single flat surface will divide the casting into cope and drag sections. Usually, a straight parting line is necessary if the pattern is to be mounted. When loose patterns are used, the mold may be made easier with a straight parting line than with a broken parting line.

Core Prints. A core print is a projection on the pattern designed to make an impression in the sand for locating and anchoring the core.

Although there are no fixed rules as to the length of core prints or how much taper they should have, practice requires that there should be sufficient bearing surface to support the weight of the core. The following table gives dimensions which have been found successful in practical application.

TABLE 3. CORE PRINT DIMENSIONS

Size of Core	Length of Core Print
Up to 1½-inch diameter	2-inch core print
From 2-inch to 5-inch diameter	At least equal to the diameter of core
Above 5 inches in diameter	6-inch core print (minimum)

In general, the length of a core print should equal or slightly exceed its diameter or width. When a core has prints in the cope, the cope prints should provide a "closing clearance" so as to avoid the possibility of crushing sand from the cope when closing the mold. This clearance, however, should not be excessive, as the core will shift under pressure from the molten metal. If it is possible for a core to be set "upside down" or "wrong end to," locating or indexing lugs (tell-tales) should be provided to prevent this.

A good practice for constructing core prints is shown in figure 35. It results in castings with fewer fins at the parting line. Fins tend to produce cracks, and require extra time to clean off the casting. Larger core prints provide better core location and support in the mold. In addition, they reduce the tendency for cracks to form in the cored openings from core fins.

The location, size, and type of vent holes, to allow gases to escape, should be indicated on the core prints and in the core box by means of strips or projections, or by some other appropriate means.

Chaplets. When the design of the core is such that additional support over and above that given by the core prints is needed, it is necessary to use chaplets. These chaplets are pieces of metal especially designed to support the core. Detailed description of chaplets and their use will be found in Chapter 5, "Making Molds." Their use is to be avoided wherever possible, particularly on pressure castings. If chaplets are necessary, their location and size should be indicated on the pattern and core box by raised sections such as shown in figure 36. This additional metal in the mold cavity serves two main purposes; first, it accurately locates the best chaplet position and insures that the location will be consistently used; second, it provides an additional mass of metal to aid in the fusion of the chaplet, which is necessary to obtain pressure tightness.

Shrinkage Rules. The patternmaker uses rules which are somewhat longer than the numbers indicate. The size of such a rule allows for shrinkage of the casting. A 1/4-inch shrink rule, for instance, is 12 1/4-inches long, although the markings would indicate that it is only 12 inches long.

The shrinkage rule to be used in constructing a pattern must be selected for the metal which will be used in the casting. It must be remembered that the shrinkage rule will also vary with the casting design. For example, light and medium steel castings of simple design and no cores require a 1/4-inch rule, whereas for pipes and valves where there is a considerable resistance offered to the contraction of the steel by the mold and cores, a 3/16-inch rule will be adequate. Shrinkage allowances for various metals and mold construction are listed in table 4.

Machining Allowances. The machining allowance or finish is usually made on a pattern to provide for extra metal on the casting surfaces that are to be machined. Some castings do not require finish since they are used in the rough state just as they come from the final cleaning operation. Most castings are finished only on certain surfaces, and no set rule can be given as to the amount of finish to be allowed. The finish is determined by the machine shop practice and by the size and shape of the casting. A casting may become distorted from stresses during the casting process or

TABLE 4. PATTERN SHRINKAGE ALLOWANCES

Casting Alloys	Pattern Dimension (inches)	Type of Construction	Contraction (inches per foot)
Gray Cast Iron	Up to 24	Open construction	1/8
	From 25 to 48	Open construction	1/10
	Over 48	Open construction	1/12
	Up to 24	Cored construction	1/8
	From 25 to 36	Cored construction	1/10
	Over 36	Cored construction	1/12
Cast Steel	Up to 24	Open construction	1/4
	From 25 to 72	Open construction	3/16
	Up to 18	Cored construction	1/4
	From 19 to 48	Cored construction	3/16
	From 49 to 66	Cored construction	5/32
Aluminum	Up to 48	Open construction	5/32
	49 to 72	Open construction	9/64
	Up to 24	Cored construction	5/32
	Over 48	Cored construction	9/64 to 1/8
	From 25 to 48	Cored construction	1/8 to 1/16
Brass			3/16
Bronze			1/8 to 1/4

during heat treatment. This distortion is also a factor in determining machining allowance. In table 5 are listed some finish allowances which may be used as a guide.

Many times, the exact finish allowances can be found by referring to the original blueprints of the part to be cast.

Draft. Draft is the amount of taper given to the sides of projections, pockets, and the body of the pattern so that the pattern may be withdrawn from the mold without breaking the sand away. This also applies to core boxes. The breaking of sand due to a lack of taper is shown in figure 37. The same pattern with correct taper is shown in figure 38. When a straight piece, such as the face of a flange or a bushing, is made, the amount of draft is usually 1/8-inch per foot. In green sand molding, interior surfaces will require more draft than the exterior surfaces, because of the lower strength of isolated volumes of sand. The draft is dependent on the shape and size of the casting and should at all times be ample. The actual draft to be used is usually determined by consultation between the patternmaker and the molder. Proper and improper drafts are shown in figure 39.

Distortion Allowance. Many times a casting is of a design which results in cooling stresses that cause distortion in the finished casting. The design may also be such that it cannot be corrected in the design. In such a case, the experience of the molder and patternmaker must be relied upon to produce a good casting. Distortion allowances must be made in a pattern and are usually determined by experience. Recorded information on castings of this type is very useful in determining distortion allowances on future work.

TABLE 5. GUIDE TO PATTERN MACHINE FINISH ALLOWANCES

Casting Alloy	Pattern Size (inches)	Bore (inches)	Finish
Cast iron	Up to 12	1/8	3/32
	13 to 24	3/16	1/8
	25 to 42	1/4	3/16
Cast steel	Up to 12	3/16	1/8
	13 to 24	1/4	3/16
	25 to 42	5/16	5/16
Brass, bronze, and aluminum	Up to 12	3/32	1/16
	13 to 24	3/16	1/8
	25 to 36	3/16	5/32

A typical casting which would require distortion allowances is a simple yoke casting shown in figure 40. Part (a) shows the casting as it was designed. The yoke made to this design is shown in part (b) with the arms widened out. The arrows indicate the direction of the cooling stresses which produced contraction in the cross member. Part (c) shows the pattern as made with distortion allowances, and part (d) shows the finished part. In part (d) the arrows again show the direction of the cooling stresses which were used to produce a straight yoke.

MAKING THE PATTERN

Skilled patternmakers are available aboard repair ships to make patterns. Construction of patterns, therefore, is not discussed in detail in this manual. Detailed information on patternmaking can be found in the patternmakers' manuals aboard ship.

If broken parts are to be used as a pattern, extreme care must be taken to insure proper alignment of the parts when they are joined or placed for molding. The surfaces should be as smooth as possible and the size of the casting should be increased wherever necessary to compensate for contraction. The use of Celastic for this purpose is described in the section on "Maintenance, Care, and Repair."

For applications where the quantity of castings required is small and the designs are quite simple, gypsum cement can be used as pattern material with success. See Pattern Materials.

One disadvantage in the use of this material is that it is fragile and is likely to be damaged in handling, molding, and storage. Internal support can be provided through the use of arbors, rods, wire frames, intermixed hair, while external reinforcements can be provided by surface coating. Typical patterns produced in gypsum cement are shown in figure 41. Making of a simple pattern is shown in figure 42.

FINISHING AND COLOR CODING

Shellac is usually used to fill the pores in wood patterns or to seal plaster patterns. The patterns are rubbed smooth to eliminate the possibility of sand adhering to the pattern because of a rough surface. Plaster patterns may be metal sprayed to produce a hard, smooth surface. After the surface of the patterns has been properly prepared, various parts are painted for identification.

The color code used for identifying different parts of a pattern is as follows:

1. Surfaces to be left UNFINISHED are painted BLACK.

2. Surfaces to be MATCHED are painted RED.

3. Seats of, and for, LOOSE PIECES are marked by RED STRIPES on a YELLOW BACKGROUND.

4. CORE PRINTS and SEATS for LOOSE CORE PRINTS are painted YELLOW.

5. STOPOFFS are indicated by DIAGONAL BLACK STRIPES on a YELLOW BASE.

MAINTENANCE, CARE, AND REPAIR

The patterns normally made aboard repair ships are used for a few castings and then they must be stored. It is important that storage space be provided which is as free of moisture as possible. This precaution will maintain the patterns in good condition and prevent warping and cracking. The storage of patterns should be in properly constructed racks wherever possible. This will keep pattern damage down to a minimum.

A record should be kept of all patterns which are on hand. These records should contain a complete description of the pattern, pattern numbers, class of ship, size of part, and drawing and piece number. Such records are useful in locating a pattern for future use. They may also be used to provide a pattern for a similar casting which may be required. Time can be saved by slightly altering a pattern already on hand or using the pattern as designed and making alterations in the machining operations. Any permanent pattern changes, no matter how small, should be noted on the pattern record, and if possible on the blueprint of the casting.

Many times a core box has to be repaired or altered slightly. Sheet lead or sheet brass of varying thicknesses may be used. Celastic may also be used to repair a pattern or core box. Minor repairs to the pattern or core box may easily be made by a molder, but any repair of a major nature should be by a patternmaker. After any repair is made, the pattern should be checked to make sure that it conforms to the drawing. A periodic check of patterns or core boxes and minor repair of them will go a long way toward keeping the patterns in good usable condition and prevent major repairs later.

Directions for Applying Celastic.

1. Clean the surface where it is to be applied.

2. Cut pieces to the size required or a number of pieces to cover the required area.

3. Immerse the Celastic in the solvent (methyl ethyl ketone) until it becomes very pliable and sticky. In this state, it can be applied to the pattern and will shape very easily, even on irregular contours, by pressure from the fingers.

4. After the solvent has evaporated, the Celastic will adhere firmly to the pattern and the outer surface will be relatively hard. It may then be sanded and lacquered to a smooth surface.

WARNING: Celastic shrinks in thickness after dipping and drying, and proper allowance must be made. If a greater thickness is desired on any surface, one or more pieces may be applied to the original layer of the Celastic. Two small metal pans should be available for submerging the Celastic; any solvent left in the pan may be returned to the bottle.

CALCULATION OF CASTING WEIGHT

The calculation of casting weights is important in the operation of any foundry. For that reason, some information on the methods and practices used is given.

It is obviously quite simple to calculate the weight required to pour a casting if the defective part is to be used as a pattern, or if it is on hand. Since risers and gates are usually round (and should be) in their cross section, it is easy to calculate their weight and add it to the weight of the casting.

Another simple method that can be used in cases where a small pattern of solid wood construction with no cores is to be used consists in weighing the patterns and multiplying this figure by the following:

For steel	17.0
For cast iron	16.5
For bronze	18.5
For aluminum	5.0

To this figure, the weights of heads and gates are added.

Caution must be used in following this practice; if the pattern is not of solid construction or if it is not made of white pine, an erroneous answer will be obtained. Sugar pine and mahogany have a greater density and a lower factor must be used to calculate the casting weights. Where neither of these methods is possible, it is necessary to break down the design into simple sections—such as rounds, squares, and plates—and calculate the weight of each section by determining its volume in cubic inches, multiplying this figure by the following weights per cubic inch, and then obtaining the total:

	Pounds per cubic inch
Cast steel	0.284
Aluminum	0.098
Cast iron	0.260
Compositions G and M	0.317
Manganese bronze	0.303

This method is demonstrated in the case of the designs shown in figures 43 and 44.

In table 6 are areas and volumes for calculating weights of castings. This table shows the various shapes and formulas which are useful in calculating casting weights.

SUMMARY

Making a pattern is the job of a skilled patternmaker, but a knowledge of the factors involved in patternmaking is useful to the molder. Many times a defective casting can be traced to not enough draft, improper parting line, or insufficient core prints. A molder who is able to recognize a defect caused by improper pattern work or a pattern requiring repair can save himself a lot of time by having the pattern corrected.

The factors discussed in this chapter are not intended to supply all the answers relating to patternmaking. The molder should use this information to guide him in maintaining his patterns and recognizing when they are in need of attention.

AREAS AND VOLUMES FOR CALCULATING WEIGHTS OF CASTINGS

Rectangle and Parallelogram

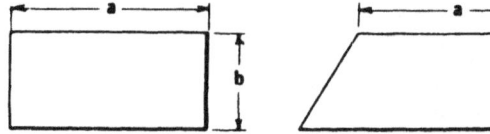

Area = ab

Triangle

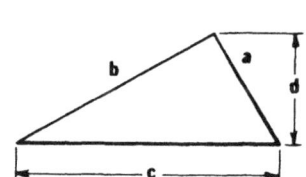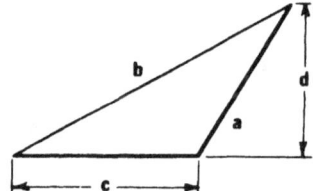

Area = ½ cd.

Area = $\sqrt{s\,(s-a)\,(s-b)\,(s-c)}$ when
s = ½ (a + b + c)

Example: a = 3", b = 4", c = 5"

$$s = \frac{3'' + 4'' + 5''}{2} = 6''$$

Area = $\sqrt{6\,(6-3)\,(6-4)\,(6-5)}$ = 6 sq. in.

Regular Polygons

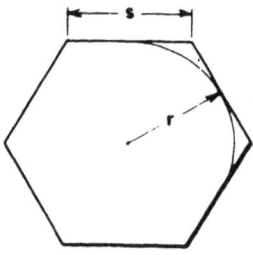

n = Number of sides, s = Length of one side, r = Inside radius

Area = ½ nsr

Number of Sides	Area
5	1.72047 s^2 = 3.63273 r^2
6	2.59809 s^2 = 3.46408 r^2
7	3.63395 s^2 = 3.37099 r^2
8	4.82847 s^2 = 3.31368 r^2
9	6.18181 s^2 = 3.27574 r^2
10	7.69416 s^2 = 3.24922 r^2
11	9.36570 s^2 = 3.22987 r^2
12	11.19616 s^2 = 3.21539 r^2

Trapezium

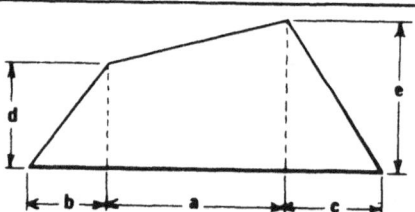

Area = ½ $\left[a\,(e + d) + bd + ce\right]$

Example: a = 10", b = 3", c = 5", d = 6", e = 8"

Area = ½ $\left[10\,(8 + 6) + (3 \times 6) + (5 \times 8)\right]$ = 99 sq. in.

Square

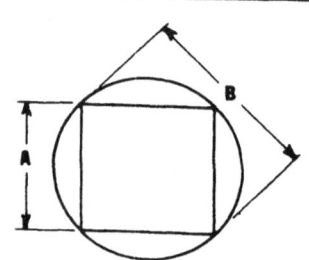

The diagonal of a square = A × 1.414

The side of a square inscribed in a given circle is: B × .707.

Circle

θ (the Greek letter Theta) = angle included between radii

π (pi) = 3.1416, D = Diameter, R = Radius, C = Chord,

h = Height of Arc, L = Length of Arc.

Circumference = $\pi D = 2\pi R = 2\sqrt{\pi \times \text{Area}}$

Diameter = $2R$ = Circumference $\div \pi = 2\sqrt{\dfrac{\text{Area}}{\pi}}$

Radius = $\tfrac{1}{2}D$ = Circumference $\div 2\pi = \sqrt{\dfrac{\text{Area}}{\pi}}$

Radius = $\dfrac{\left(\dfrac{C}{2}\right)^2 + h^2}{2h}$

Area = $\tfrac{1}{4}\pi D^2 = 0.7854\, D^2 = \pi R^2$

Chord = $2\sqrt{h(D-h)} = 2R \times \text{sine } \tfrac{1}{2}\theta$

Height of Arc, $h = R - \sqrt{R^2 - \left(\dfrac{C}{2}\right)^2}$

Length of Arc, $L = \dfrac{\theta}{360} \times 2\pi R = 0.0174533\, R\theta$

$\tfrac{1}{2}\theta$ (in degrees) = $28.6479\dfrac{L}{R}$

Sine $\tfrac{1}{2}\theta = \dfrac{C}{2} \div R$

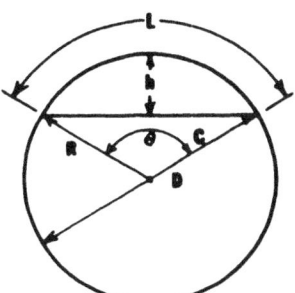

Sector of a Circle

Area = $\tfrac{1}{2}LR$

Example: L = 10.472", R = 5"

Area = $\dfrac{10.472}{2} \times 5 = 26.180$ sq. in.

or Area = $\pi R^2 \times \dfrac{\theta}{360} = 0.0087266\, R^2\theta$

Example: R = 5", θ = 120°

Area = $3.1416 \times 5^2 \times \dfrac{120}{360} = 26.180$ sq. in.

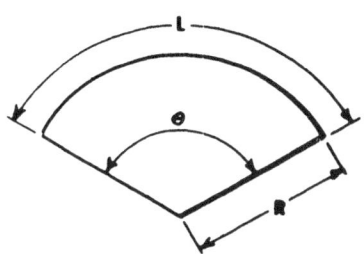

Segment of a Circle

Area = $\pi R^2 \times \dfrac{\theta}{360} - \dfrac{C(R-h)}{2}$

Example: R = 5", θ = 120°, C = 8.66", h = 2.5"

Area = $3.1416 \times 5^2 \times \dfrac{120}{360} - \dfrac{8.66(5-2.5)}{2} = 15.355$ sq. in.

Length of arc $L = 0.0174533\, R\theta$

Area = $\tfrac{1}{2}\left[LR - C(R-h)\right]$

Example: R = 5", C = 8.66", h = 2.5", θ = 120°

L = $0.0174533 \times 5 \times 120 = 10.472$"

Area = $\tfrac{1}{2}\left[(10.472 \times 5) - 8.66(5-2.5)\right] = 15.355$ sq. in.

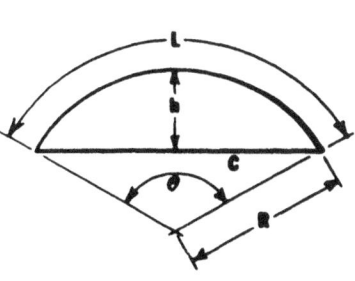

Circular Ring

Area = $0.7854 (D^2 - d^2)$, or $0.7854 (D-d)(D+d)$

Example: $D = 10"$, $d = 3"$

Area = $0.7854 (10^2 - 3^2) = 71.4714$ sq. in.

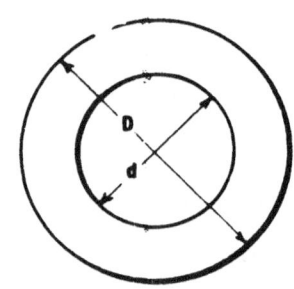

Spandrel

Area = $0.2146 R^2 = 0.1073 C^2$
Example: $R = 3$
Area = $0.2146 \times 3^2 = 1.9314$

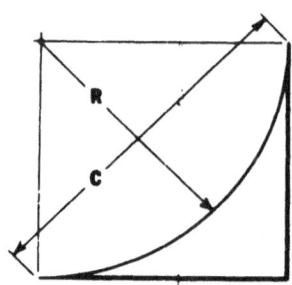

Parabolic Segment

Area = $\frac{2}{3} sh$
Example: $s = 3$, $h = 4$
Area = $\frac{2}{3} \times 3 \times 4 = 8$

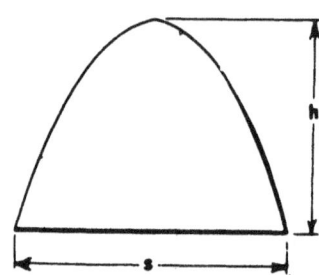

Ellipse

Area = $\pi ab = 3.1416\, ab$
Example: $a = 3$, $b = 4$
Area = $3.1416 \times 3 \times 4 = 37.6992$

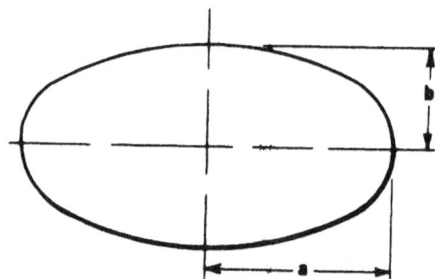

Irregular Figures

Area may be found as follows:
Divide the figure into equal spaces
as shown by the lines in the figure.
  (1) Add lengths of dotted lines.
  (2) Divide sum by number of spaces.
  (3) Multiply result by "A."

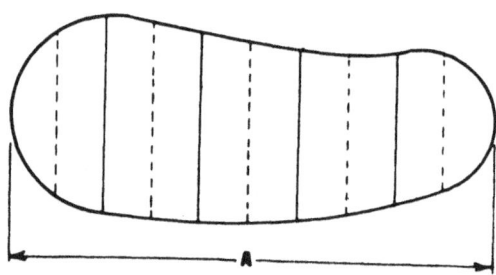

Ring of Circular Cross Section

Area of Surface = $4\pi^2 Rr$ = 39.4784 Rr
Area of Surface = $\pi^2 Dd$ = 9.8696 Dd
Volume = $2\pi^2 Rr^2$ = 19.7392 Rr^2
Volume = $\frac{1}{4}\pi^2 Dd^2$ = 2.4674 Dd^2

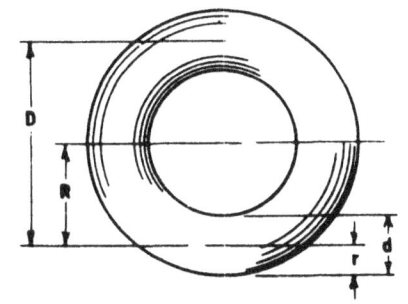

Sphere

Surface = $4\pi r^2$ = 12.5664 r^2 = πd^2
Volume = $\frac{4}{3}\pi r^3$ = 4.1888 r^3
Volume = $\frac{1}{6}\pi d^3$ = 0.5236 d^3

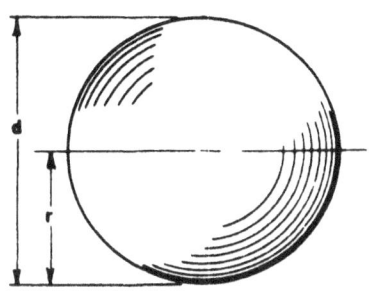

Segment of a Sphere

Spherical Surface = $2\pi rh$ = $\frac{1}{4}\pi(c^2 + 4h^2)$ = 0.7854 $(c^2 + 4h^2)$
Total Surface = $\frac{1}{4}\pi(c^2 + 8rh)$ = 0.7854 $(c^2 + 8rh)$
Volume = $\frac{1}{3}\pi h^2 (3r - h)$ = 1.0472 $h^2 (3r - h)$
    or
Volume = $\frac{1}{24}\pi h (3c^2 + 4h^2)$ = 0.1309 $h(3c^2 + 4h^2)$

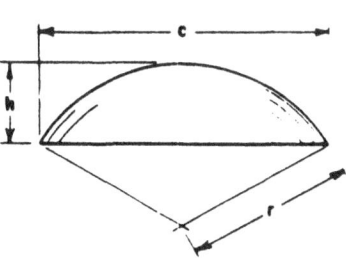

Sector of a Sphere

Total Surface = $\frac{1}{2}\pi r(4h + c)$ = 1.5708 $r(4h + c)$
Volume = $\frac{2}{3}\pi r^2 h$ = 2.0944 $r^2 h$

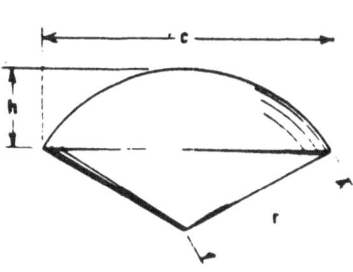

Cylinder

Cylindrical Surface = πdh = $2\pi rh$ 6.2832 rh
Total Surface = $2\pi r(r + h)$ 6.2832 (r + h)
Volume = $\pi r^2 h$ = $\frac{1}{4}\pi d^2 h$ 0.7854 $d^2 h$

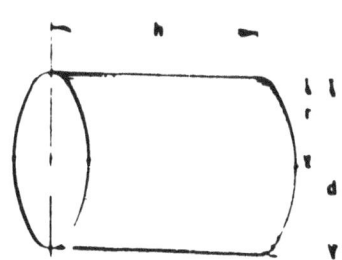

Pyramid

A = area of base
P = perimeter of base
Lateral Area = ½ Ps
Volume = ⅓ Ah

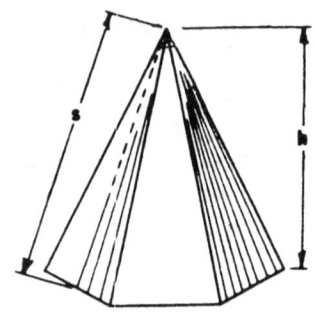

Frustum of a Pyramid

A = area of base
a = area of top
m = area of midsection
P = perimeter of base
p = perimeter of top
Lateral Area = ½ s (P + p)
Volume = ⅓ h (a + A + $\sqrt{aA}$)
Volume = ⅙ h (A + a + 4 m)

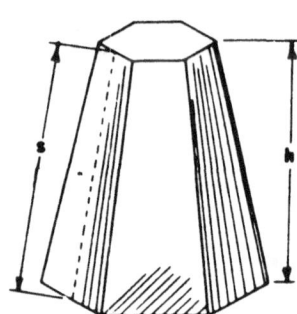

Cone

Conical Area = πrs = πr $\sqrt{r^2 + h^2}$

Volume = ⅓ $\pi r^2 h$ = 1.0472 $r^2 h$ = 0.2618 $d^2 h$

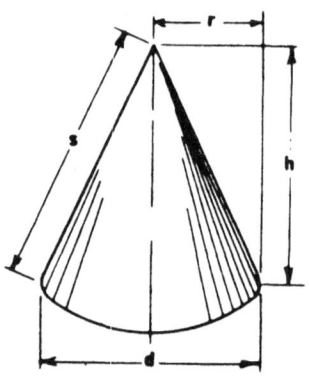

Frustum of a Cone

A = area of base
a = area of top
m = area of midsection
R = D ÷ 2; r = d ÷ 2
Area of Conical Surface = ½ πs (D + d) = 1.5708 s (D + d)
Volume = ⅓ h (R^2 + Rr + r^2) = 1.0472 h (R^2 + Rr + r^2)
Volume = ¹⁄₁₂ h (D^2 + Dd + d^2) = 0.2618 h (D^2 + Dd + d^2)
Volume = ⅓ h (a + A + $\sqrt{aA}$) = ⅙ h (a + A + 4 m)

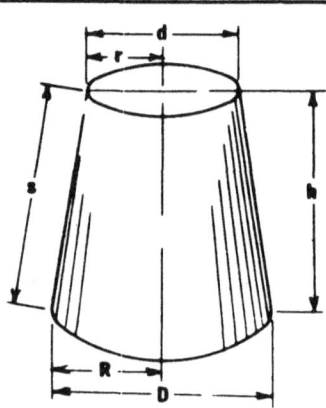

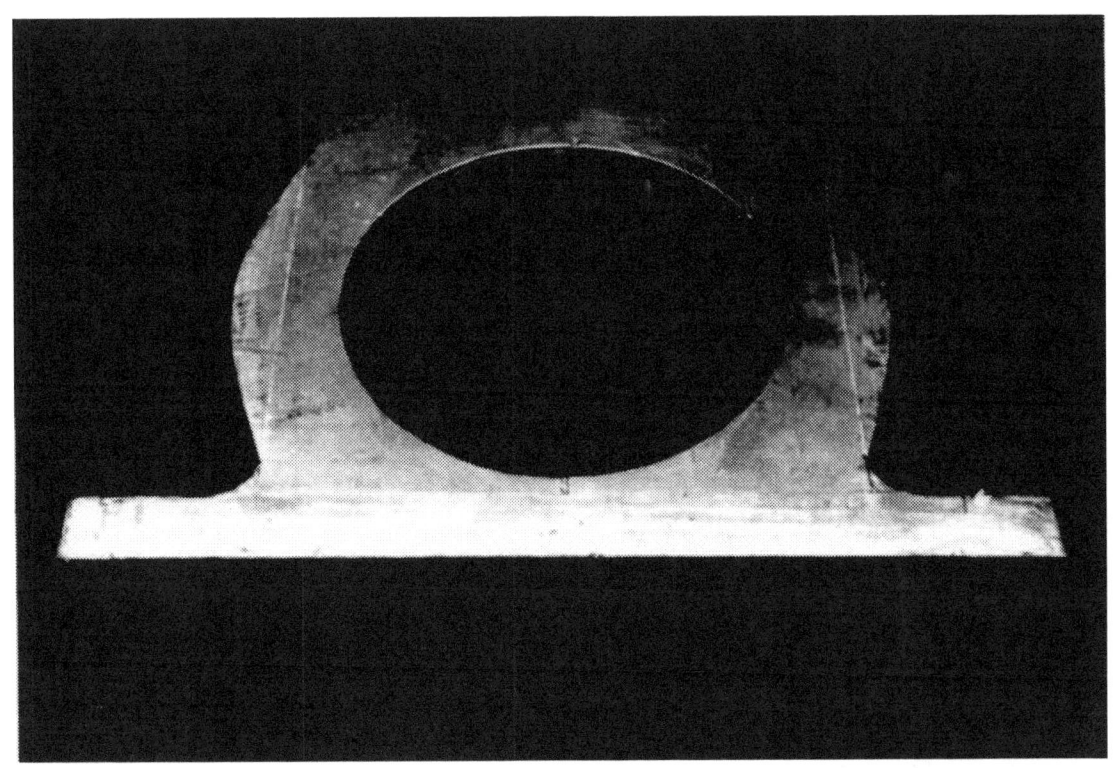

Figure 33. One piece pattern

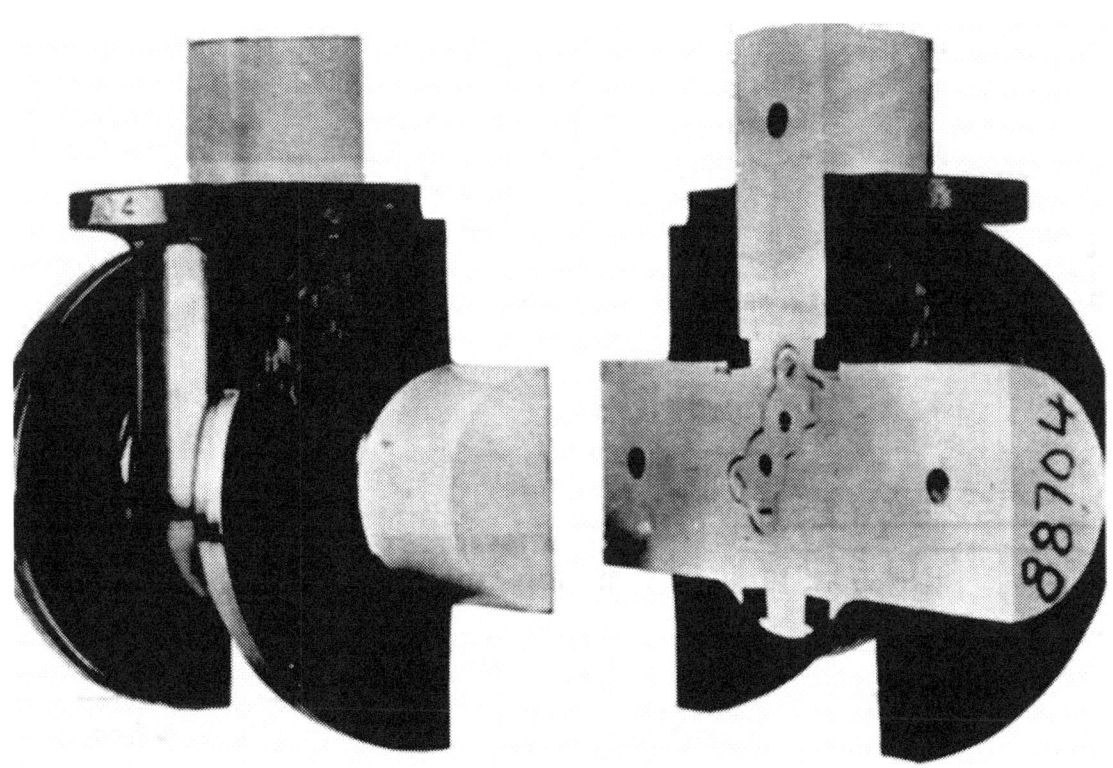

Figure 34. Split pattern.

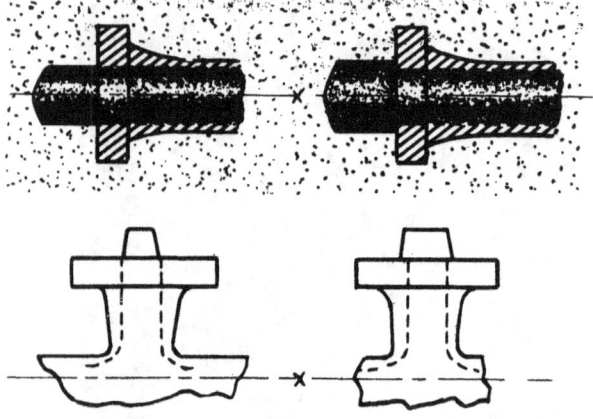

Figure 35. Core print construction.

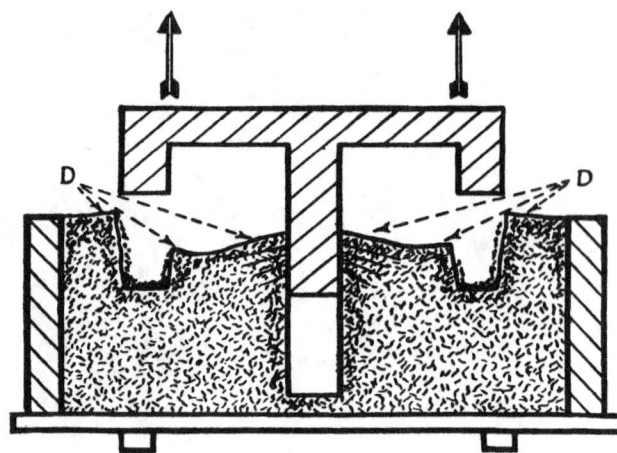

Figure 37. Mold broken due to a lack of taper.

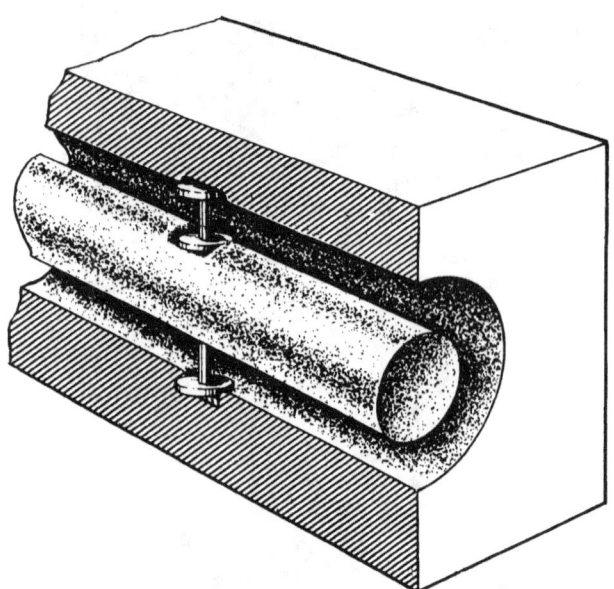

Figure 36. Chaplet location with pads.

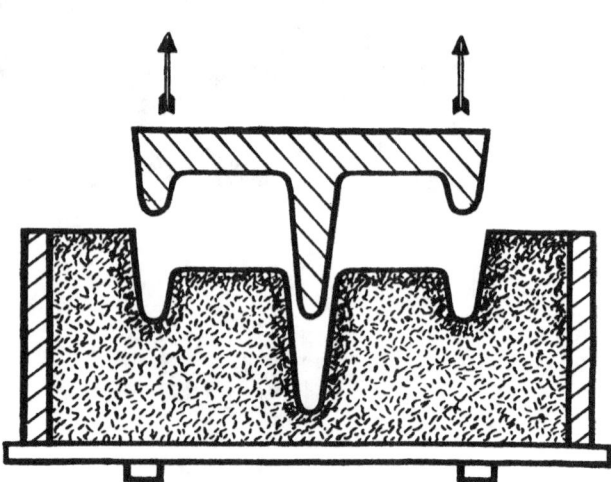

Figure 38. Clean pattern draw with correct taper.

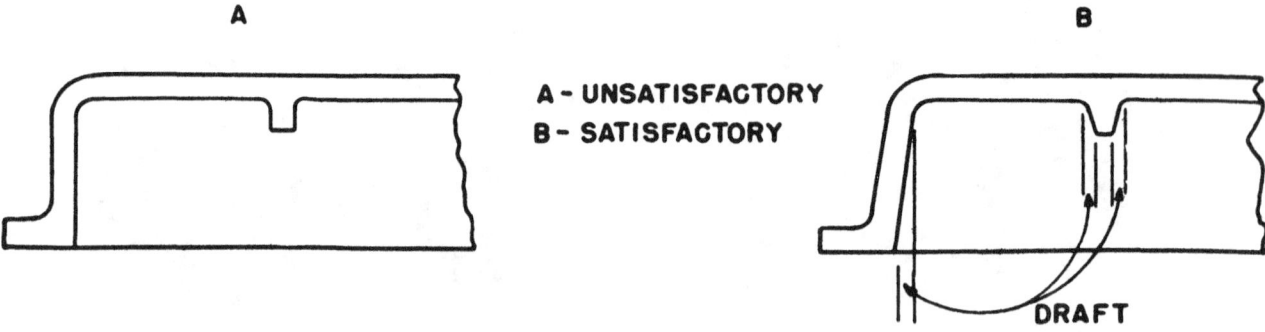

A - UNSATISFACTORY
B - SATISFACTORY

Figure 39. Pattern draft.

- 36 -

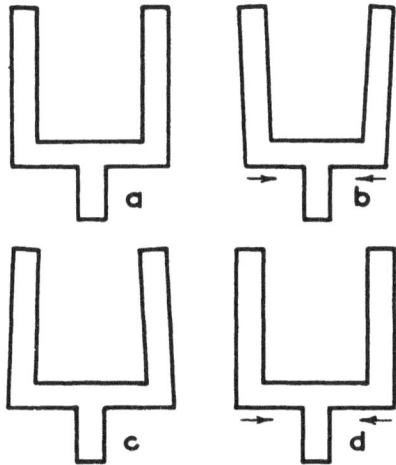

Figure 40. Distortion allowance in a simple yoke pattern.

Figure 41. Plaster patterns and core boxes.

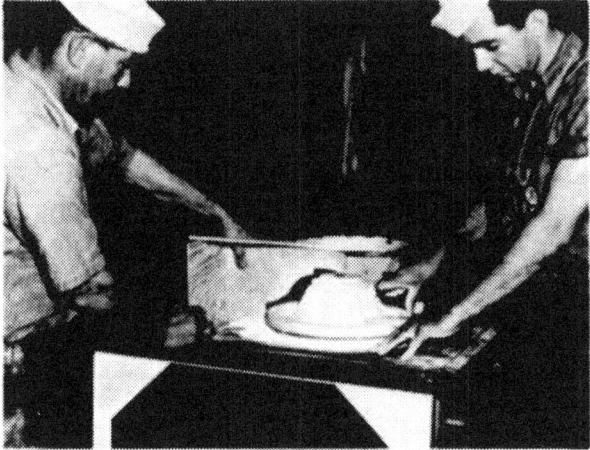

Figure 42. Making a simple plaster pattern.

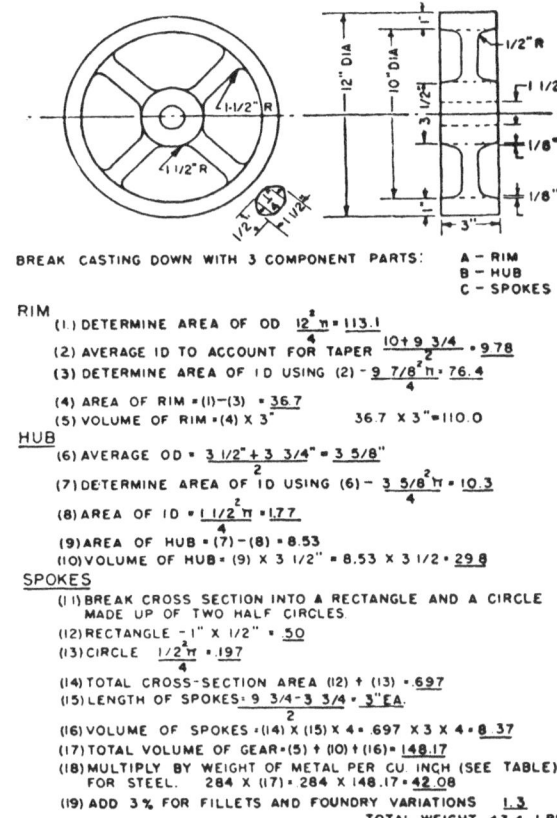

BREAK CASTING DOWN WITH 3 COMPONENT PARTS: A – RIM
B – HUB
C – SPOKES

RIM
(1) DETERMINE AREA OF OD $\frac{12^2 \pi}{4} = 113.1$
(2) AVERAGE ID TO ACCOUNT FOR TAPER $\frac{10 + 9\ 3/4}{2} = 9.78$
(3) DETERMINE AREA OF ID USING (2) = $\frac{9\ 7/8^2 \pi}{4} = 76.4$
(4) AREA OF RIM = (1) − (3) = 36.7
(5) VOLUME OF RIM = (4) X 3"   36.7 X 3" = 110.0

HUB
(6) AVERAGE OD = $\frac{3\ 1/2" + 3\ 3/4"}{2} = 3\ 5/8"$
(7) DETERMINE AREA OF ID USING (6) = $\frac{3\ 5/8^2 \pi}{4} = 10.3$
(8) AREA OF ID = $\frac{1\ 1/2^2 \pi}{4} = 1.77$
(9) AREA OF HUB = (7) − (8) = 8.53
(10) VOLUME OF HUB = (9) X 3 1/2 = 8.53 X 3 1/2 = 29.8

SPOKES
(11) BREAK CROSS SECTION INTO A RECTANGLE AND A CIRCLE MADE UP OF TWO HALF CIRCLES.
(12) RECTANGLE − 1" X 1/2" = .50
(13) CIRCLE $\frac{1/2^2 \pi}{4} = .197$
(14) TOTAL CROSS-SECTION AREA (12) + (13) = .697
(15) LENGTH OF SPOKES = $\frac{9\ 3/4 - 3\ 3/4}{2} = 3"$ EA.
(16) VOLUME OF SPOKES = (14) X (15) X 4 = .697 X 3 X 4 = 8.37
(17) TOTAL VOLUME OF GEAR = (5) + (10) + (16) = 148.17
(18) MULTIPLY BY WEIGHT OF METAL PER CU. INCH (SEE TABLE) FOR STEEL. .284 X (17) = .284 X 148.17 = 42.08
(19) ADD 3% FOR FILLETS AND FOUNDRY VARIATIONS = 1.3
TOTAL WEIGHT 43.4 LBS.

Figure 43. Calculating casting weight.

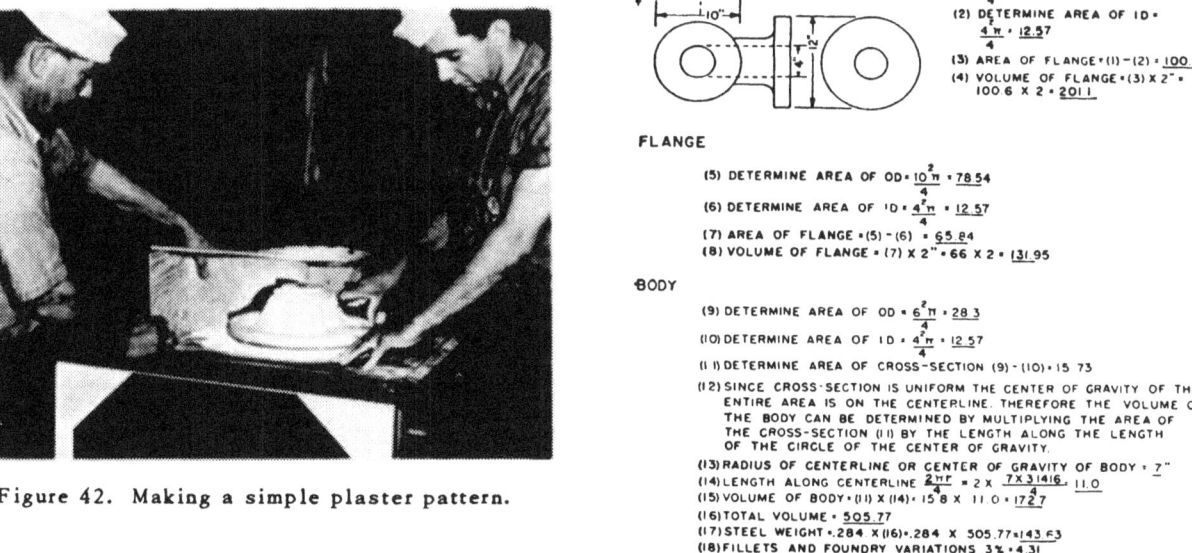

BREAK CASTING DOWN INTO THE COMPONENTS. FLANGE
FLANGE
BODY OF FITTING

FLANGE
(1) DETERMINE AREA OF OD = $\frac{12^2 \pi}{4} = 113.097$
(2) DETERMINE AREA OF ID = $\frac{4^2 \pi}{4} = 12.57$
(3) AREA OF FLANGE = (1) − (2) = 100.5
(4) VOLUME OF FLANGE = (3) X 2" 100.6 X 2 = 201.1

FLANGE
(5) DETERMINE AREA OF OD = $\frac{10^2 \pi}{4} = 78.54$
(6) DETERMINE AREA OF ID = $\frac{4^2 \pi}{4} = 12.57$
(7) AREA OF FLANGE = (5) − (6) = 65.84
(8) VOLUME OF FLANGE = (7) X 2" = 66 X 2 = 131.95

BODY
(9) DETERMINE AREA OF OD = $\frac{6^2 \pi}{4} = 28.3$
(10) DETERMINE AREA OF ID = $\frac{4^2 \pi}{4} = 12.57$
(11) DETERMINE AREA OF CROSS-SECTION (9) − (10) = 15.73
(12) SINCE CROSS-SECTION IS UNIFORM THE CENTER OF GRAVITY OF THE ENTIRE AREA IS ON THE CENTERLINE. THEREFORE THE VOLUME OF THE BODY CAN BE DETERMINED BY MULTIPLYING THE AREA OF THE CROSS-SECTION (11) BY THE LENGTH ALONG THE LENGTH OF THE CIRCLE OF THE CENTER OF GRAVITY.
(13) RADIUS OF CENTERLINE OR CENTER OF GRAVITY OF BODY = 7
(14) LENGTH ALONG CENTERLINE $\frac{2 \pi r}{4} = 2 \times \frac{7 \times 3.1416}{4} = 11.0$
(15) VOLUME OF BODY = (11) X (14) = 15.8 X 11.0 = 172.7
(16) TOTAL VOLUME = 505.77
(17) STEEL WEIGHT .284 X (16) = .284 X 505.77 = 143.63
(18) FILLETS AND FOUNDRY VARIATIONS 3% = 4.31
(19) TOTAL WEIGHT = 147.94 LBS.

Figure 44. Calculating casting weight.

Chapter IV
SANDS FOR MOLDS AND CORES

The principal molding material used in foundries is silica sand. Silica sand is readily available, low in cost, and possesses properties that enable it to withstand the effects of molten metals.

The primary function of any molding material is to maintain the shape of the casting cavity until the molten metal is poured and until the casting solidifies. The properties of silica sand that make it useful as a molding material are its refractoriness and its ability to be formed into complicated shapes easily. Its refractoriness enables it to withstand the intense heat from molten metals. Its ability to be formed into shapes is attained by the action of naturally occurring clay (clay that is quarried with the sand) or added clay, additional binders, and water. The binder maintains the sand in place until the casting is poured and solidified.

The three major parts of a molding sand are: (1) the sand grains, which provide the necessary refractory properties; (2) the bonding material, which may be a naturally occurring clay in the sand or an added material such as bentonite or cereal; and (3) water, which makes possible the bonding of the sand grains by the binder to make the sand a useful molding material.

MOLDING SANDS

Because storage space aboard repair ships is limited, it is to the molder's advantage to stock only a few types of foundry sands. From this point of view, the use of an all-purpose sand is advantageous in that only one facility for new sand is required for all of the metals cast aboard ship. Many times, it may be impossible to obtain the all-purpose sand required, and a locally available sand will have to be used. In such instances, the various properties of the substitute sand will have to be determined before the sand is used in the foundry. All of the sand properties discussed in the section, "Sand Properties," apply to natural sands as well as to synthetic and all-purpose sands.

NATURAL SANDS

Natural sands contain only the clay that is already associated with them when mined. Such a sand is often used as it is received, with only moisture added to obtain the desired properties. Albany sand is a typical example of a natural sand. A naturally bonded sand has the advantages of maintaining its moisture content for a long period of time, having a wide working range for moisture, and permitting easier patching and finishing of molds. One disadvantage of natural sand is that its properties vary and are not so consistent as desired. Additions of bentonite are sometimes made to natural sands. Such a sand is called "semisynthetic."

SYNTHETIC SANDS

Sands that fall under the designation of "synthetic" sands are not actually synthesized from the various elements. They are made by mixing together the various individual materials that make up a molding sand. (See glossary, Synthetic Sand.) A more appropriate name would be "compounded" sands. However, the name synthetic has become established in the foundry industry, through usage, to designate a sand of this type.

Synthetic sands consist of a naturally occurring sand with a very low clay content, or a washed sand (all of the natural clay removed), and an added binder, such as bentonite. Synthetic sands have the following advantages over naturally bonded sands: (1) more uniform grain size, (2) higher refractoriness, (3) mold with less moisture, (4) require less binder, (5) the various properties are more easily controlled, and (6) less storage space is required, since the sand can be used for many different types of castings.

ALL-PURPOSE SANDS

Sands that are used for a variety of casting sizes and types of metals are called "all-purpose" sands. In commercial practice, different sands are used to cast different metals and different sizes of castings of the same metal, but in a shipboard foundry, the limitation of storage space makes the practice of maintaining many special sands impossible. A synthetic sand used as a base for an all-purpose sand has the requirements for a molding sand for shipboard use. Naturally, some advantages will have to be sacrificed in using one sand for making all types of castings. The major factor that will be sacrificed in this respect is that of surface finish. However, the principal purpose of a shipboard foundry is to produce serviceable castings. Surface finish is often not a major requirement. As an example, a coarse-grained sand suitable for steel castings will produce rough surface finishes on lighter nonferrous castings made in the same sand. This is a minor disadvantage for an all-purpose sand when compared to its advantages for repair-ship use.

SAND PROPERTIES

There are a great many properties of sand which are of interest to the production foundryman. Among the most important are: (1) green permeability, (2) green strength, (3) dry strength, (4) moisture content, (5) clay content, and (6) grain fineness. These will be discussed in greater detail. The other properties include hot strength, sintering point, deformation, and collapsibility. The six properties selected as most important are those with which repair-ship molders should be most familiar. These are also the properties that can be determined by the use of the sand-test equipment aboard ship.

GREEN PERMEABILITY

Green permeability is that property of a molding sand that permits the passage of air, gases, or steam through the sand. The openings between the sand grains in a mold give sand its permeability. There are four factors that control the permeability of foundry sand: (1) fineness of the sand grains, (2) shape of the sand grains, (3) the amount and type of binder, and (4) the moisture content. Permeability is expressed as a number that increases with an increasing openness of the sand.

Grain Fineness. Grain fineness is an indication of the grain size of the sands. It is expressed as a number that tells a molder if he has a fine sand, made up largely of very small sand grains, or a coarse sand, composed mainly of large sand grains. A detailed description of grain-fineness number is given under "Methods for Testing Sands," in this chapter.

The general effect of grain size on permeability is shown by figure 45. Data for this curve were obtained by screening a given sand through a series of test screens and then making a permeability test on the sand retained on each screen. The permeability of the coarse sand is very high. As the sand grains become smaller, the permeability decreases rapidly. This decrease is due to the smaller voids or openings between the individual sand grains for the fine sand. Coarse sand grains have the same general size relation to fine sand grains as basketballs have to marbles.

The permeabilities of four typical foundry sands, ranging from coarse to fine, are shown in figure 46. The numbers shown on the graph are the grain-fineness numbers. The coarse sand, having a greater amount of large sand grains and large voids between the grains, has a high permeability. The other sands, having a greater amount of small sand grains and small voids, have lower permeabilities.

Shape of the Sand Grains. There are two primary shapes of sand grains, angular and rounded. There are many degrees of roundness or angularity between the two extremes. Angular grains can be compared to crushed stone. There are sharp edges and corners on the grains. The rounded sand grains have the appearance of beach pebbles that have been rounded by the action of the sea. Sharp angular sand grains cannot pack together as closely as rounded sand grains. As a result, sand with angular grains have a higher permeability than sands with rounded grains. The effect of grain shape on the permeability of molding sand is shown in figure 47. (The word "sharp," incidentally, when applied to molding sands has nothing to do with grain shape. A sharp sand is simply a sand very low in clay content.)

Binder. The amount and type of binder also have an effect on the permeability of foundry sand. The effect of increasing amounts of bentonite on permeability is shown in figure 48. The permeabilities are shown for moisture contents of 2 and 4 percent. With 2 percent moisture, the sand shows a rapid decrease in permeability with increased bentonite content. Sands containing 4 percent moisture show a fairly constant permeability after 4 percent bentonite is reached. This type of information indicates that 4 percent of moisture in this particular sand would produce the best permeability over a range of bentonite contents. The type of binder also affects permeability, as shown in figure 49.

Moisture Content. The effect of moisture content on permeability was shown in figures 46 and 47. Low permeability at very low moisture content is caused by the dry clay particles filling the spaces between the sand grains. Figures 46 and 47 both show an increase in permeability to a maximum value, and then a decrease with further additions of water. The increase in permeability is produced when the moisture causes the clay particles to agglomerate or stick together. This action is similar to the addition of water to dust to form a firm piece of soil. When water is added in excess of the amount to produce this sticking together, the excess water begins to fill in the holes between the sand grains and as a result, the permeability goes down. This action is similar to the addition of water to a firm soil to produce mud.

GREEN STRENGTH

Green strength is the strength of molding sand just after it has been tempered. (Refer to glossary, "temper.") It is the strength which is required for the handling of the sand during the molding operation and, if a mold is poured soon after it is completed, it is the strength which must maintain the shape of the mold.

Green strength is expressed as the number of pounds per square inch required to crush a standard specimen. The same factors that control permeability also control the green strength of foundry sand. They are (1) grain fineness, (2) shape of the sand grains, (3) the amount and type of binder, and (4) the moisture content. Mulling practice or mixing practice also affect the green strength of sand. This is discussed in detail in the section on "Mixing."

Grain Fineness. The smaller the size of the sand grains in a given amount of molding sand, the greater will be the area of contact between the many grains. As a result, the green strength of the finer sand is high. A coarse sand, on the other hand, will have a much smaller contact area for the same amount of sand, and the green strength is lower. This is illustrated in figure 50. The green strength increases as the sand changes from a coarse sand to a fine sand. Figure 51 shows the variation in green strength for four different sands. The sand with the highest fineness number (108) is the finest sand and has the highest green strength for a given moisture content. The other sands become progressively weaker as they become coarser.

Grain Shape. The area of contact between the sand grains is also affected by the shape of the grains. Round grains pack together much more closely than sharp, angular grains and, as a result, have a stronger bond than the angular sand. A comparison of the green strengths of round and angular sands is made in figure 52.

Binder. Green strength is affected directly by the amount of binder which is added. The more binder used, the higher will be the green strength, as shown in figure 53. The type of binder used (clay, cereal, dextrine, or rosin) also affects the green strength of molding sand. The effect of bentonite and fireclay on the green strength is shown in figure 54.

Moisture Content. The effect of moisture on green strength is similar to its effect on permeability. The green strength increases with the first additions of water, reaches a maximum strength, and then starts to decrease. This is illustrated in figures 47 and 53.

DRY STRENGTH

The dry strength of sand mixtures is generally affected in the same way as green strength by grain fineness, grain shape, and moisture content. Different binders, however, can affect dry strength and green strength differently. For example, in comparison with western bentonite, southern bentonite produces a high green strength and a low dry strength. Southern bentonite is widely used for its low dry strength and the resulting easy shakeout of castings.

When cereal and dextrine are added to bentonite, the bonded mixtures give a higher dry strength. For more information on the behavior of different binders, see the next section.

BINDERS

Binders are the materials added to molding sands to hold the individual sand grains together to provide a usable molding material.

Green strength, dry strength, and permeability are the properties of the sand which are directly affected by the amount and type of binder. The change in permeability with a change in bentonite content is shown in figure 48. Figure 49 shows the effect of bentonite and fire clay on permeability.

The change in green strength with a change in the bentonite content is shown in figures 53 and 55. In figure 55, it can be seen that for any given amount of bentonite, there is not a large change in green strength with a change in the moisture content. If the moisture content is maintained at a given value, the green strength can be changed over a considerable range by adjusting the amount of bentonite. Figure 54 shows the effect of fire clay and bentonite on green strength. This shows the advantage of bentonite over fire clay as a binder. The green strength due to fire clay decreases rapidly with increased moisture, while the green strength due to bentonite decreases much less for the same moisture contents.

The effects of blending western and southern bentonite on the green strength and dry strength of a sand with an AFS Fineness Number of 50 to 60 are shown in figure 56. There is a rather uniform decrease in dry strength with a changeover from western to southern bentonite. The green strength increases slightly from 100 percent western bentonite through the various mixtures and then increases rapidly as the 100 percent southern bentonite bond is used. This shows the difference in properties that result from the use of the two different bentonites, or mixtures of the two bentonites. The low dry strength of southern bentonite is especially advantageous when a sand mixture having good collapsibility is required, for instance, when casting alloys that are apt to hot tear easily.

Other binders (such as cereal, dextrine, and rosin) are often used as additives to augment or modify the clay binders. The cereal binders are wheat and corn flours. A corn-flour binder slightly improves the green strength and makes a decided improvement in the dry strength. Wheat-flour, on the other hand, contributes very little to green strength, but improves the collapsibility of a sand. It is important to realize that the effects of all cereal binders are not the same in influencing the properties of molding sands. Dextrine binders are a form of sugar and produce a much higher dry strength than do cereal binders. However, dextrines also cause a reduction in the green strength of the sand mixture. Molasses can be used as a substitute for dextrine, but its influence on sand properties is not so great as that of dextrine. Rosin binders are commercial by-products that are used principally as core binders or in sand mixes for dry sand molds. A rosin-bonded sand has a very hard surface when baked, but has the disadvantage that it absorbs moisture on standing. Because of this characteristic, molds and cores made with rosin-bonded sands should be used as soon after baking as possible.

GRAIN FINENESS

The effect of fineness of a foundry sand is discussed under the various other properties. Briefly, a fine sand will have a higher strength and lower permeability, for a given moisture and binder content, than will a coarse sand.

OTHER PROPERTIES

In a molding sand, hot strength and collapsibility are two properties which are important to the foundryman. Hot strength is the strength that a molding sand has when it is at the pouring temperature of the various molten metals. Hot strength is necessary in a sand mixture to retain the shape of the mold before solidification of the metal starts. Hot strength should not be confused with retained strength, which is the strength of molding sand after it has been heated and permitted to cool to room temperature. Collapsibility is the property that permits a sand mold or core to crumble when it is subjected to the forces exerted by a contracting casting. The determination of hot strength and collapsibility is impossible with the sand-testing equipment aboard ship, but general determinations of these properties can be made by observation. The two properties of hot strength and collapsibility go hand in hand, and one cannot be discussed without the other. The ideal foundry sand would have a high hot strength and good collapsibility, but this combination is difficult to attain, except through very close control of sand processing. The hot strength and collapsibility of the sand can be checked by observing the condition of the sand when shaking out a casting. If the sand is difficult to remove from deep pockets, then the sand lacks adequate collapsibility. A hot tear in a casting is an indication of too high a hot strength, and also a lack of collapsibility.

In this discussion of sand properties, it is obvious that all of the various factors affecting the properties of molding sand are dependent on each other. This interdependence of properties must be kept in mind constantly, especially when trying to determine the cause of casting defects due to sand. The apparently obvious cause of a defect may not be the actual factor causing that defect, and in many cases it is a combination of sand properties that leads to a defect.

REBUILDING OF SANDS

The binder in foundry sands is burned out by the heat of the molten metal. As a result, the green strength of the sand becomes lower and the permeability decreases as the sand is reused. The permeability decreases because of the increase in fines in the sand. The use of sand-testing equipment periodically to measure these properties of molding sands enables the molder to make appropriate additions to the sand before it has deteriorated to the point where it must be discarded. If a continuous check is made, corrections can be made by the addition of small amounts of binder, and more uniform day-to-day properties can be maintained.

Additions of new binder may be as little as one-third to one-half percent of the sand by weight if additions are made frequently and are made as shown necessary by test information. The actual amount of binder required will depend on the type of binder and on the manner in which it is added. The effect of fire clay and bentonite as binders is shown in figures 49 and 54. Note that the fire clay gives a much weaker bond than bentonite, and would require a larger addition to attain the same strength as a bentonite-bonded sand.

MIXING

When rebonding sands, the use of a muller is necessary to obtain the maximum benefits. A much larger percentage of binder is required if the sand is mixed manually with a shovel.

Mulling Sand. To obtain the maximum properties from a molding sand, a muller should be used for the mixing of all foundry sands. It is especially important that core sands and facing sands be mixed in a muller, but the mixing of all sands in a muller provides a more uniform day-to-day operation. The use of a muller to mix and rebond sands is essential to good sand control, and shows up in the production of better castings.

The literature supplied with the mullers aboard repair ships should be consulted for proper operating instructions. The best results are obtained by mixing the dry sand and dry bond for at least one minute. This operation distributes the bond evenly throughout the sand. A part of the temper water is then added, the sand mixed for a suitable period of time, the balance of the temper water added, and mixing completed. The total mixing time after the water additions should be approximately as shown in table 7.

TABLE 7. MIXING TIMES USED IN SOME OF THE COMMON TYPES OF MULLER MIXERS

Type of Mixer	Size of Batch, cu ft	Mixing Time for Facing Sand, minutes	Mixing Time for Backing Sand, minutes
Clearfield	4	5	3
Mulbaro	3	5	3
Simpson	5½	5	3
Speedmullor	3	1½	1

A mixing time longer than those listed in table 7 does not increase the green strength. This is shown in figure 57. It is good practice to make a series of tests for green strength after various mulling times to determine the time needed to attain the maximum green strength.

Mulling of sand distributes the clay and other binders over the individual sand grains by a kneading and smearing action. Such distribution of the binder is impossible to achieve by manual operation, no matter how thoroughly it is done. In addition to distributing the binder uniformly, the mulled sands require a smaller amount of binder than does a hand-mixed sand. The increased amount of binder required in a hand-mixed sand also results in a lower permeability than in a mulled sand of the same green strength.

Manual Mixing. Situations may arise when mulling of sand is impossible and manual mixing of the sand will have to be done. When such mixing is necessary, it should be done preferably the day before the sand is to be used.

The binder should be added to the sand heap in small amounts in the dry condition and mixed thoroughly after each addition. After the binder has been added and dry mixing completed, the temper water should be added a little at a time with a sprinkling can while the sand is being mixed. On completion of the mixing operation, the sand should be passed through a three or four-mesh riddle and permitted to stand (or temper) for at least a few hours. Preferably, a hand-mixed sand should be covered with wet burlap bags and permitted to stand overnight.

ALL-PURPOSE SAND

The all-purpose sand that is used in Navy foundries is a "compounded" or "synthetic" sand that has been developed by the Naval Research Laboratory. A wide range of properties can be attained in the molding sand with a minimum of bonding materials such as bentonite, cornstarch, and dextrine. Sand properties for an all-purpose sand having an AFS Fineness Number of 63 are discussed in the following section. Properties of sands having higher or lower AFS Fineness Numbers (finer or coarser sands) will generally vary as described in the section on Sand Properties. (See figures 45, 46, 50, and 51.)

PROPERTIES OF A 63 AFS FINENESS NUMBER SAND

The principal properties (green strength, permeability, and dry strength) of a 63 AFS Fineness sand are shown in figures 58 and 59. This graphical method of presenting the information is used so that the interrelation of the various properties can be easily seen.

The relationships between green compressive strength, moisture content, bentonite content, and permeability are shown in figure 58. The green strength of the sand increases with increased amounts of bentonite. Notice that for each bentonite content, there is a rapid increase in green strength with the first additions of moisture, and then a gradual decrease in green strength as the moisture content is increased. The broken lines in figure 58 show the various permeabilities that are obtained for the various bentonite and moisture contents.

The relationships between green compressive strength, moisture content, bentonite content, and dry strength are shown in figure 59. In figure 59, the broken lines show the dry strengths that are obtained with the various bentonite and moisture contents. Figures 58 and 59 provide information on the direction in which changes can be made to correct the sand properties, and also give information on the particular combinations of binder and moisture to use in a new mix to obtain certain desired properties.

As an example, assume that a sand was prepared with 4 percent bentonite and 4 percent moisture, and that it had a green compressive strength of 4.5 p.s.i. and a permeability of 95. Assume that this sand is found to be unsatisfactory because the green strength is too low, and it is desired to increase the green strength without changing the permeability. Reference to figure 58 shows that this change could be made by increasing the bentonite content to 5 percent

- 43 -

and reducing the moisture content to 3 percent. This new combination of bentonite and moisture contents would provide a sand that has a green strength of 7 p. s. i., with the permeability still at 95. From figure 59, it can be seen that this change in moisture and binder contents will probably cause a decrease in dry strength of only 10 p. s. i., reducing the dry strength from 110 to 100 p. s. i.

As a second example, assume that a sand was prepared with 4 percent bentonite and 4.5 percent moisture. This sand would probably have a green compressive strength of 4.5 p. s. i., permeability of 90, and a dry strength of 120 p. s. i. Assume that this sand is found to cause difficulties in shake-out or to cause hot tearing in the casting. This would indicate that the dry strength might possibly be too high. References to figure 59 shows that by keeping the bentonite content at 4 percent, but decreasing the moisture content to 3 percent, the dry strength will be decreased to approximately 90 p. s. i. This change in moisture would produce only a small increase in the green strength from 4.5 to 5 p. s. i., and increase the permeability from 85 to 105.

When referring to these figures, it must be remembered that only bentonite was considered as a binder. Other materials can also be added as binders to improve green strength or dry strength. The effects of these other binders were discussed in the section on binders.

Another word of caution on figures 58 and 59. These figures should not be used as an indication of properties for all other sands that may have a similar fineness number and the same type of binder. Figures 58 and 59 were based on information obtained from a particular sand, and are used here mainly to show a method of presenting sand-property information in a condensed and usable form.

The green compressive strength of sands of the various grain class numbers that are to be used in shipboard foundries will vary generally as shown in figure 60. This figure should not be taken to mean that there is a sharp separation between the properties of the different classes of sands. There will be some overlapping of the indicated areas because of differences in sand-grain distributions within sands having the same fineness numbers.

It is recommended that a test series (such as that required to produce the information for figures 58 and 59) be made on each new shipment of sand before it is used in the foundry. Conducting such a series of tests and putting the information in graphical form would be a useful and informative way of conducting shipboard instruction periods. The information is developed by making a series of sand mixtures having different bentonite (or other binder) contents and different moisture contents. As an example, a series of 2 percent bentonite sand mixes with 1/2, 1, 2, 3, 4, 5, and 6 percent moisture can be tested for green strength, permeability, and dry strength. A second series of sand mixes containing 3 percent of bentonite and the same moisture contents can be tested to obtain the same properties. This procedure is then repeated for the remaining bentonite contents. The final information is then plotted to produce graphs similar to those shown in figures 58 and 59.

When a new shipment of sand is received aboard ship, a few spot tests can be made to determine how the new lot of sand compares with the previous lot. If the properties are reasonably close, the charts developed for the previous sand may be used for the new sand. However, if there is a significant difference in the physical properties, a complete series of tests should be conducted on the new lot of sand to develop a complete picture of the properties of the new sand.

MOLDING SAND MIXES

Listed in the following tables are various examples of sand mixes that may be used as a starting point in preparing all-purpose sand for use aboard repair ships.

TABLE 8. SAND MIXES FOR GRAY IRON CASTINGS

Sand			Materials, percent by weight					Properties		Casting Weight, lb
Type	Grain Class	Fineness Number	Sand	Bentonite	Cereal	Other	Water	Green Strength p.s.i.	Permeability	
Green	4	70-100	89.4	5.3		5.3 Fireclay	2.8	8.3	110	1-30
	4	70-100	94.0	4.1	0.2	1.7 Sea coal	4.4-5.5	10.2	76	150-800
Skin dried	4	70-100	45.5							
	3	100-140	45.5	3.9	0.6	4.5	3.5-4.0	8.0	70-80	60 and over

- 44 -

TABLE 9. SAND MIXES FOR STEEL CASTINGS

Sand			Materials, percent by weight					Properties		Casting Weight, lb
Type	Grain Class	Fineness Number	Sand	Bentonite	Cereal	Other	Water	Green Strength, p.s.i.	Perme- ability	
Green Facing Sand	5	50-70	94.0	5.0	1.0		3.0 4.0	7.5 - 9.0	120	to 500
Green Backing Sand	Used heap		97.5	1.8	0.7		2.5- 3.5	5.0 - 7.0	120	to 500
Skin dried	5	50-70	95.5	3.0	1.5		4.0- 4.5	5.5 - 6.5	90-120	100 and over

TABLE 10. SAND MIX FOR ALUMINUM CASTINGS

Sand			Materials, percent by weight					Properties		Casting Weight, lb
Type	Grain Class	Fineness Number	Sand	Bentonite	Cereal	Other	Water	Green Strength, p.s.i.	Perme- ability	
Green	4	70-100	95.0 97.0	5.0 3.0			5.0- 5.5	5.0-10.0	50-100	to 200

TABLE 11. SAND MIXES FOR COPPER-BASE ALLOYS

Sand			Materials, percent by weight					Properties		Casting Weight, lb
Type	Grain Class	Fineness Number	Sand	Bentonite	Cereal	Other	Water	Green Strength, p.s.i.	Perme- ability	
Green	4	70-100	95.0	4.0	1.0		4.0	6.0 - 7.0	60-70	to 2000
	3	100-140 Used heap	20.0 75.0	5.0			4.0	7.0–12.0	30-50	to 2000
	4	70-100	80.0	4.0	1.0	15.0 Silica Flour	5.5	7.0-12.0	40-80	Special purpose

The sand mixes given in the preceding tables are given only as a guide. The properties obtained with the all-purpose sands aboard ship will probably vary somewhat from those listed.

CORES

Cores used aboard repair ships are usually baked sand cores. Other types (such as green sand cores) have limited use and are not discussed here. Baked sand cores should have the following properties:

1. Hold their shape before and during the baking period.

2. Bake rapidly and thoroughly.

3. Produce as little gas as possible when molten metal comes in contact with the core.

4. Be sufficiently permeable to permit the easy escape of gases formed during pouring.

5. Have hardness sufficient to resist the eroding action of flowing molten metal.

6. Have surface properties which will prevent metal penetration.

7. Be resistant to the heat contained in the metal at its pouring temperature.

8. Have hot strength that is sufficient to withstand the weight of the molten metal at the pouring temperature and during the beginning stages of solidification.

9. Have good collapsibility so the core won't cause hot tears or cracks in the casting.

10. Absorb a minimum amount of moisture if the mold is required to stand a considerable period of time before pouring. This also holds true if storage of cores is necessary.

11. Retain its strength properties during storage and withstand breakage during handling.

CORE PROPERTIES

In addition to the special properties listed in the preceding section, the properties discussed for molding sands in the section "Sand Properties," also apply to core sands.

There are three major factors which influence the properties of cores. They are (1) baking time and temperature, (2) type of core binder, and (3) collapsibility.

Baking Time and Temperature. The best combination of baking time and temperature varies with: (1) the type of binder used, (2) the ratio of oil to sand, and (3) the type of core ovens used. Figure 61 shows the dependence of the baked strength on baking time and temperature. It will be noted that the same strength was attained in one hour when baking at 450°F., as was attained in six hours at a baking temperature of 300°F. It is always good practice to make a series of tests on the effect of baking time and temperature on the baked strength of cores before using a new core mix. Such information will provide the shortest baking time to obtain a given strength for that mix. This type of investigation will also provide information on the baking characteristics of a core oven.

In the baking of oil sand cores, two things occur. First, the moisture is driven off. Following this, the temperature rises, causing drying and partial oxidation of the oil. In this way, the strength of the core is developed.

For proper baking of oil-sand cores, a uniform temperature is desired. This temperature should not be over 500°F. nor under 375°F. If linseed-oil cores are baked at a moderate temperature of 375°F. or 400°F., they will be quite strong. If the same cores are baked quickly at 500°F., they will be much weaker. Continuing the baking of the cores to the point where the bonding material decomposes must be avoided, as this causes the cores to lose strength.

The size of the core must be considered in drying. The outer surface of a core will bake readily and will be the first part to develop maximum strength. If the temperature is maintained, the inside will continue to bake until it finally reaches maximum strength, but by that time, the outer surface of a large core may be overbaked and low in strength. The tendency for this to happen in large cores can be partly overcome by filling the center of the core with highly permeable material with a low moisture and bond content, by the use of well-perforated core plates, and by using low baking temperatures. It is not only a matter of heating the center of the core, but also of supplying it with oxygen. Thus, there is need for free circulation of air around and through the core while baking.

The most skillful and careful preparation of metal and mold can easily be canceled by careless technique, and the necessity for proper baking cannot be overemphasized. If cores are not properly baked, the following is likely to happen to the casting:

1. Excessive stress, possibly cracks, caused by the core continuing to bake from the heat of the metal, thus increasing in strength at the time the metal is freezing and contracting.

2. Unsoundness caused by core gases not baked out.

3. Entrapped dirt due to eroded or spalled sand caused by low strength in the core.

When overbaked, the loss of strength of the core results in excessive breakage in handling or during casting, and cutting or eroding of the core surface.

To establish a full appreciation of the problems of drying cores, a series of 3-inch, 5-inch, and 8-inch cube cores should be made without rods and baked at temperatures of 400°F., 425°F., 450°F., 475°F., and 500°F. for varying predetermined times. After being taken out of the oven and cooled, they should be cut open with a saw to determine the extent to which they are baked. This simple test will aid in determining the proper times and temperatures to use for various cores in a given oven and under given atmospheric conditions.

Practice is necessary to determine accurately when a core is baked properly. A practical method is to observe the color of the core. When it has turned a uniform nut brown, it is usually properly baked. A lighter color indicates insufficient baking, and a darker color indicates overbaking.

Type of Core Binder. The type of core binder is important from the viewpoint of gas-generating properties as well as the strength the binder will develop. Figure 62 shows the volume of core gas generated from a linseed-oil compound and an oil-pitch mixture. The volume generated by both is the same for the first minute, but then the generation of gas from linseed oil decreases rapidly, while the generation of gas by the oil-pitch mixture decreases at a much lower rate. A core oil having gas-generating characteristics similar to those of linseed oil is preferred, since core gas is generated for a much shorter length of time and the possibility of defects due to core gas is lessened.

Combinations of several binders can be used to obtain a better overall combination of green strength, baked strength, and hot strength than can be obtained with the individual binders. For example, in a sand mixture containing core oil and cereal binder, the cereal binder contributes most of the green strength, whereas, the core oil contributes most of the baked strength. This is the reason for using combinations of cereal binder with oil binders. The strength attained from a cereal-oil combination is shown in figure 63. Notice that the strength obtained by the combination is higher than the total strengths of the individual binders.

Collapsibility. The sand-testing equipment used aboard ship does not permit the high-temperature testing of cores for collapsibility. A rule-of-thumb practice must be followed in determining this property. Close observation must be made in shaking out a casting to determine if the core mix had good collapsibility. A core that is still very hard during shakeout is said to lack collapsibility. If a crack should later be observed in the cored area of the casting, the core sand mixture definitely is too strong at high temperatures and the sand mix should be corrected. One remedy is to add about 2 percent of wood flour to the mixture.

CORE SAND MATERIALS

Standard Materials. Core sand mixtures are made from clean, dry silica sands and various binders. The fineness of the sand is determined by the size of the core and the metal being poured. One important point in the mixing of core sand mixtures is to have the sand dry before any materials are added.

The materials used for binders are primarily corn flour, dextrine, raw linseed oil, and commercial core oils. Corn flour and dextrine are cereal binders. Dextrine greatly increases the strength of baked cores and is used in small amounts with other binders. Dextrine-bonded cores have the disadvantage that they absorb moisture very easily and, therefore, should not be stored for any length of time before being used. Corn flour is used to give the core green strength and hold it together until it is baked. The cereal binders are used in combination with core oil to produce the desired strength. They are rarely used by themselves.

Cereal binders have the following advantages that make them very useful binder materials: (1) good green strength, (2) good dry bond, (3) effective in angular sand, (4) core oil is not absorbed as in naturally bonded sands, (5) quick drying, and (6) fast and complete burn out. Core oils are used to provide a hard strong core after baking. They have the following advantages over other types of binders: (1) ability to coat the individual sand grains evenly with a reasonable amount of mixing, (2) generate a small amount of smoke and gas, (3) work clean in the core boxes, and (4) give cores good strength.

Substitute Materials. Aboard repair ships, the situation may arise where the standard core materials are unavailable. In such cases, substitute materials must be used. Substitute materials should be used only as an emergency measure. Molasses and pitch are two materials which can be obtained easily for use as core materials. Molasses should be mixed with water to form a thin solution known as "molasses water." In this condition, it is added to the core mix as part of the temper water during the mulling operation. Pitch is seldom used alone. Used with dextrine, it imparts good strength to a core mix. Sea coal in small amounts is used with pitch to prevent the pitch from rehardening after it has cooled from the high temperatures caused by the molten metal.

If new washed silica sand is not available, reclaimed backing sand may be used for facing, if properly bonded. Some beach or dune sands, relatively free from crustaceous matter and feldspar, some fine building sands, and some natural sand deposits containing clay may be used. If bentonite is not available, portland cement, fireclay, or some natural clays may be used. The corn flour may be replaced with ordinary wheat flour. Sugar or molasses will take the place of dextrine. Wherever substitutes must be used, the amount of organic materials and clay should be kept to a minimum and the amount of good clean sand grains to a maximum.

Other Core Materials. Silica flour and wood flour are added to core mixtures to get special properties. Silica flour is usually added to prevent metal penetration and erosion of cores by molten metal. Silica flour must be used carefully and not used in excess. Excessive use may lead to hot tears because of too high a hot strength. Wood flour is not a binder but a filler material. Its use is that of softening or weakening a core so that it has better collapsibility.

MIXING

Core sands should be mixed in a muller or some other type of mechanical mixer to obtain the maximum properties from the various binders. Many of the binders are added in very small amounts, and only a thorough mixing operation can distribute the binder uniformly throughout the sand. Manual mixing with a shovel requires the addition of much more binder to obtain the desired properties, and results are not consistent. Manual mixing of core sands is to be discouraged.

Mulling time has the same effect on core sand as on molding sand, as is shown in figure 57. The proper mulling time should be determined for each mix used. In the mixing of core sands, the additions are made in the following order with the mixer running: (1) sand, (2) dry ingredients, (3) run the mixture dry for a short time, (4) add liquids, and (5) continue to mix for the desired period of time. Laboratory tests have shown that if the core oil is added to the sand before the water and mixed for a short period of time, more consistent core properties will be obtained. If cereal binder is used, the batch should not be mixed too long before adding the liquids. Excessive mixing of the sand with cereal binders without the liquids will cause the batch to become sticky, and a longer length of time will be needed to bring the core mix to its proper condition.

CORE SAND MIXES

The following tables suggest various representative core mixes. They are given primarily as a guide to obtain good core mixes for work aboard repair ships.

TABLE 13. CORE SAND MIXES FOR GRAY IRON CASTINGS

Sand			Materials, percent by weight								Use in Castings
Type	Grain Class	Fineness Number	Sand	Bentonite	Core Oil	Cereal	Silica Flour	Molasses Water(3:1)	Other	Water	
New	4	70-100	98.0		1.5	0.5				5.0	General castings

TABLE 14. CORE SAND MIXES FOR STEEL CASTINGS

Sand			Materials, percent by weight								Use in Castings
Type	Grain Class	Fineness Number	Sand	Bentonite	Core Oil	Cereal	Silica Flour	Molasses Water(3:1)	Other	Water	
New	5	50-70	88.0	0.5	1.0	0.5	10.0			5.0	100 to 1000 lb
New	4	70-100	98.0		1.5	0.5				5.0	General small castings

TABLE 15. CORE SAND MIXES FOR ALUMINUM CASTINGS

Sand			Materials, percent by weight								Use in Castings
Type	Grain Class	Fineness Number	Sand	Bentonite	Core Oil	Cereal	Silica Flour	Molasses Water(3:1)	Other	Water	
New	4	70-100	98.5	0.2	1.0	0.3				5.0	General castings
New	4	70-100	99.0		0.5	0.5				5.0	Thin sections

TABLE 16. CORE SAND MIXES FOR COPPER-BASE ALLOYS

Sand			Materials, percent by weight								Use in Castings
Type	Grain Class	Fineness Number	Sand	Bentonite	Core Oil	Cereal	Silica Flour	Molasses Water(3:1)	Other	Water	
New	5	50-70	95.0		1.5	0.5				4.0	100 lbs and over
New	4	70-100	98.0		1.5	0.5				3.5	General purpose

TABLE 17. CORE SAND MIX FOR COPPER CASTINGS

Sand			Materials, percent by weight								Use in Castings
Type	Grain Class	Fineness Number	Sand	Bentonite	Core Oil	Cereal	Silica Flour	Molasses Water(3:1)	Other	Water	
New	4	70-100	98.0		1.7	0.3				4.0	

CORE PASTE AND FILLER

A very good core paste for use in joining core sections may be made from 3 percent bentonite, 6 percent dextrine, and 91 percent silica flour. The ingredients should be mixed dry, and water added to produce a mixture the consistency of soft putty.

A filler to seal the cracks between parts of the core may be made from 3 percent bentonite, 3 percent dextrine, and 94 percent silica flour. The ingredients are mixed dry, then water is added to make a mixture with the consistency of stiff putty. This material is pressed into the cracks between the core sections to prevent metal penetration.

MOLD AND CORE WASHES

Core and mold washes may be needed in some cases to prevent erosion of the sand and metal penetration into the sand. The following mix contains the same bonding material as the molding sands, with silica flour replacing the sand, and with sodium benzoate added to prevent the mixture from becoming sour.

	Weight percent
Silica flour	64.0
Bentonite	1.5
Dextrine	3.0
Sodium benzoate	0.2
Water	31.3

The dry material should be mixed thoroughly in a closed container. The water is then added, and the mixture stirred thoroughly. The mixture is sprayed onto the green core like paint and then baked, or it may be brushed on the dried core or mold. It must be allowed to dry thoroughly in air or be baked in an oven, and should be used only when absolutely necessary.

In most cases, the green or air-dried sand mixtures will produce excellent casting surfaces without use of the wash.

In brass castings, where erosion and penetration are problems, a core wash made from a silica base is satisfactory. A plumbago wash is useful for bronze castings. A core wash for use with high-lead alloys and phosphor bronzes, may be made from a paste of plumbago and molasses water. Such a treatment should be followed by a thin coating of the regular core wash.

METHODS FOR TESTING SAND

The testing of foundry sands should not be a series of tests for obtaining a great deal of meaningless information. Regular sand testing along with records of the results is the one way of establishing the cause of casting defects due to sand. Regular sand testing results in a day-to-day record of sand properties, and indicates to the molder how the sand properties behave. Proper interpretation of the results of sand tests permits the molder to make corrections to the sand before it is rammed up in molds, thereby not only saving time but also preventing casting losses.

TEST-SPECIMEN PREPARATION

A sample of sand, at least one quart in volume, should be taken from various sections of the sand heap and from a depth of at least six inches. The sand should be riddled through a 1/4 inch mesh riddle or the size of riddle used in the foundry. The same riddle size should be used for all sand tests.

Enough tempered sand is weighed out to make a rammed sample 2 inches high. The proper amount of sand can be determined by trial and error. The sand is then placed in the specimen tube, which rests on the specimen-container pedestal. The tube with the pedestal is then placed under the rammer, as shown in figure 64. Care should be taken to keep the tube upright so as not to lose any of the sand.

The rammer is lowered gently into the specimen tube until the rammer is supported by the sand. The rammer is raised slowly by the cam to the full 2-inch height, and permitted to fall. This is repeated until a total of three rams have been applied. The top of the rammer rod should be between the 1/32 inch tolerance marks for control work. If the end of the rammer rod is not within the tolerance, the specimen must be discarded and a new test specimen made. If the specimen is of the correct height, the rammer rod should be lifted carefully to clear the specimen tube, and the specimen tube removed from the pedestal. The type of rammer supplied for shipboard use is shown in figure 64.

PERMEABILITY

The permeability of foundry sand is determined by measuring the rate of flow of air under a standard pressure through a standard specimen 2 inches high by 2 inches in diameter. The specimen is prepared as described in the previous section, "Test-Specimen Preparation." The equipment for the permeability determination is shown in figure 65. The sand specimen, still in the tube, is placed in the mercury well with the sand specimen in the top position. The air chamber is raised to its proper position, released, and permitted to drop. When the water column in the manometer becomes steady, the permeability scale, which is on the curved part of the indicator, is rotated until the edge of the scale is opposite the top of the water column. The reading on the scale at this point is the permeability for control purposes. It is good practice to take permeability readings on three different specimens from the same lot of sand and to average the readings. (The test, as described, measures green permeability.)

GREEN STRENGTH

Green compressive strength is the property most useful in foundry sand control in repair ship foundries. The specimen is prepared as described under "Test-Specimen Preparation," and then stripped from the tube with the stripping post. The specimen used for permeability test is suitable if not damaged in previous test. The sand specimen is placed between the compression heads on the lower part of the test apparatus shown in figure 66. The face of the sand specimen which was the top face when the specimen was rammed should be placed against the right-hand compression head.

Care should be taken to seat the specimen carefully in the compression head. A small magnetic rider is placed on the scale against the compression head, and the arm is raised by the motor-driven mechanism or by hand. If hand operation is used, care must be taken to maintain a slow and uniform speed of operation, since the rate of motion of the arm affects the test results. When the specimen breaks, the motor automatically reverses itself and returns to its bottom position. With hand operation, the arm is returned manually when the specimen is seen to break. The magnetic rider will remain at the position attained by the arm when the break occurred. The green compressive strength is read from the back of the rider on the appropriate scale. The testing equipment must be maintained in good operating condition at all times, and sand from the broken specimens must be completely removed from the equipment after each test. Pay special attention to keeping grains of sand and dirt out of the bearings. Use only dry lubricants, such as graphite, on sand-testing equipment.

DRY STRENGTH

For test specimen for determining dry strength is prepared as described in the section "Test-Specimen Preparation." After the specimen is made and removed from the stripping post, it should be placed on a flat rigid plate and dried for at least two hours. Drying is done at a temperature between 220°F. and 230°F. The specimen is removed from the oven and cooled to room temperature in a container that will prevent moisture pickup by the dry specimen. The specimen is then tested in the same manner as described for obtaining green strength in the section "Green Strength." The specimen should be tested as soon as it has cooled to room temperature, and should not be permitted to stand for any appreciable length of time before testing.

MOISTURE

The moisture content of molding sands is determined with the apparatus shown in figure 67. A representative 50-gram sample of tempered sand is placed in the special pan, which is then placed in the holder. The timer switch is set for 3 minutes. Setting the timer automatically starts the dryer, which runs for the set time interval. After drying is complete, the pan is removed and weighed. The loss in weight multiplied by two is the percent of moisture in the tempered sand.

CLAY CONTENT

A representative sample is obtained from the sand which is to be tested for clay content. The sand is then thoroughly dried and a 50-gram sample taken. The sample is placed in the jar shown in figure 68 with 475 cc of distilled water and 25 cc of standard sodium hydroxide solution. The standard sodium hydroxide solution is made by dissolving 30 grams of sodium hydroxide in distilled water and diluting to 1000 cc. The jar containing the sand sample and solution is assembled with the stirrer and stirred for five minutes. (The assembled sand-washing equipment is shown in figure 69.) The stirrer is then removed from the jar and any adhering sand washed into the jar. The jar is then filled with distilled water to a depth of six inches from the bottom of the jar. The contents of the jar should be well stirred by hand and then allowed to settle for 10 minutes. The water is then siphoned off to a depth of 1 inch. Distilled water is then added again to a depth of 6 inches, the solution agitated and allowed to settle for 10 minutes a second time. The water is siphoned off a second time to a depth of 1 inch. Water is added a third time, the solution agitated and permitted to settle for a 5-minute period, after which the water is siphoned off again. Distilled water is added to a depth of 6 inches, the solution agitated, permitted to settle for 5 minutes, siphoned off to a depth of 1 inch, and the procedure repeated until the solution is clear after the 5-minute settling period. The glass cylinder is then removed from the base of the jar so as to leave the sand in the base. The sand is dried thoroughly in the base. The dry sand is weighed. The weight lost multiplied by two is the percent of AFS clay in the sand.

If clay determinations are made on used sand, the result will not be a true clay content, since sea coal and other additives will be removed along with the clay. The determination would then give a false value.

GRAIN FINENESS

Grain fineness is expressed as the grain fineness number and is used to represent the average grain size of a sand. The number is useful in comparing sands. Grain fineness numbers, however, do not tell the molder the distribution of grain sizes, and the distribution does affect the permeability and potential strength of the sands. Two sands may have the same grain fineness number and still differ widely in permeability, due to differences in their grain-size distribution. Clay content and the shape of sand grains also influence the sand properties, and may differ in sands having the same grain fineness number.

The sample for determining the grain fineness number should be washed of all clay as described under "Clay Content," and thoroughly dried. A 50-gram sample of the sand is then screened through a series of standard sieves. The sand remaining on each screen should be carefully weighed and recorded. The grain fineness number is then calculated as shown in table 18.

TABLE 18. CALCULATION OF GRAIN FINENESS NUMBER

Size of Sample: 50 grams
Clay Content: 5.9 grams - 11.8 percent
Sand Grains: 44.1 grams - 88.2 percent

Screen	Amount Retained on Screen		Multiplier	Product
	Grams	Percent		
6	None	0.0	3	0
12	None	0.0	5	0
20	None	0.0	10	0
30	None	0.0	20	0
40	0.20	0.4	30	12
50	0.65	1.3	40	52
70	1.20	2.4	50	120
100	2.25	4.5	70	315
140	8.55	17.1	100	1710
200	11.05	22.1	140	3094
270	10.90	21.8	200	4360
Pan	9.30	18.6	300	5580
Total	44.10	88.2		15243

$$\text{Grain Fineness Number} = \frac{\text{Total Product}}{\text{Total Percent of Retained Grain}} = \frac{15243}{88.2} = 173$$

A better method for comparing sands is to compare them by the actual amounts retained on each screen. A method for plotting this type of information is shown in figure 70. Two sands have been plotted for grain distribution. Notice that although both sands have the same grain fineness number, the size distribution of the grains is different.

SUMMARY

The need for proper sand control through the use of sand-test equipment cannot be stressed too strongly. There is only one way to determine the properties of molding sands and core sands, and that is to make tests. Day-by-day testing of foundry sands provides the molder with information which enables him to keep the molding sand in proper condition. The recording of these test results, along with appropriate comments as to the type of castings made and any defects which may occur, can help the molder to determine the causes of casting defects, and point the way toward corrective measures.

As a summary to the chapter on foundry sand, the various factors affecting sand are tabulated below with the results produced in the sand.

Molding Sand

Factor	Variation	Effect
Grain fineness	Sand too fine	Permeability reduced, green strength increased. Possible defects: blisters, pinholes, blowholes, misruns, and scabs.
	Sand too coarse	Permeability increased, green strength decreased. Possible defects: rough casting surface and metal penetration.
Binder	Too much binder	Accompanied by too little moisture, results in decreasing permeability, increasing green strength. Possible defects: hot cracks, tears, and scabs.
	Too little binder	Low green strength and high permeability. Possible defects: drops, cuts, washes, dirt, and stickers.
Moisture content	Too high	Permeability and green strength decreased. Possible defects: blows, scabs, cuts, washes, pin holes, rat tails, and metal penetration.
	Too low	Permeability and green strength too low. Possible defects: drops, cuts, washes, and dirty castings.

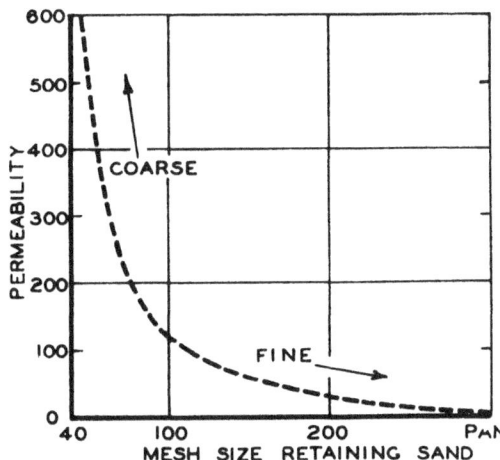

Figure 45. Permeability as affected by the grain size of sand.

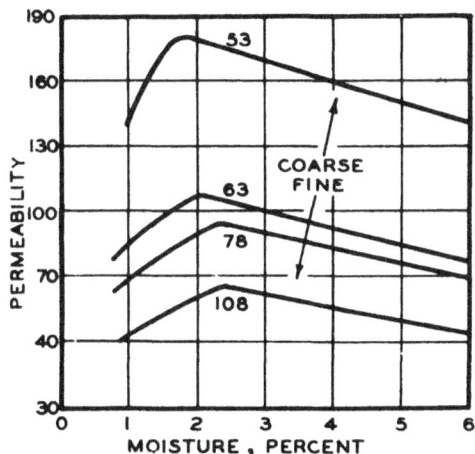

Figure 46. Permeability as affected by sand fineness and moisture.

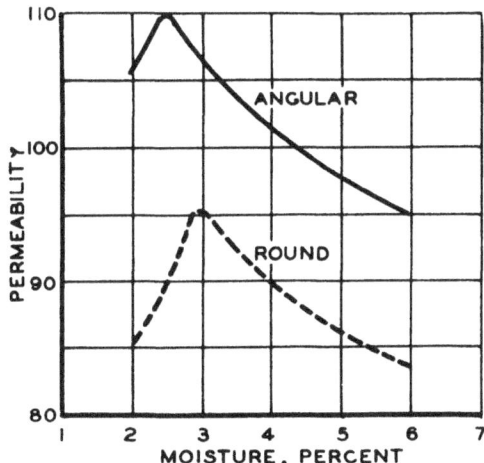

Figure 47. The effect of sand grain shape on permeability.

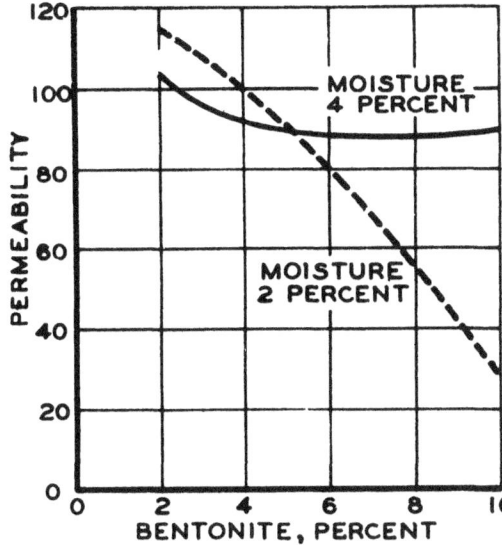

Figure 48. Permeability as affected by the amount of binder.

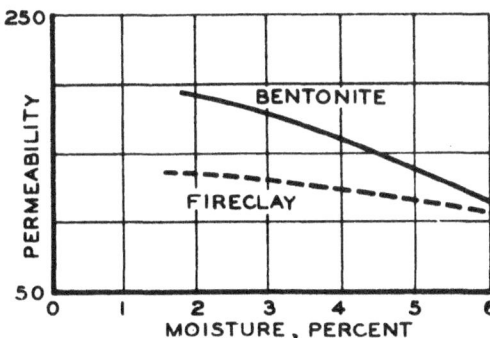

Figure 49. The effect of bentonite and fireclay on permeability.

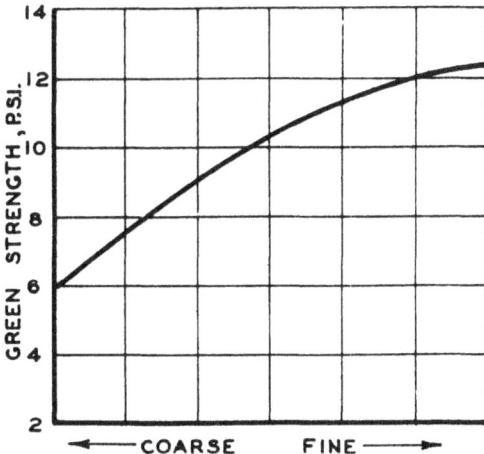

Figure 50. Green strength as affected by the fineness of sand.

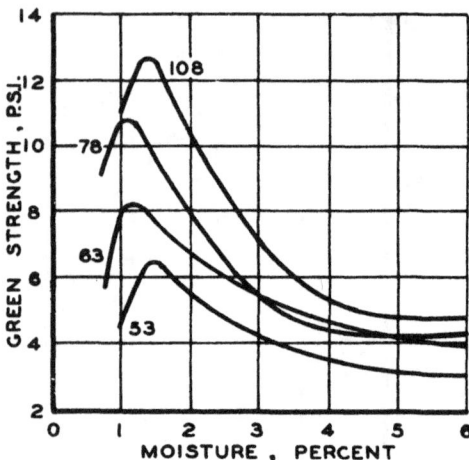

Figure 51. Green strengths of sands with varying fineness numbers.

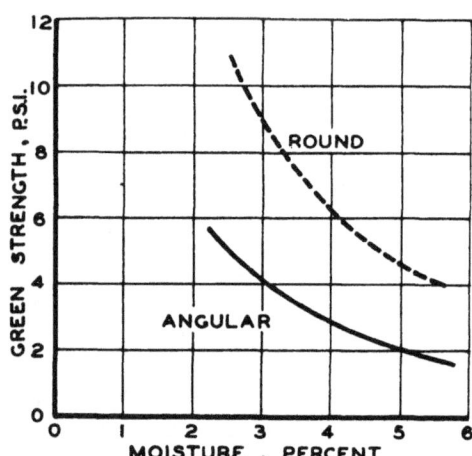

Figure 52. Green strength as affected by the shape of sand grains.

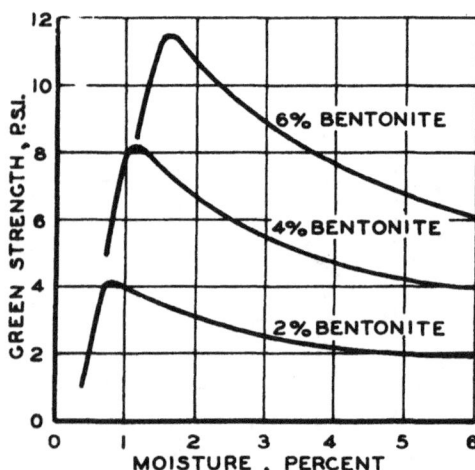

Figure 53. Green strength as affected by moisture and varying bentonite contents.

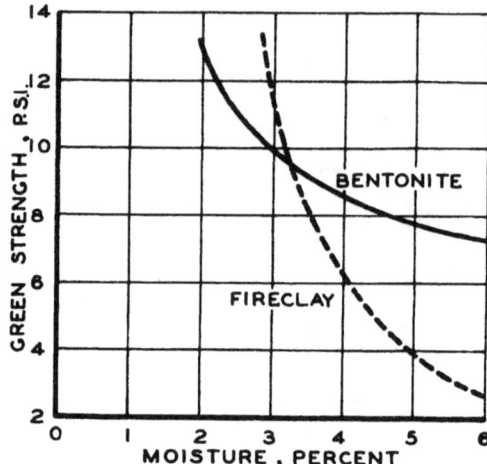

Figure 54. The effect of bentonite and fireclay on green strength of foundry sand.

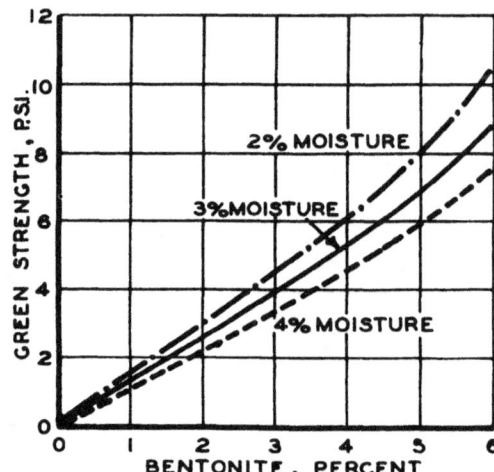

Figure 55. The effect of bentonite on sands with various moisture contents.

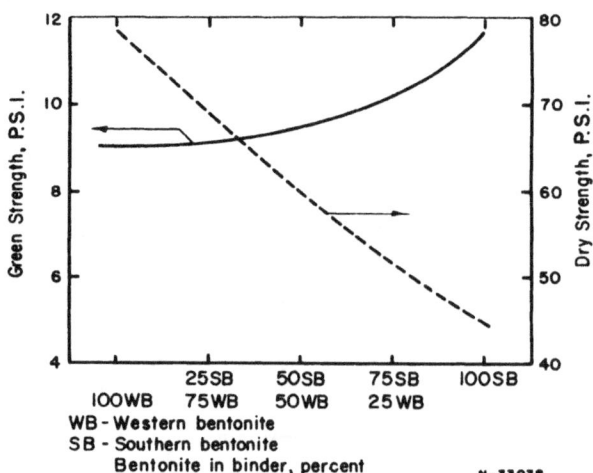

Figure 56. The effect of western and southern bentonite on green strength and dry strength.

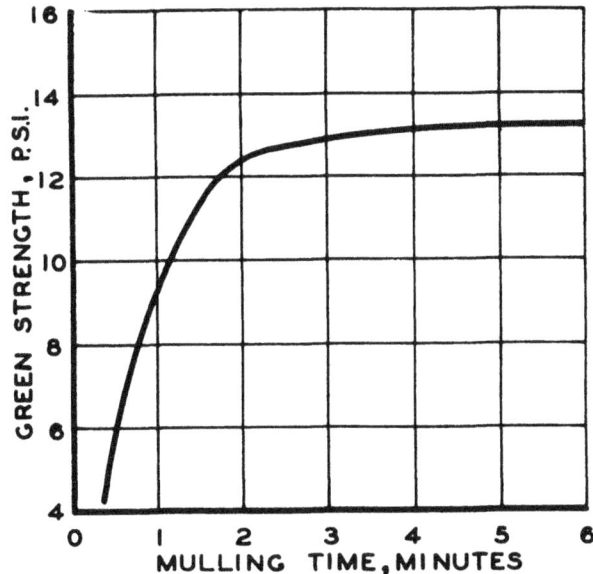

Figure 57. Green strength as affected by mulling time.

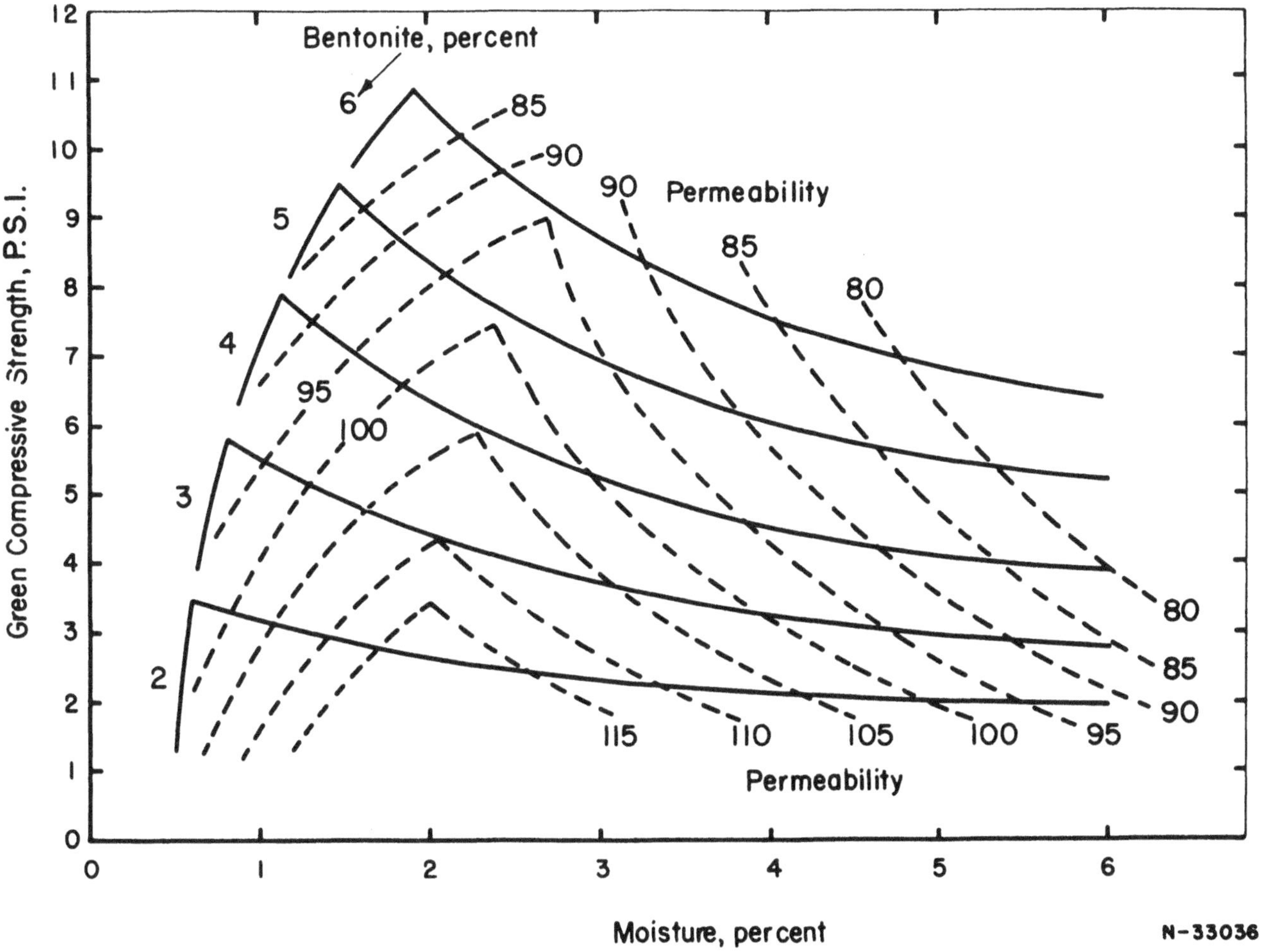

Figure 58. Relationship between moisture content bentonite content, green compressive strength, and permeability for an all-purpose sand of 63 AFS fineness number.

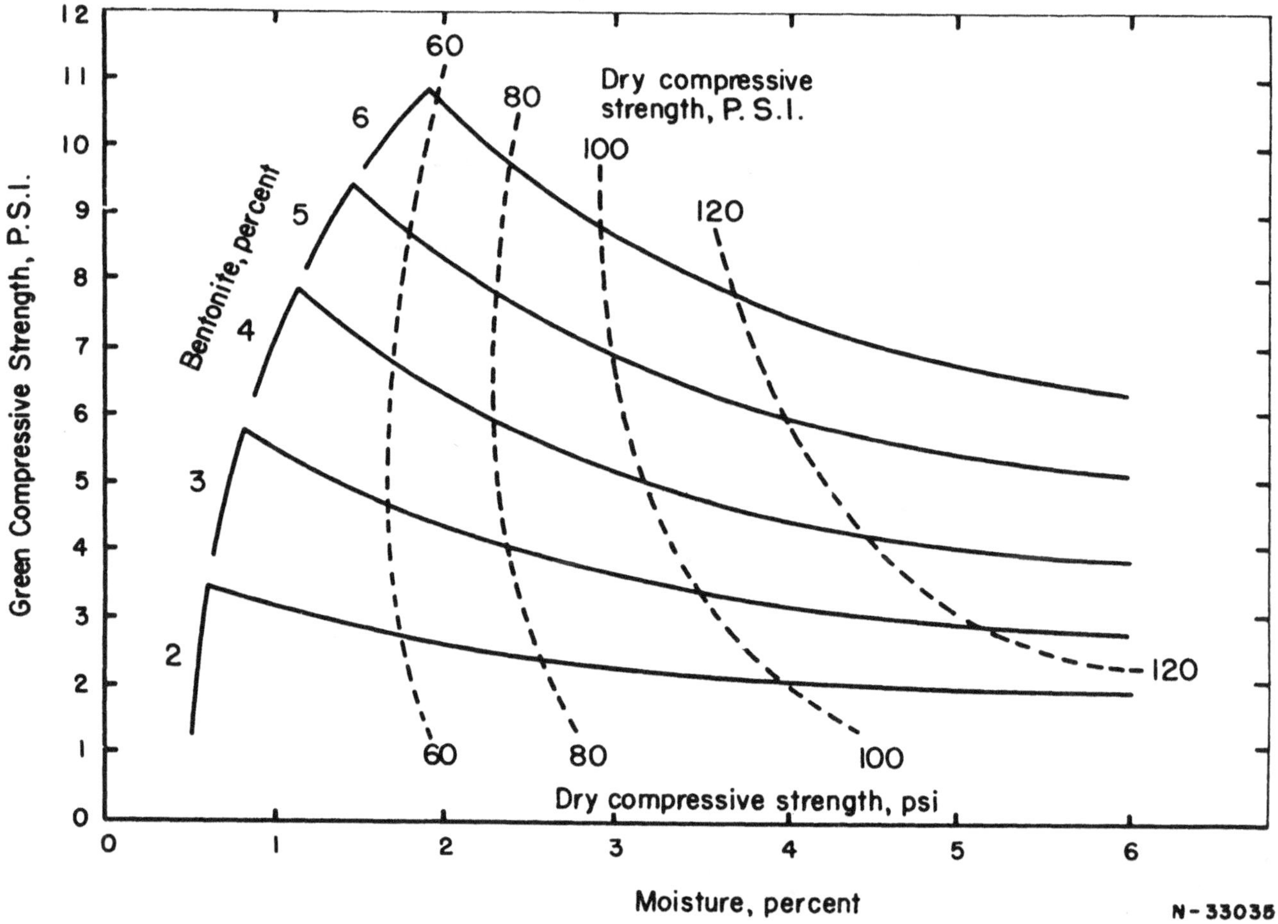

Figure 59. Relationship between moisture content, bentonite content, green compressive strength, and dry strength for an all-purpose sand of 63 AFS fineness number.

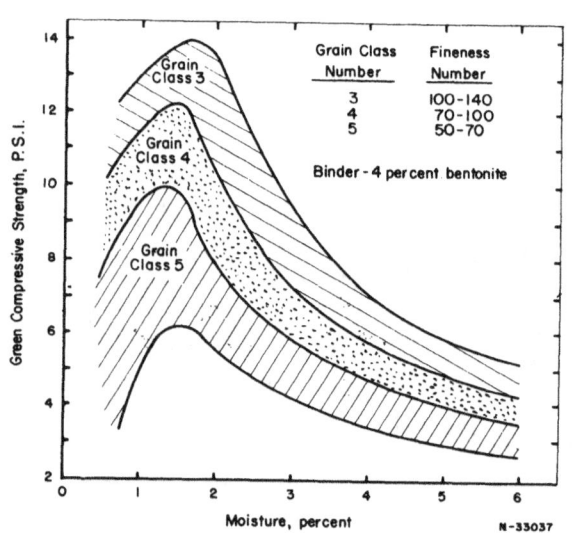

Figure 60. General green compressive strengths for sands of different grain class numbers.

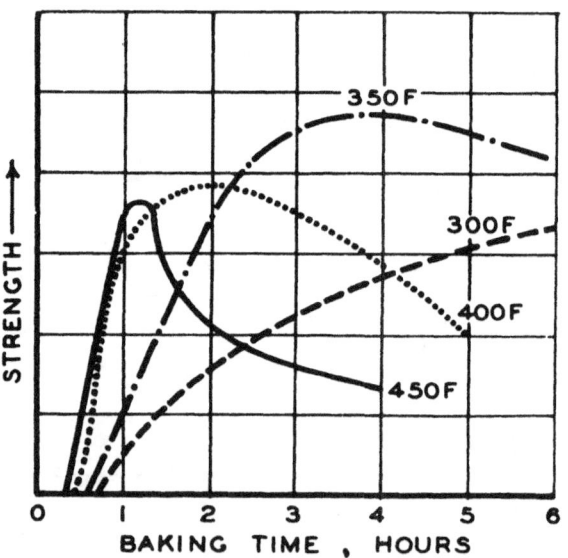

Figure 61. Strength of baked cores as affected by baking time and baking temperatures.

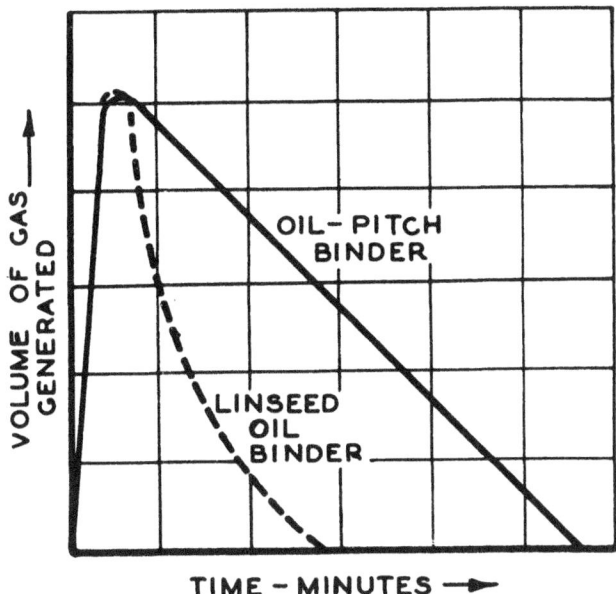

Figure 62. Core gas generated by two different core binders.

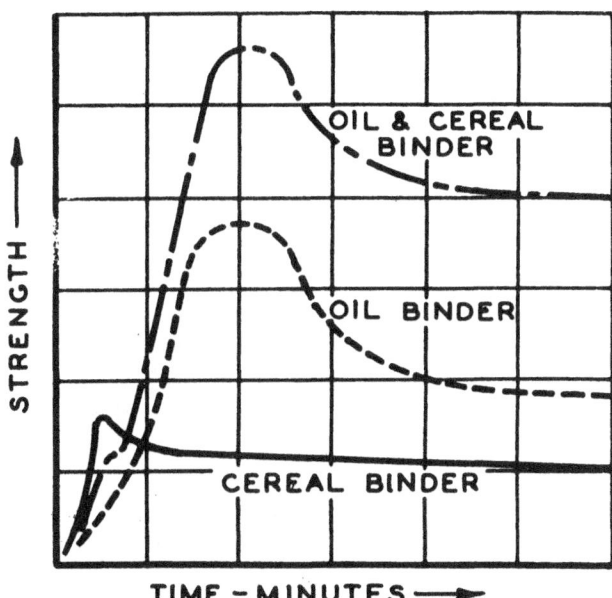

Figure 63. The effect of single binders and combined binders on the baked strength of cores.

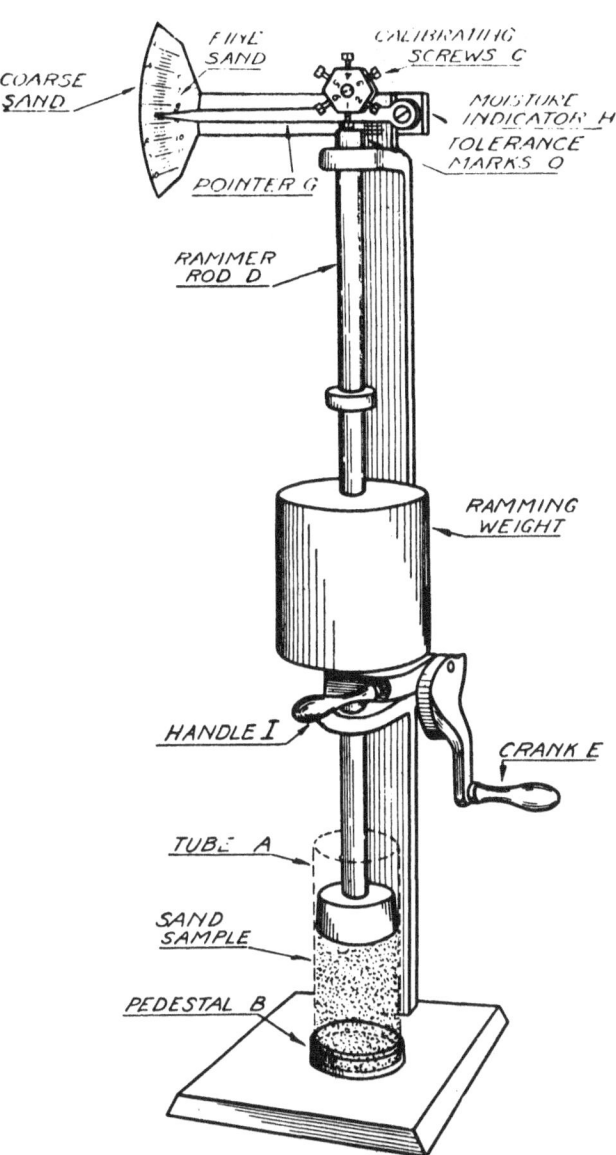

Figure 64. Rammer used for test specimen preparation.

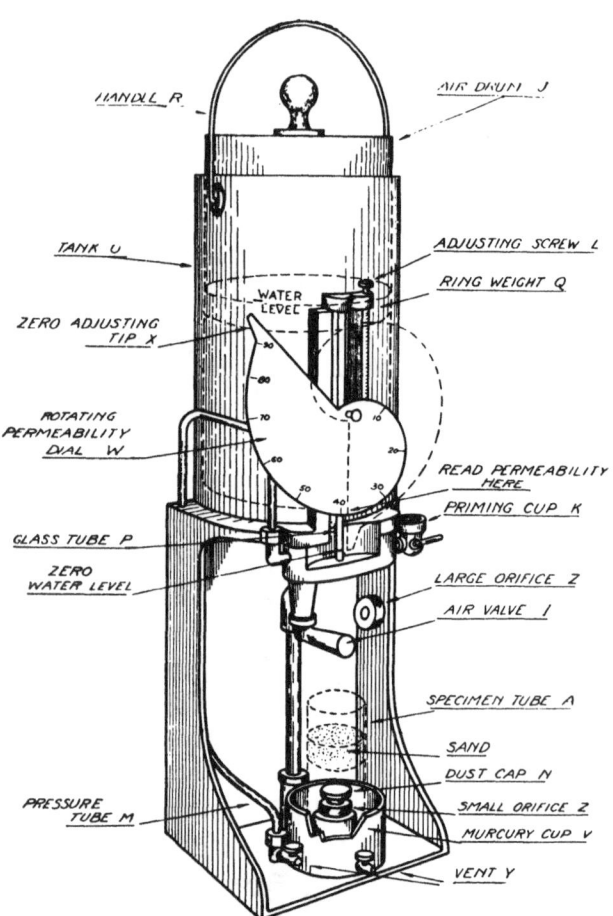

Figure 65. Permeability test equipment.

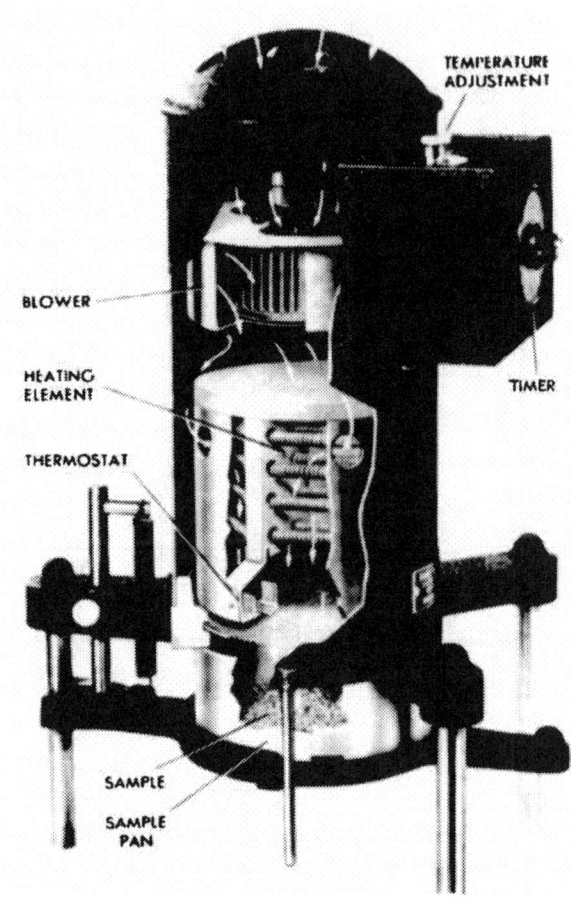

Figure 67. Equipment for drying sand specimens for moisture determination.

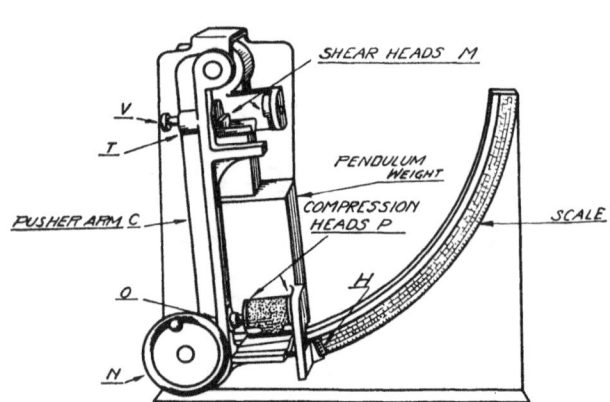

Figure 66. Strength testing equipment.

Figure 68. Jar and stirrer for washing sand.

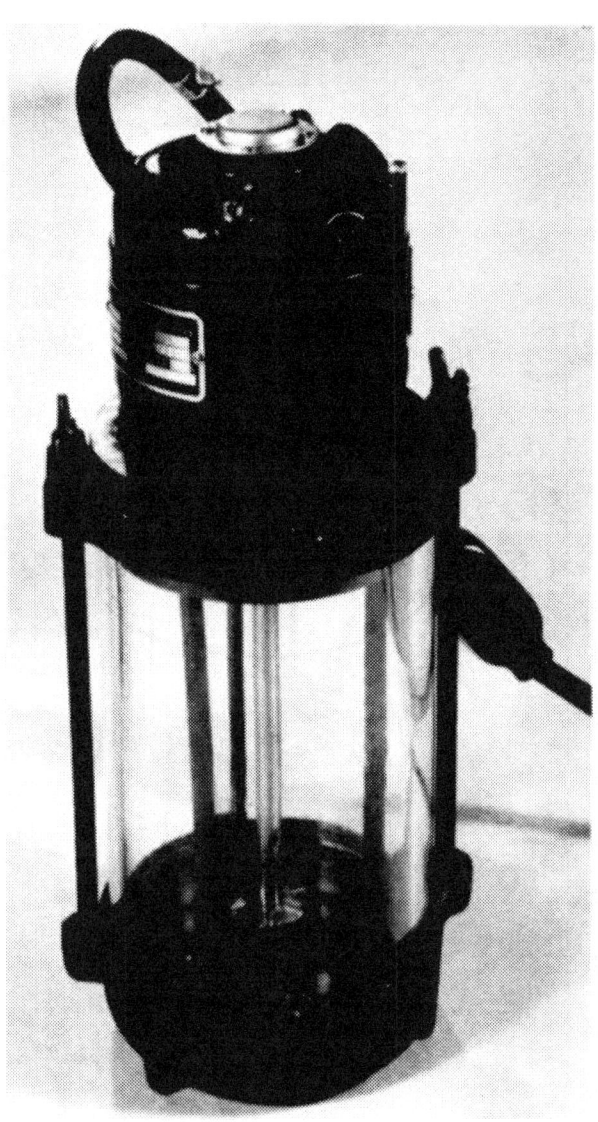

Figure 69. Sand washing equipment assembled.

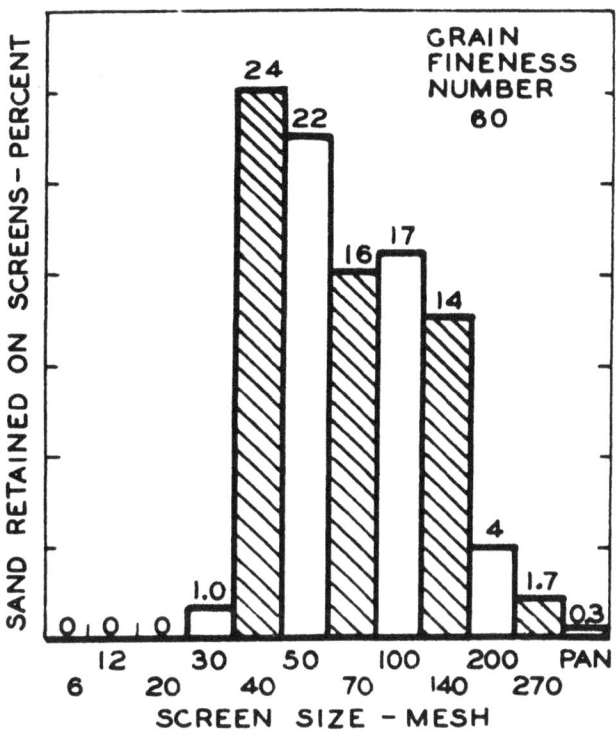

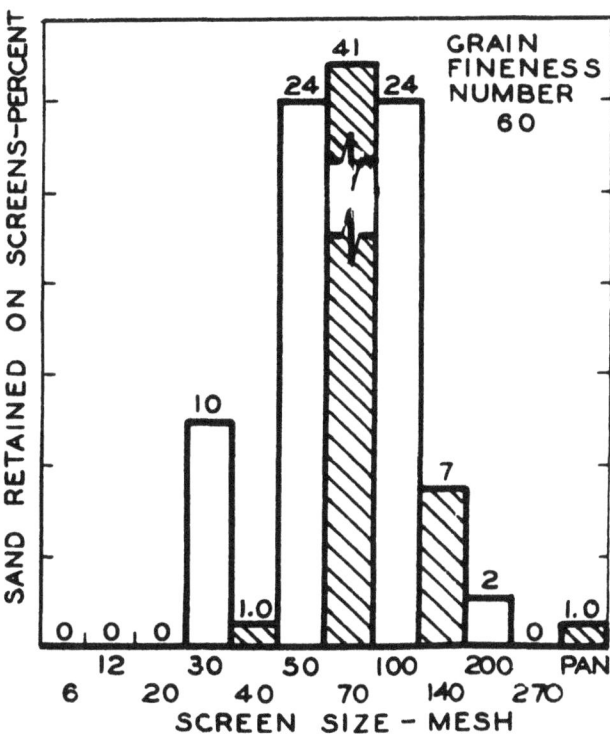

Figure 70. The difference in sand grain distribution for two foundry sands having the same grain-fineness number.

Chapter V
MAKING MOLDS

Castings are made by pouring molten metal into refractory molds and allowing the metal to solidify. The solidified metal will retain the shape of the mold cavity and can be removed from the mold when the metal is solid. A mold is made by shaping a suitable sand mixture around a pattern of the desired form. A metal or wood box (flask) is used to retain the sand. The pattern is then removed from the sand, leaving a cavity in the sand into which the molten metal can be poured.

The molder's skill is the basic skill of the foundry. He must know how to prepare molds with the following characteristics:

1. Strong enough to hold the weight of the metal.

2. Resistant to the cutting action of the rapidly moving metal during pouring.

3. Generate a minimum amount of gas when filled with molten metal.

4. Constructed so that any gases formed can pass through the body of the mold itself rather than penetrate the metal.

5. Refractory enough to withstand the high temperature of the metal, so it will strip away cleanly from the casting after cooling.

6. Collapsible enough to permit the casting to contract after solidification.

The refractory material normally used by foundries is silica sand bonded with clay. The material usually provided for the variety of castings made aboard repair ships is a washed and graded silica sand mixed with clay and cereal bond as described in Chapter 4, "Sands for Molds and Cores."

MOLDING TOOLS AND ACCESSORIES

The basic molding tools and accessories used by the molder and coremaker are described below and shown in figures 71 and 72.

FLASKS

Flasks are wood or metal frames in which the mold is made. They must be rigid so that distortion does not take place during ramming of the mold or during handling. They must also resist the pressure of the molten metal during casting. The pins and fittings should be continually checked for wear and misalignment to avoid mismatched or shifted molds.

The use of steel flasks is preferred, but cases will arise requiring a size of flask not available. Under such circumstances, a flask may be constructed of wood. It should be husky enough to stand wear and tear. If it is planned to use the flask for several molds, allowance should be made for some burning of the wood, which will often occur when the metal is poured.

A flask is made of two principal parts, the cope (top section) and the drag (bottom section). When more than two sections of a flask are necessary, either because of the size or design of the casting, intermediate flask sections, known as cheeks, are used.

HAND TOOLS

RIDDLES are used for sifting the sand over the surfaces of the pattern when starting a mold. The size of the riddle is given by the number of meshes to the inch. A No. 8 riddle has eight meshes per inch, a No. 4 riddle has four meshes per inch, etc. The particular riddle used depends on the kind and character of casting to be made; castings with fine surface detail require finer sand and a finer riddle.

RAMMERS are used for tamping the sand around the pattern in the flask. For the heavier class of molding, they are made of iron. Sometimes they are made with a wooden handle with a cast iron butt on one end and a cast iron peen on the other. The small rammers used in bench work are usually made of maple, although sometimes they are made of cast iron or aluminum.

STRIKES are used to scrape the extra sand from the top of the cope or drag after ramming. They are usually a thin strip of metal or wood. They should have one straight edge and should be light but sturdy.

CLAMPS are used for holding together the cope and drag of the completed mold or for clamping together the mold-board and the bottom-board on either side of the drag when the latter is rolled over. They are of many styles and sizes. Some are adjustable and are tightened on the flask by means of a lever. Other types use wedges to secure them on the flask. The WEDGES are usually of soft wood, but for the heavier work are either of hard wood or iron.

BELLOWS are used to blow excess parting materials from the pattern and also to blow loose sand and dirt from the mold cavity. Compressed air hoses have almost replaced bellows for this purpose.

TROWELS are of many different styles and sizes to suit the individual taste of the molder and the particular requirements of the job. The trowel is used for making joints and for finishing, smoothing, and slicking the flat surfaces of the mold.

VENTS - Thin, rigid steel strips are used for making vents. Hacksaw blades are suitable for this purpose. Rods are also used for vents but they often cause a shrinkage depression at their base on the casting.

BOSHES or SWABS are made of hemp, tasselled to a point at one end and bound with twine at the other to hold it together. They are used for placing a small amount of water on the sand around the edge of the pattern before the pattern is rapped for drawing from the mold. Boshes will hold considerable water and the amount which they deliver to the sand can be regulated by the pressure the molder applies when squeezing them. Boshes are also used to apply wet blacking to dry-sand molds when they are to be blacked before the mold is dried.

SOFT BRUSHES are used to brush the pattern and the joint of the mold. The hard brush is used to spread beeswax or tallow on metal patterns and to brush and clean out between the teeth of gears and similar patterns.

CAMEL'S HAIR BRUSHES are used to brush dry blacking on the face of the mold.

RAPPING and CLAMPING BARS are usually bars of steel about 3/4 inch in diameter and 2 feet long. They are pointed at one end to enter rapping plates in a pattern and are flattened and turned up at the other end for convenience in tightening clamps on a flask.

DRAW SCREWS are eye-bolts threaded on one end. They are used for drawing large wooden patterns from the sand by screwing into holes drilled for that purpose in the rapping plate. They are also used for drawing metal patterns where pointed spikes could not be used.

DRAW SPIKES are steel rods which are sharpened at one end for driving into a wooden pattern to rap and draw it and are principally used in bench work for drawing small patterns.

LIFTERS are used for removing loose sand from deep cavities in molds. They are of different lengths and sizes, one end being turned at right angles to the stem; this portion is called the heel. The straight, flattened portion is known as the blade and is used to slick the sides of the mold where they cannot be reached by the trowel or slicker. The heel is also used to slick the bottom of deep recesses after the sand has been removed.

SLICKERS are formed with blades of varying widths, sometimes with one end of the tool turned to form a heel somewhat similar to the lifter. It is used for lifting loose sand from shallow parts of the mold, for patching, and to form corners to the proper shape. This tool is widely used by molders.

SPOON SLICKERS have spoon-shaped ends and are used to slick rounded surfaces in a mold. They are usually made with one end larger than the other.

The DOUBLE ENDER has a slicker at one end and a spoon at the other. They are usually made to the molder's order and are used on small molds.

CORNER TOOLS are used to slick the corners of molds where a slicker or the heel of a lifter is not satisfactory. Corner tools are made with different angles for special work.

Various specialized tools such as flange tools, pipe tools, and hub tools are also used.

WOODEN GATE-PINS or SPRUES are round tapered pins used to form the sprue or downgate through which metal is poured into the mold. The size depends on the size of the mold.

GATE CUTTERS are pieces of sheet brass bent to a semicircle on one edge. They are used to cut the ingate in the drag leading from the base of the sprue to the mold cavity.

SPRUE CUTTERS are cylindrical metal tubes used to cut the sprue in the cope when the sprue-stick is not used. Tapered sprue cutters are available for making the more desirable tapered sprue. They must be pressed down from the cope side before stripping the mold from the pattern.

CALIPERS are used more often by the coremaker than the molder. The molder uses them to verify the sizes of cores in order to insure proper fit in the core print and also to obtain the length of smaller cores. The calipers in this case are set at the proper dimension and the core filed to fit. This is important in dry-sand work to prevent crushing of the mold if the core is too large when the mold is closed.

CUTTING NIPPERS are used to cut small wires to the desired length for use in cores or molds.

FACING NAILS

Facing nails are used: (1) to reinforce mold surfaces and to prevent washing of the mold face, (2) to mechanically lock the sand on the face with that deeper in the body of the mold, and (3) to act as a means for slightly accelerating solidification at internal corners of the casting. These nails are similar to "roofing nails," having a flat, thin head of large diameter and shanks of various lengths. Caution must be exercised to see that no galvanized, rusty, oily, or dirty nails are used. The use of anything but clean dry nails will result in defective castings.

GAGGERS

Gaggers are used to give support to hanging masses of sand which would break under their own weight unless they were supported. Gaggers should be cleaned and are coated with clay before use to provide a better bond with the sand. Care must be taken in the placing of gaggers in the mold so that they are not too close to a mold surface, where they would cause a chilling of the metal where it is not wanted. Many times, a casting defect can be traced to a gagger located too near to a mold surface.

CHAPLETS

Chaplets are metal supports used to hold a core in place when core prints are inadequate. They are too often used to compensate for poor design, improper pattern construction, or bad core practice. In all castings (especially in pressure castings), chaplets are a continual source of trouble and should be avoided whenever possible. Figures 73, 74, and 75 show typical chaplets. It is absolutely necessary that they be clean. Rust, oil, grease, moisture, or even finger marks, cause poor fusion or porosity. Sandblasting immediately before use is a good practice if no other protection is used. Copper and nickel plating is a good method of protecting chaplets from rusting but does not eliminate the need for absolute cleanliness. Their size must bear a direct relationship to the type and section of metal in which they are to be used. Soft-steel chaplets are used in iron and steel, and copper chaplets in brass and bronze castings. Chaplets should be the same composition as the casting, if possible. The strength of the chaplet must be enough to carry the weight of the core until sufficient metal has solidified to provide the required strength, but it should be no heavier than necessary. The use of an oversize chaplet results in poor fusion and often causes cracks in the casting. A chaplet which can be made in the machine shop for emergency use is shown in figure 76. Chaplets should not have any sharp, internal corners because metal will not fill a sharp internal groove.

It is well to consider the forces which a chaplet must resist. In all metals except aluminum and the light alloys, a core tends to float when molten metal is poured into the mold. It is buoyed up by a force equal to the weight of the displaced molten metal. A core with dimensions of 12 inches by 12 inches by 12 inches, or one cubic foot, will weigh approximately 100 pounds. Immersed in molten gray iron, which weighs 450 pounds per cubic foot, the core will tend to remain in place until it has displaced 100 pounds of iron, and then it will tend to float. In order to keep it submerged (displacing 450 pounds of cast iron) it will be necessary to exert 350 pounds of force on it (450 - 100 = 350). It takes no more force to keep it submerged at greater depths than just below the surface. A greater head does not increase the lifting effect, although it does increase the pressure on the core.

The ratio of 100 to 350, or 1 to 3.5, holds good for cores of any size, so we can make the rule that the force resulting from the tendency of a sand core to lift in cast iron is roughly 3.5 times its weight; for steel, 3.9 times; for copper, 4.5 times; etc.

Where chaplets are used on large cores with extensive surface areas exposed to the metal, the usual practice is to use ordinary chaplets in the drag (since they are only required to hold the core in place until the metal is poured around them) and to use stem chaplets in the cope. Stem chaplets, instead of bearing on the mold face, pass through the mold body and are brought to bear against a support placed across the top of the flask. They are thus able to withstand very high forces, such as imposed when large cores tend to float on the metal. Figure 77 illustrates this method. It also shows a useful method for increasing the load-carrying ability of the green sand mold. A dry-sand core is used as a chaplet support in the mold. A dried oil sand core will safely support a load of 70 to 90 p. s. i. while the strength of green sand is 5 or 6 p. s. i.

Metal wedges or shims must be used to hold the stem chaplet down because the force of the molten metal acting on the core and transmitted through the stem of the chaplet will force it into a wooden wedge and thus allow the core to rise.

Table 19 for calculating the load-carrying capacity of chaplets of various sizes is given below.

TABLE 19. CHAPLET LOAD-CARRYING CAPACITIES

Double Head			Stem Chaplet		
Diameter of Stem, inch	Size of Square Head, inch	Safe Load, lbs	Diameter of Head, inch	Thin Metal Section, lbs	Heavy Metal Section, lbs
3/16	3/4	45	3/4	45	22
3/8	1-1/2	180	1-1/4	180	90
5/8	2-1/2	500	1-3/4	500	250

CHILLS

A detailed description of the use of chills will be found in Chapter 7, "Gates, Risers, and Chills."

Chills used in making molds are internal chills and external chills. Internal chills are set so they project into the mold cavity. They are expected to fuse with the solidifying metal and become a part of the casting. Extreme care should be taken to make sure the chills are clean. Any grease, finger marks, film, or dirt will prevent good fusion between the chill and the casting. External chills are rammed up with the mold to anchor them firmly in the sand. They also should be rust free and clean when used without special treatment. Many times, when external chills fuse to a casting, the condition can be overcome by coating the chill surface with a thin coat of shellac, or other adhesive material, applying a very thin layer of fine, dry sand, and then drying the chill to drive off any moisture. Many commercial chill coating materials are available also. Torch drying of coated chills in the mold should be avoided because moisture from the flame will condense on the chill. Moisture will condense on cold chills in green sand molds if the molds are closed and allowed to stand for an appreciable time before pouring.

CLAMPS AND WEIGHTS

Clamps and weights are used to hold the cope and drag sections of a mold together and to prevent lifting of the cope by the force of the molten metal. It is safe practice to use a weight on small molds, but when the molds are of considerable size, both weights and clamps should be used. The use of insufficient weights is a common cause of defective castings.

TYPES OF MOLDS

The types of molds which are made aboard repair ships are (1) green-sand molds, (2) dry-sand molds, and (3) skin-dried molds. These three types of molds differ mainly in their sand mixture content.

GREEN-SAND MOLDS

Molds made from tempered sand (see chapter on foundry terminology) and not given any further treatment are called green-sand molds. Green-sand molds are used for normal foundry work aboard ship. They have the necessary green strength and other properties which make them suitable for a great variety of castings. Green sand gives less resistance to contraction of a casting than does dry sand, and thereby tends to prevent hot cracks in the casting. Green-sand molds are the easiest to make.

DRY-SAND MOLDS

Dry-sand molds, as the name implies, are molds made with tempered sand and then thoroughly dried by baking. Dry-sand molds are used when a mold of high strength is needed, or when low moisture content is important. Dry-sand molds are not recommended for complicated castings unless special care is taken to obtain sand mixtures which will give good collapsibility, so as to prevent hot cracks or tears.

SKIN-DRIED MOLDS

Skin-dried molds are green-sand molds which have been dried only on the mold surface by the use of a torch or some other source of heat. Skin-dried molds are used where a mold surface low in moisture content is necessary. The mold surface is usually sprayed with additional special binding materials and then dried by the use of a torch. This type of mold combines the firm sand face obtained from a dry-sand mold with the collapsibility of a green-sand mold in the backing sand. In general the sand used for skin-dried molds has a moisture content higher than for a green-sand mold and dry-sand molds require a still higher original moisture content.

MOLDING LOOSE-PIECE PATTERNS

Loose-piece patterns are in one piece or are split to make molding easier. Molding with a split pattern will be described here. Molding with a single-piece pattern (and the use of broken parts) usually involves the cutting of a parting line and will be described under the section, "False-Cope Molding," later in this chapter.

In making a mold, a flask should be selected so that sufficient room is allowed between the pattern and flask for risers and the gating system. There must also be enough space over and under the pattern to prevent any break-outs of the metal during pouring or straining of the mold. Many castings are lost, or require extra cleaning, and many injuries to personnel are caused by the use of undersized flasks. It is better to err on the side of safety and choose too large a flask, rather than to use a flask that is too small. In addition to the safety factor, an undersized flask makes positioning of the gages and risers difficult. Gates and risers placed too close to a steel flask will be chilled by the flask and will not perform their function properly. Safe practice in the selection of a flask is shown in figure 78.

For a split pattern, such as that for a pump housing, a smooth ram-up board and a bottom board are needed. The ram-up board should be of sufficient size to project an inch or two beyond the flask. A one-piece board, such as 3/4-inch plywood, is preferred. The use of such a ram-up board keeps mold finishing and slicking to a minimum.

Before use, the pattern should be checked for cleanliness and the free working of any loose pieces which must seat securely. Any chills which will be required should be clean and on hand ready for use. The chills should be checked to make sure that they fit the pattern correctly and have the proper means for anchoring them.

When using a split pattern, the drag part of the flask is turned upside down and placed on the ram-up board. If the flask is not too large, the ram-up board and drag can be placed on the cope of the flask. The drag pattern is placed with the parting surface down on the ram-up board along with any pieces used for the gating and risering system. Figure 78 shows a pump-housing pattern set in the drag with the parts of the gating system. The facing sand is then riddled to a depth of about one inch on the pattern and the ram-up board. Riddling of the sand is absolutely necessary for good reproduction of the pattern. The riddled sand is then tucked into all pockets and sharp corners and hand packed around the pattern as shown in figure 79.

Backing sand is then put into the flask to cover the facing sand to a depth of three or four inches. The backing sand should be carefully rammed into any deep pockets as shown in figure 80. The remainder of the mold is then rammed with a pneumatic or hand rammer, care being taken to avoid hitting or coming too close to the pattern. The mold must be rammed uniformly hard in order to obtain a smooth, easily cleaned casting surface and to avoid metal penetration into the sand, swelling, break-outs, or other casting defects. When this ramming is completed, five or six more inches of sand are added at a time and rammed until the flask is filled to a point about one inch above the top of the flask.

Next, the excess sand is "struck off," by means of a straight edge or strike as shown in figure 81. Instead of striking the sand in one motion, it is often easier to loosen the sand by a series of short strokes and then remove it with one motion. When the struck-off surface is smooth, a scattering of a small amount of loose sand on the struck surface helps to give better contact with the bottom board. The bottom board is placed into position with a slight circular motion. Good, full, and solid contact between bottom board and drag sand is important if the mold and pattern are to have adequate support when they are rolled over. The drag section is then clamped between the bottom board and the ram-up board and turned over. The ram-up board (which is now on top) is removed and the mold face cleaned and slicked. Figure 82 shows the drag of the mold ready for the cope.

A parting material is sprinkled over the mold joint and pattern. The parting material prevents the sand in the cope from sticking to the sand in the drag when the cope is rammed up. The parting material for large castings is usually fine silica sand. For medium and small castings, finely ground powders (such as talc or silica flour) are used.

The cope of the flask is set on the drag and seated firmly with the aid of the flask pins. The cope pattern, riser forms, and any other parts of the gating system are set in their proper positions. Figure 83 shows the mold with the cope pattern, sprue, whirl-gate, and cross-gate pieces set.

The facing sand is riddled into the cope and hand packed around corners and in deep pockets. At this point, any gaggers which are necessary are placed. Care should be taken not to set the gaggers too close to the pattern risers or parts of the gating system. Gaggers set too close to the mold surface will cause undesired chilling of the metal. Any mold showing exposed gaggers in the cope after the cope pattern has been drawn, should be shaken out and made over. The number of gaggers and supporting bars will depend upon the size of the casting. Large flasks will require cope bars to support the sand. Gaggers may be fastened to the cope bars. The flask is then filled with sand and rammed as in making the drag. The sand should be packed by hand around any riser forms or raised portion of the pattern and care should be taken to avoid striking any forms or patterns during ramming. The partially filled cope is shown in figure 84. Notice that the cope has been peened around the inside edge of the flask. This procedure should be followed for both cope and drag, as it serves to pack the sand tightly against the flask and to prevent the sand from dropping out during handling. Also notice that the sprue and swirl gate form are slightly below the top of the cope of the flask. The cope is filled the same as the drag with successive fillings and uniform ramming. The completed mold is then struck off. With the riser, gating, and sprue forms a little below the top of the flask, the excess sand can be struck off without disturbing them.

After the mold has been struck off, it is vented in the cope as shown in figure 85. The cope is then removed from the drag, set on its side, and rolled over to facilitate drawing the cope pattern. By tapping the runner, riser, and sprue forms lightly on the parting side of the cope, they will come free easily and can be removed.

Any cope pattern pieces are also withdrawn at this time. The pieces used for the gating system are drawn from the drag. Cutting of ingates is done before drawing the pattern if possible. Loose sand should be cleaned from the drag. The drag pattern is drawn by the use of eyebolts or draw pins as dictated by the pattern. A light rapping of the pattern and eye bolt before and at the beginning of the draw will make this operation easier. (NOTE: avoid excessive rapping.) The beginning of the pattern draw for the pump housing is shown in figure 86. The drawn pattern is shown in figure 87. Notice that this pattern was drawn with both hands. Such procedures give the molder better control over the pattern. The cope and drag are both inspected and cleaned, if necessary. Slicking of the mold should be kept to a minimum, but the pouring gate in the cope should be compacted and smoothed so as to eliminate loose sand and prevent washing out by the molten metal.

If any facing nails are required to resist washing of the mold face, they should be placed at this time. Any sharp corners or fins of sand in the mold cavity or in the gating system should be carefully removed. Any such projections will be washed out by the stream of molten metal and result in defective castings.

Once the cope and drag have been properly finished, the cores should be set in place. It is good practice to rest the arms against the body while setting the core to make the operation easier and smoother and to avoid damage to the mold. Both hands should be used for setting all but the smallest cores. The handling of the core for the pump housing is shown in figure 88. The cope and drag with the core set and ready for closing are shown in figure 89.

The mold is closed carefully by using pins to guide the cope. The cope should be lowered slowly and kept level. Any binding on the pins because of cocking of the cope should be avoided. A jerking motion caused by binding pins often causes sand to drop from the cope. This is one reason why flask equipment should be kept in top condition. After the mold is closed, it is clamped, the weights are placed, and the pouring cup or basin set over the sprue. The mold, ready for pouring, is shown in figure 90.

The proper pouring techniques are discussed in detail in Chapter 9, "Pouring Castings." Pouring of the pump housing is shown in figure 91. A block of iron was used to hold the pouring basin down. Notice that the lip of the ladle is close to the basin and that the basin is kept full of metal.

The finished pump-housing casting is shown in the two views in figure 92.

MOLDING MOUNTED PATTERNS

A mounted pattern is one which is attached to a ram-up board. It is called a match-plate pattern if the cope pattern is mounted on one side of the board and the drag pattern on the other. For a match-plate, the cope and drag patterns must be aligned perfectly.

The molding of a mounted pattern is much easier than the molding of loose patterns. Mounted patterns are usually metal and are used for quantity production, but their use is often justified when quite a few castings of one design are required. Mounted patterns may also be of wood, but these require proper care and storage facilities to prevent warping.

The advantages of a mounted pattern are that the parting-line surface can be rammed much harder than with loose-pattern molding and a vibrator can be attached to the pattern plate to make the drawing of the pattern much easier. Another important advantage is that the gating and risering system can be made a fixed part of the pattern. As a result, smooth hard surfaces will be obtained and sand-erosion problems reduced.

The sequence of operations for molding with mounted patterns is the same as for molding loose patterns. The pattern is set between the cope and drag parts of the flask and held in place by the flask pin. The drag is rammed up first, the flask rolled over, and the cope rammed up. When the mold is completed, the cope is drawn off the pattern and then the pattern drawn from the drag. Core-setting operations and closing the mold are the same as for loose-pattern molding.

FALSE-COPE MOLDING AND THE USE OF BROKEN PARTS AS PATTERNS

Some patterns do not have a straight or flat parting line that permits them to be placed solidly against a ram-up board. Broken castings or parts which are to be used as a pattern usually fall into this class. Very often castings of this type would be impossible to mold by the common cope-and-drag method. The difficulties come from the facts that: (1) the parting line is not straight, and (2) the pattern or broken part requires special support while being molded. The false-cope method provides this extra support and makes it possible to have a very irregular parting line. Essentially the method consists of molding the part or pattern roughly into a false cope that is used to support the part while the final drag section is molded. The false cope is then removed and a final cope section molded to take its place in the final mold assembly.

The cutting of an irregular parting line is probably the most important step in false-cope molding and will be described here. A small boat propeller is used as the pattern.

The propeller is set in the drag, on the ram-up board, as shown in figure 93. The facing sand is riddled onto the pattern, and the drag filled with sand and rammed in the conventional manner. The bottom board is set and the flask rolled over. The cope of the flask and ram-up board are removed. The parting line is then cut with the use of spoons and slicks. The sand must be removed to provide a gradual slope from the casting parting line to the flask parting line. A 45° slope usually is the maximum that can be tolerated and prevent sand from dropping. The completely cut parting line is shown in figure 94. The cope section of the flask is set in position, parting compound sprinkled on the drag, and the cope rammed. Extreme care must be taken in ramming to prevent damage to the drag. After the cope is completed, it is carefully removed from the drag. The drawn cope is shown with the drag in figure 95. The pattern is then drawn, and the sprue, gates, and risers cut. The mold ready for closing is shown in figure 96. The as-cast propeller is shown in figure 97.

In false-cope molding, the false cope provides a bearing surface for the pattern when ramming up the drag. It has the advantage that the finished mold is not disturbed in cutting the parting line.

The preparation of a false cope consists of molding the pattern in the cope and cutting the parting line. The sand is rammed as hard as possible to provide a good bearing surface when ramming the drag. An alternative way of preparing a false cope is to ram the cope without the pattern. The pattern is then bedded into the parting line side of the cope. The parting line can be cut into the cope or built up with additional sand, or it may be a combination of both.

The drag section of the flask is then placed into position, parting material sprinkled over the mold joint, and the drag made as described for loose-pattern molding. The flask is then rolled over and the cope drawn. Extreme care must be taken in drawing the cope. The original cope is discarded, the cope section returned to the drag, parting compound sprinkled over the mold joint and a new cope made. Extreme care must be taken in ramming the cope to prevent any damage to the drag. The cope is then carefully drawn, the pattern drawn, sprues, gates, and risers cut, and the mold closed. This type of molding provides a firm, sharp, parting line without any loose sand particles that might wash into the mold cavity.

If several castings are required from a pattern with an irregular parting line, a more permanent type of false cope, or "follow board" can be used. A shallow box, the size of the flask and deep enough for the cope section of the pattern, is made from wood. The box should be made so that it is held in place by the flask pins. The pattern should be given a light grease coating to prevent any sticking. It is then positioned in the box, cope side up, in the manner previously described. Plaster is poured around the pattern and permitted to set firmly but not hard. The false cope and follow board are then turned over together. The pattern is worked back and forth slightly so that it can be drawn easily. While the plaster is still workable, the parting line is cut and the plaster permitted to harden. After the plaster has dried completely, it may be coated with shellac to prevent any moisture pickup. Nails may be used through the sides of the frame to help support the plaster. A follow board may be made in a similar manner by building up the required backing with fireclay mixed to the consistency of heavy putty, and working it around the pattern. A fireclay match has the disadvantage that it must be kept slightly moist to keep the fireclay from cracking.

The follow board is used in place of the false cope in providing the necessary support when ramming the drag. The pattern is set in the follow board, and the drag rammed up as for molding a loose pattern. After the bottom board is set, the drag is rolled right side up and the match plate drawn, exposing the pattern in the drag with the parting line made. The cope is then placed and the molding completed as for loose-pattern molding.

SETTING CORES, CHILLS, AND CHAPLETS

In the setting of cores, it is important to check the size of the core print against the core itself. A core print is a depression or cavity in the cope or drag, or both. The print is used to support a core and, when the core is set, is completely filled by the supporting extensions on the core. A typical example of a core print in use is shown at the extreme left of the mold in figure 89. An oversize print or an undersize core will cause fins on the completed castings, which may lead to cracks or chilled sections in the core area. An oversize core or an undersize print may cause the mold to be crushed and result in loose sand in the mold and a dirty casting.

Setting simple cores in the drag should be no problem to a molder. Care should be taken in handling and setting the core. After a core has been properly set, it should be seated by pressing it lightly into the prints. Another item which should be checked is the venting of cores through the mold. Many times, the cores themselves are properly vented but the molder forgets to provide a vent through the mold for the core gases to escape.

In come instances, the cores may have to be tied to the cope. In such a case, they are attached to the cope by wires extending through the cope. The wires are wound around long rods resting on the top of the cope to provide additional support. The rods should rest on the flask to prevent crushing or cracking of the cope.

Such operations should be done with the cope resting on its side or face up. The tieing should be done with as little disturbance as possible to the rammed surface. The core should be drawn up tight to prevent any movement of the core while the mold is being closed. Before closing the mold, the cope should be checked to make sure it is free of any loose sand.

Chills are rammed in place with the mold and are described under "Molding Tools" in this chapter. Again it is emphasized that chills must be clean and dry. Even chills which have just been removed from a newly shaken-out mold should be checked before immediate reuse.

The use of chaplets was described earlier in this chapter under "Molding Tools." It must be remembered that chaplets should be used only when absolutely necessary. Preferably, another method for support (for example, core prints) should be used, if at all possible. The use of chaplets in pressure castings should be completely avoided.

CLOSING MOLDS

The most important factor in the proper and easy closing of molds is to have flask equipment in good condition. Clean pins and bushings and straight sides on the flasks are the factors that make the closing of molds an easy operation. The opening of a mold after it has been closed is sometimes recommended. This procedure may prove useful. By using an excess of parting compound, the molder can then determine, with a fair degree of certainty, any mismatch or crushing of the mold. Nevertheless, the fewer times a mold is handled, the fewer chances there are to jar it and cause sand to drop.

SUMMARY

The molding operation aboard ship depends primarily on the molder and his ability to do his job. Skill in this type of molding can be attained only through experience, but a high level of skill can be reached in a shorter length of time by following correct molding techniques. For a beginning molder, it may appear much easier to patch molds that have been made haphazardly, than to take the time to make them properly. A molding technique based on careful attention to the various details involved in making a mold is by far the best approach to attaining molding skill. As with many other trades, speed in molding comes about by itself, if proper attention is given to the basic techniques.

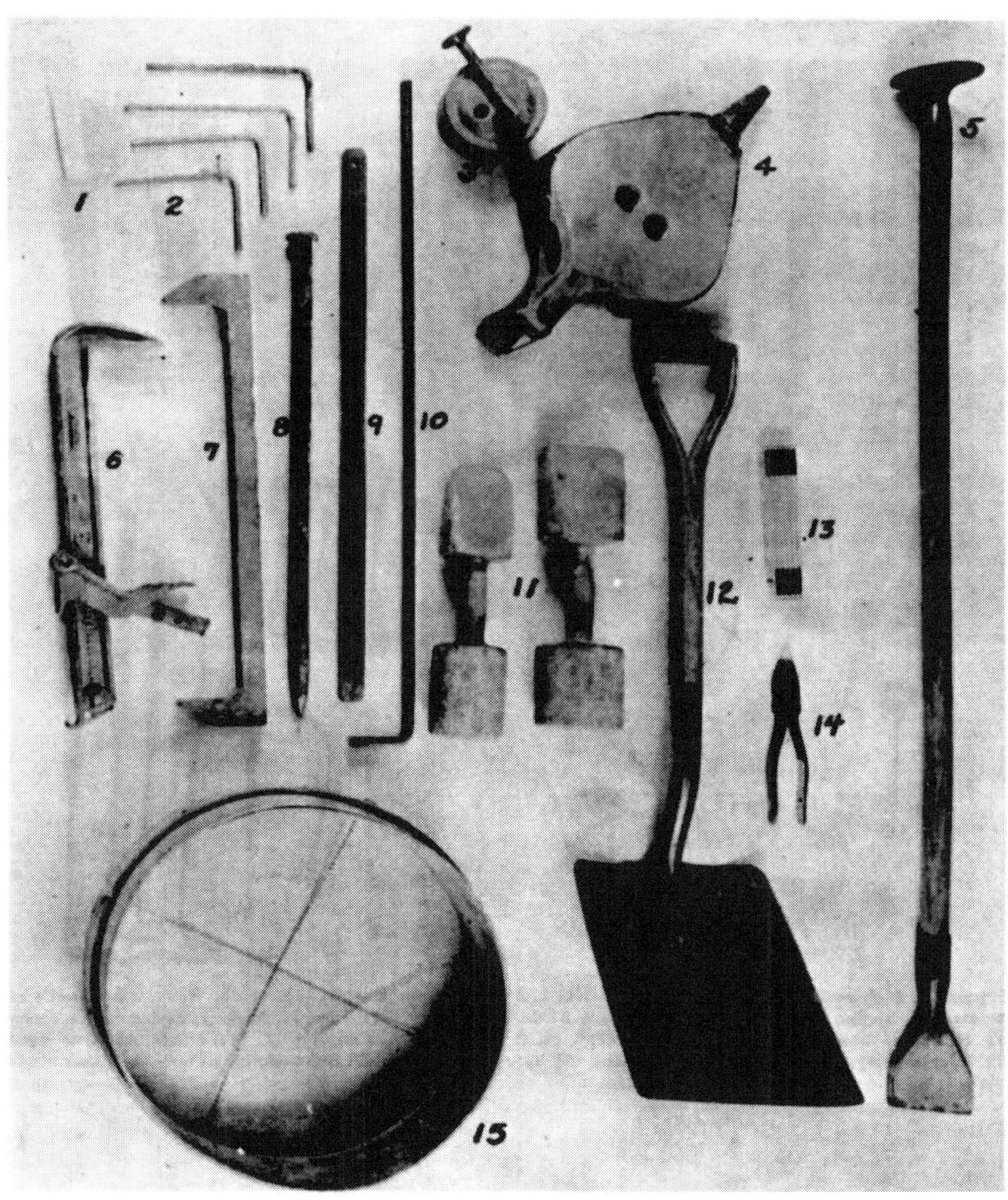

Figure 71. Molder's hand tools. 1. Wedge; 2. Gaggers; 3. Blow can; 4. Bellows; 5. Floor rammer; 6. Adjustable clamp; 7. Clamp; 8. Rapping iron; 9. Strike; 10. Rammer; 11. Bench rammers; 12. Molder's shovel; 13. Six-foot rule; 14. Cutting pliers; 15. Riddle.

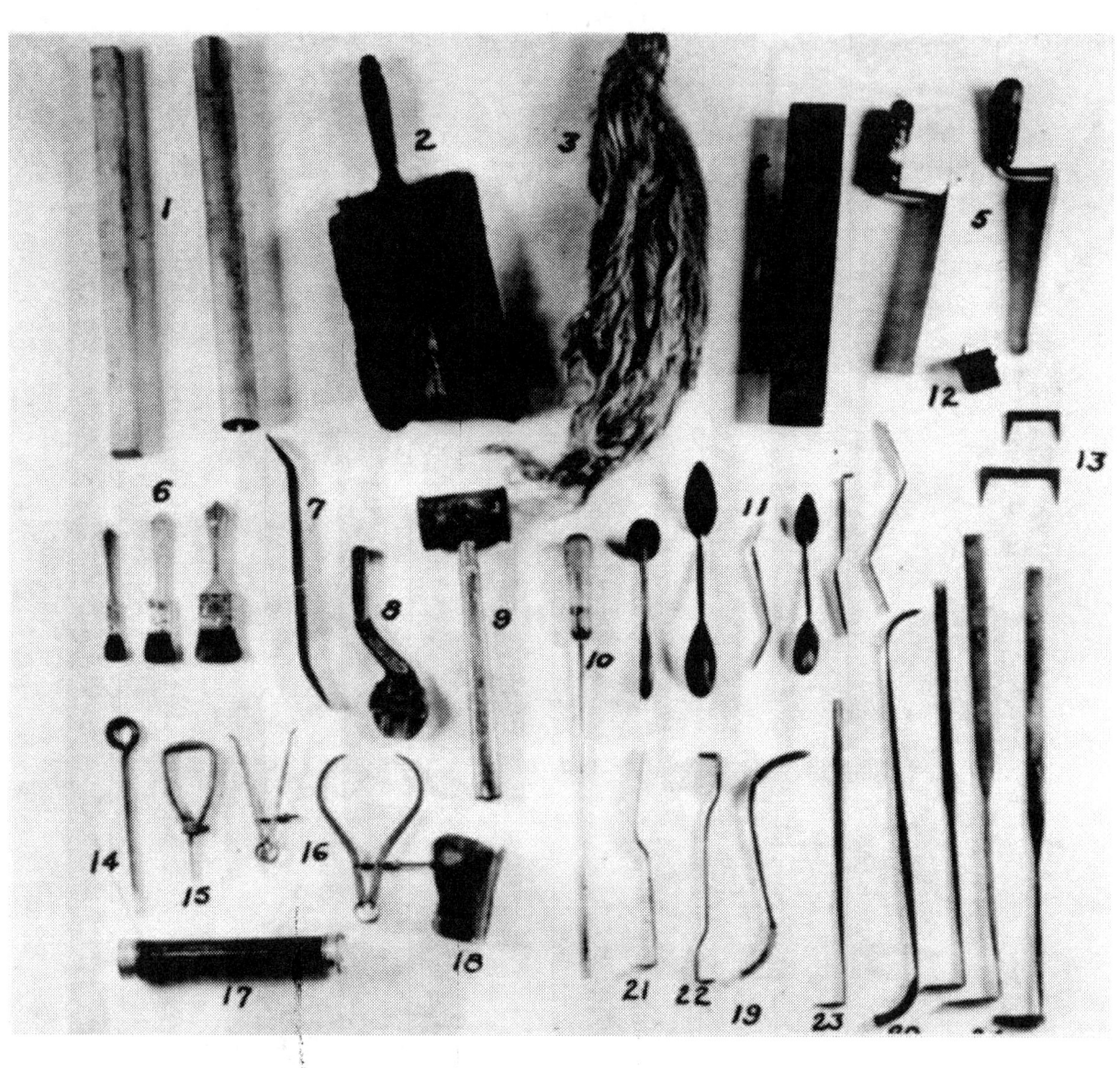

Figure 72. Additional molder's tools. 1. Gate stick; 2. Brush; 3. Bosh or swab; 4. Level; 5. Trowels; 6. Camel's hair brushes; 7. Rapping or clamping bar; 8. Wrench; 9. Rawhide mallet; 10. Vent wire; 11. Slickers, double-enders, spoons; 12. Half-round corner; 13. Dogs; 14. Draw spike; 15. Draw screw; 16. Calipers; 17. Flash light; 18. Gate cutter; 19. Circular flange tool; 20. Circular flange tool; 21. Bench lifter (bent); 22. Hub tool; 23. Lifter; 24. Lifters.

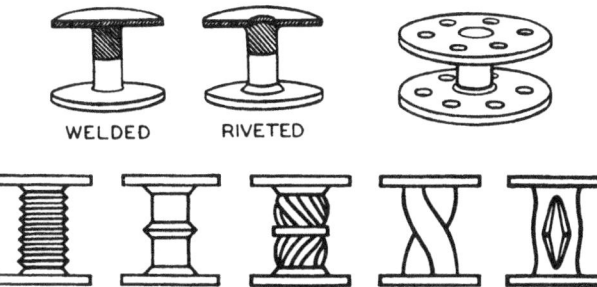

Figure 73. Double-headed chaplets.

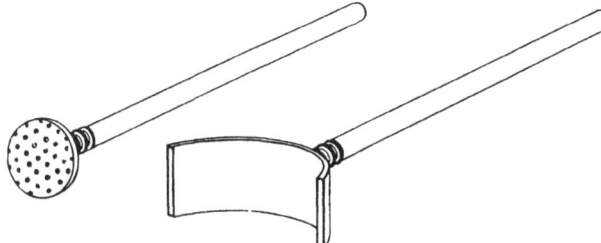

Figure 74. Stem chaplets.

Figure 75. Perforated chaplets.

Figure 76. Recommended chaplet design for emergency use.

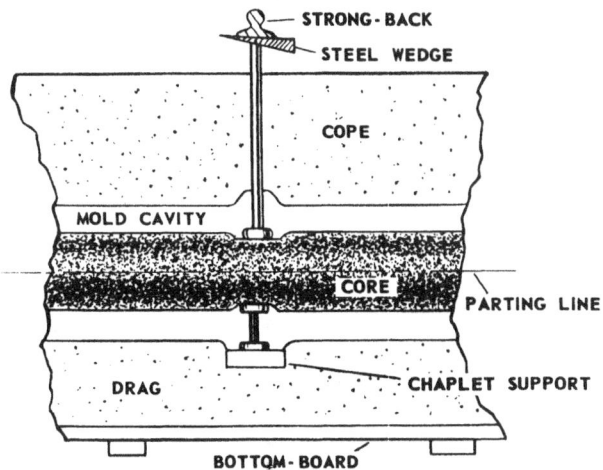

Figure 77. Anchoring cores with chaplets.

Figure 78. Pattern set in drag with gating system parts.

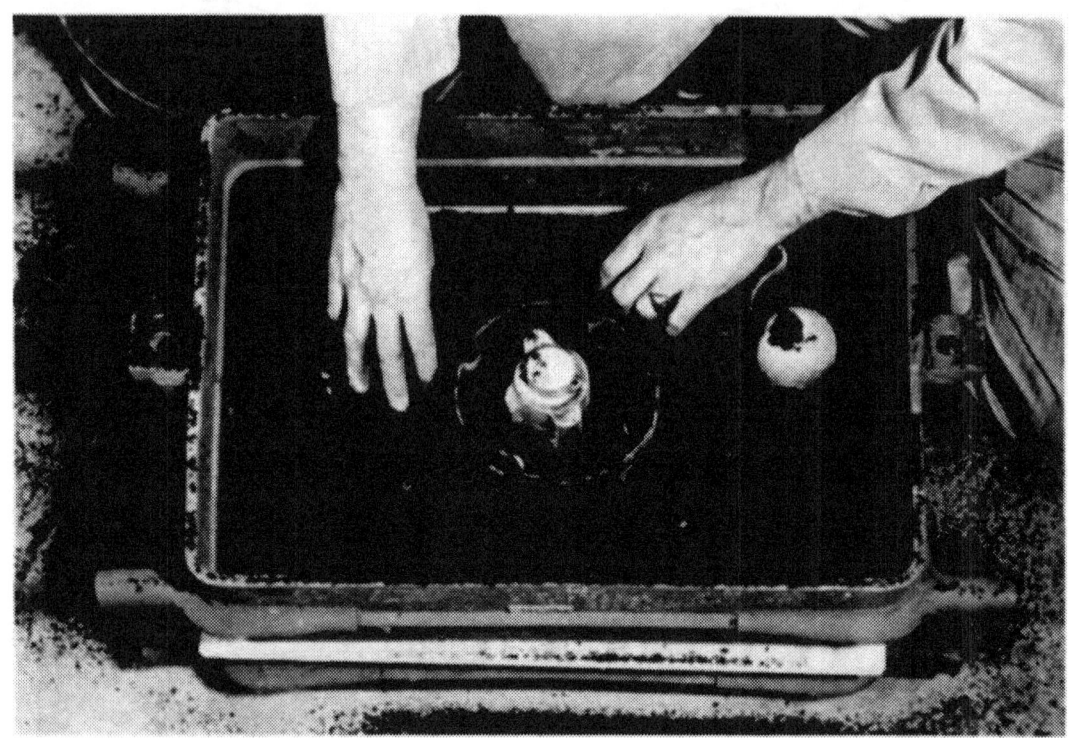
Figure 79. Hand packing riddled sand around the pattern

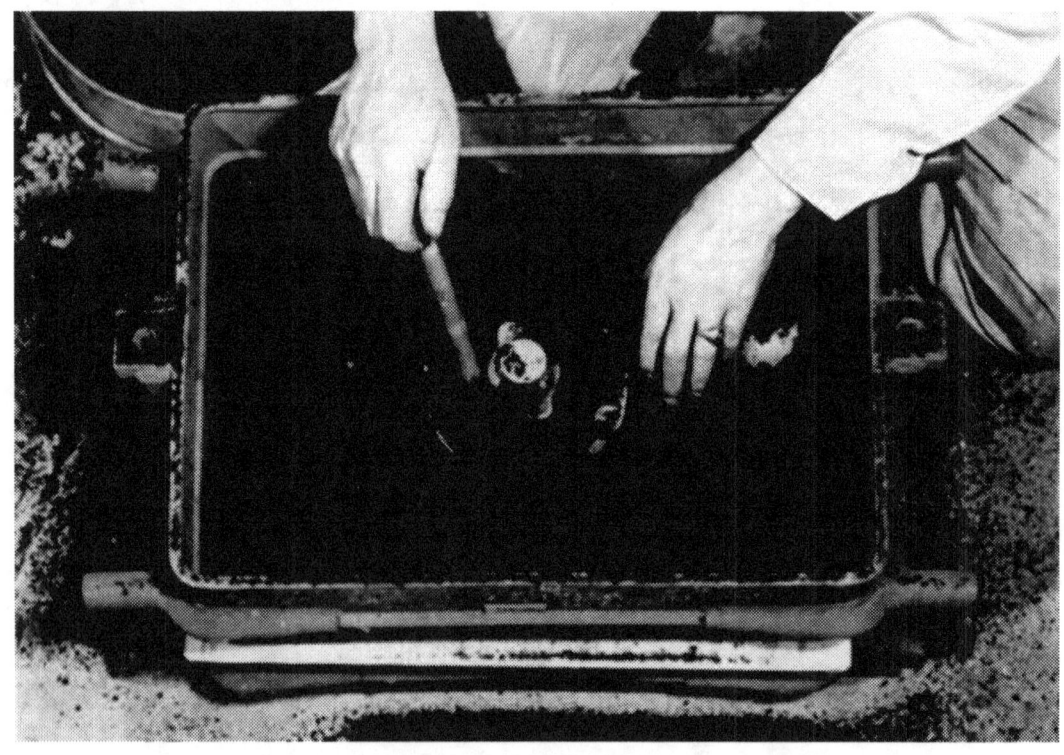
Figure 80. Ramming a deep pocket.

Figure 81. Striking off the drag.

Figure 82. Drag ready for the cope

Figure 83. Cope with pattern and gating pieces set

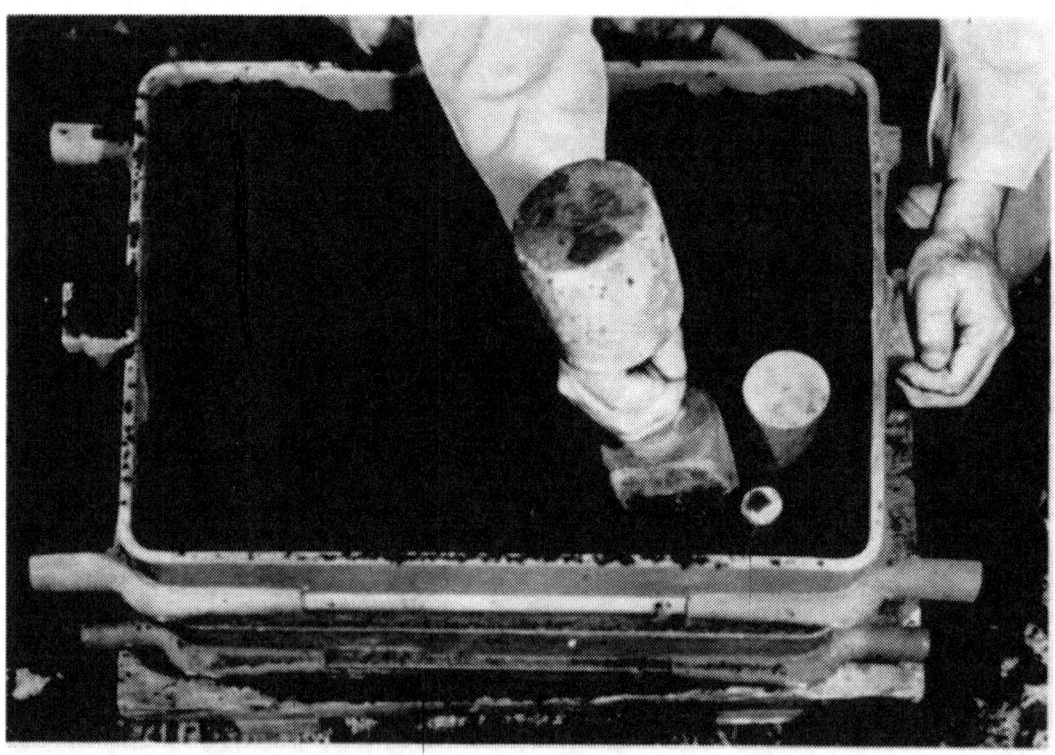

Figure 84. Ramming the partially filled cope.

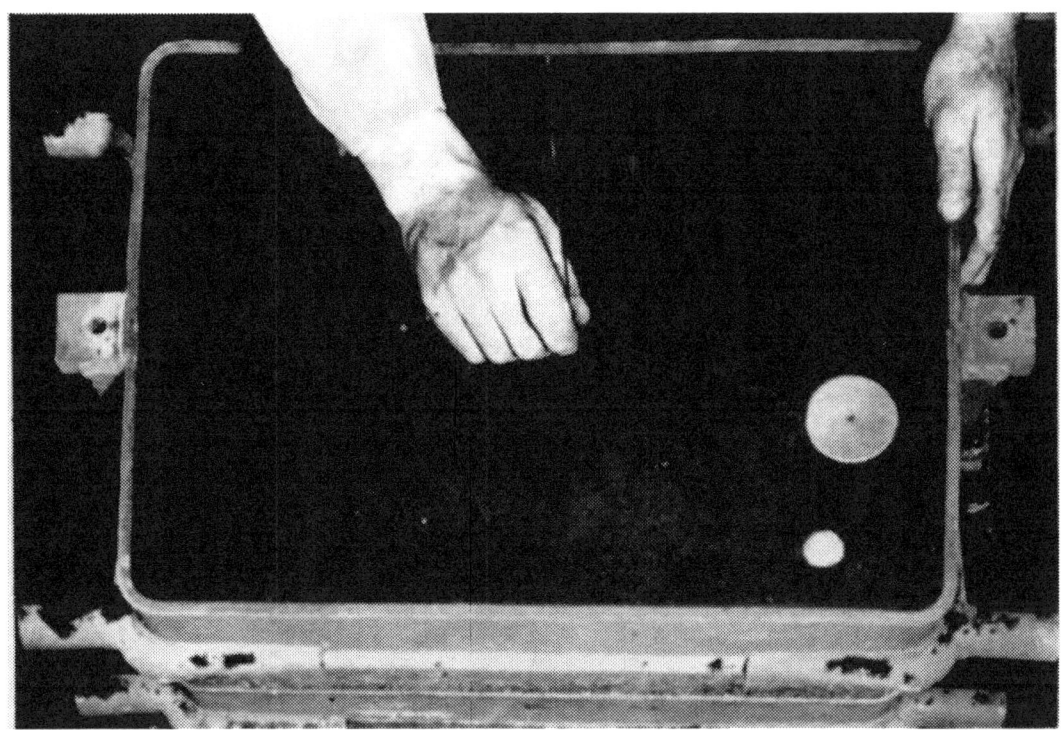

Figure 85. Venting the cope.

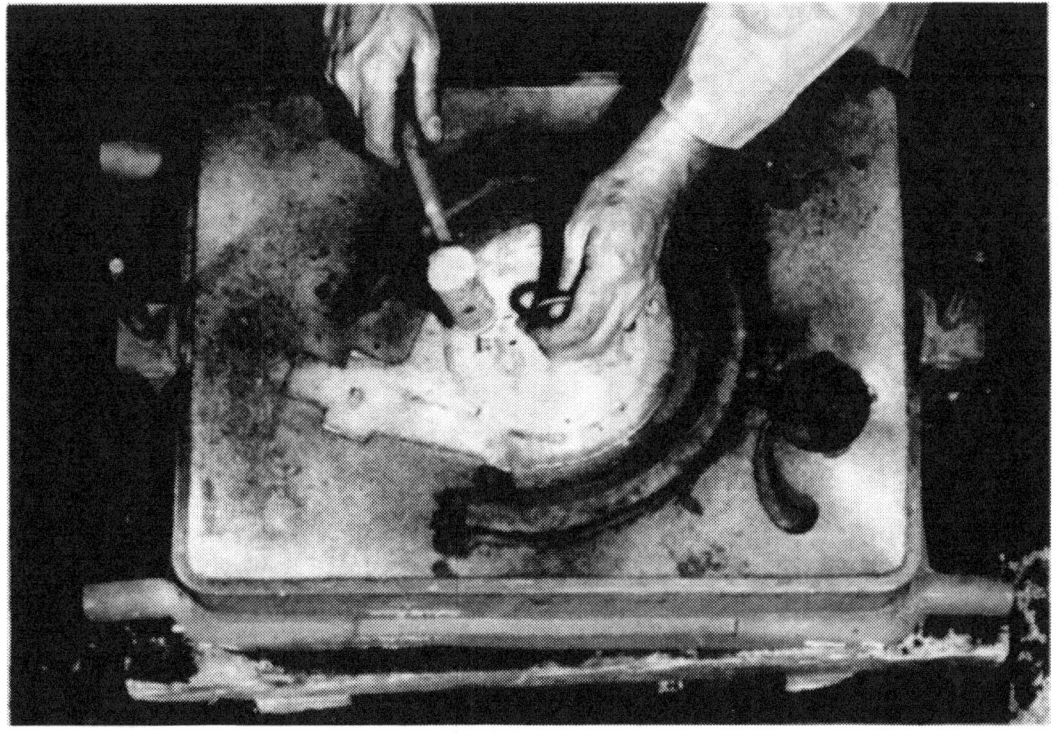

Figure 86. Start of the pattern draw

Figure 87. Pattern completely drawn

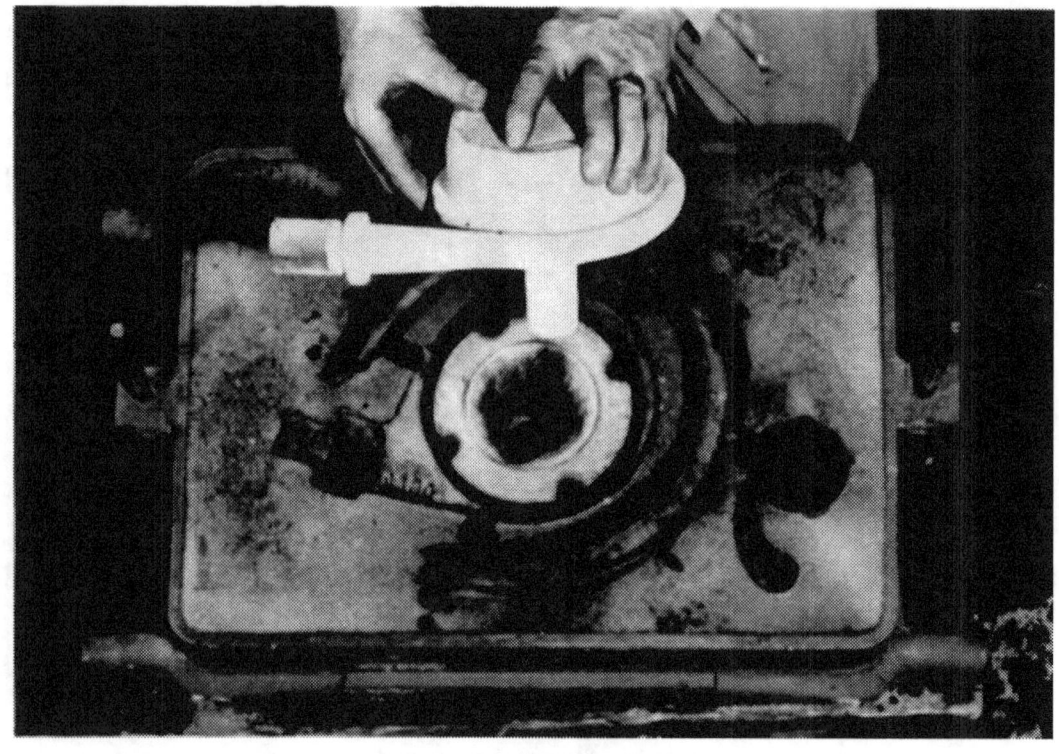

Figure 88. Setting the core.

Figure 89. Cope and drag ready for closing

Figure 90. Clamped mold with weights and pouring basin

Figure 91. Pouring the mold.

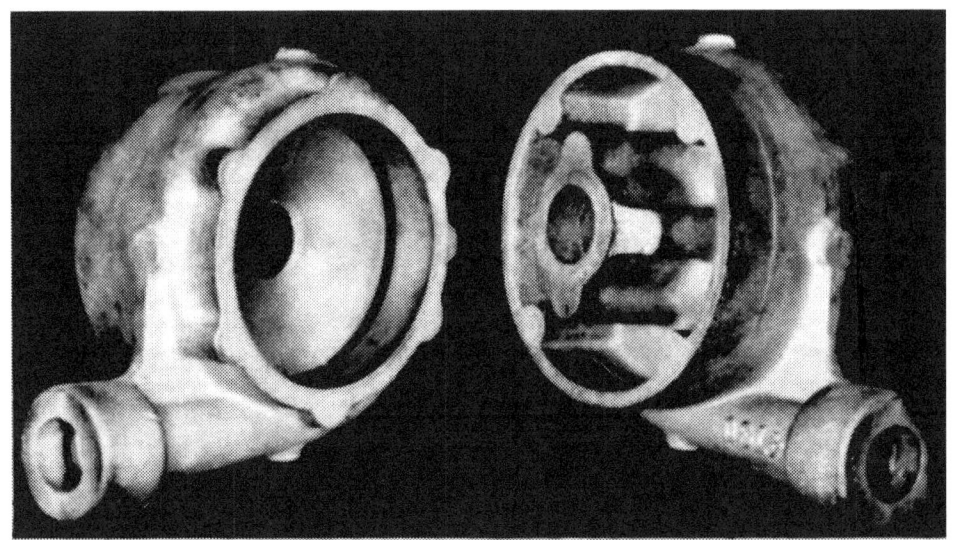

Figure 92. Finished pump housing casting

Figure 93. Propeller set in the drag.

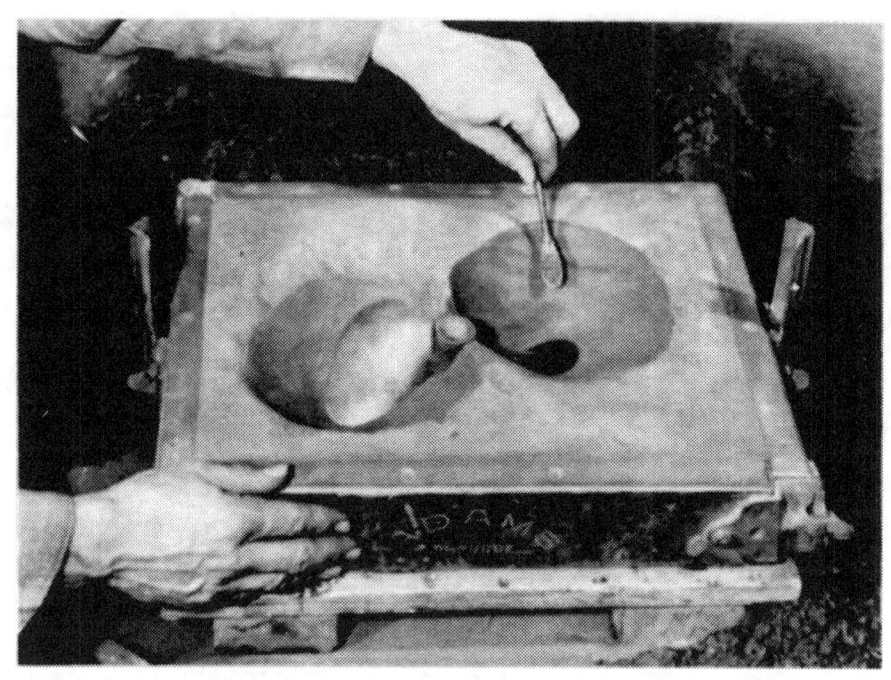

Figure 94. Propeller in the drag with parting line cut

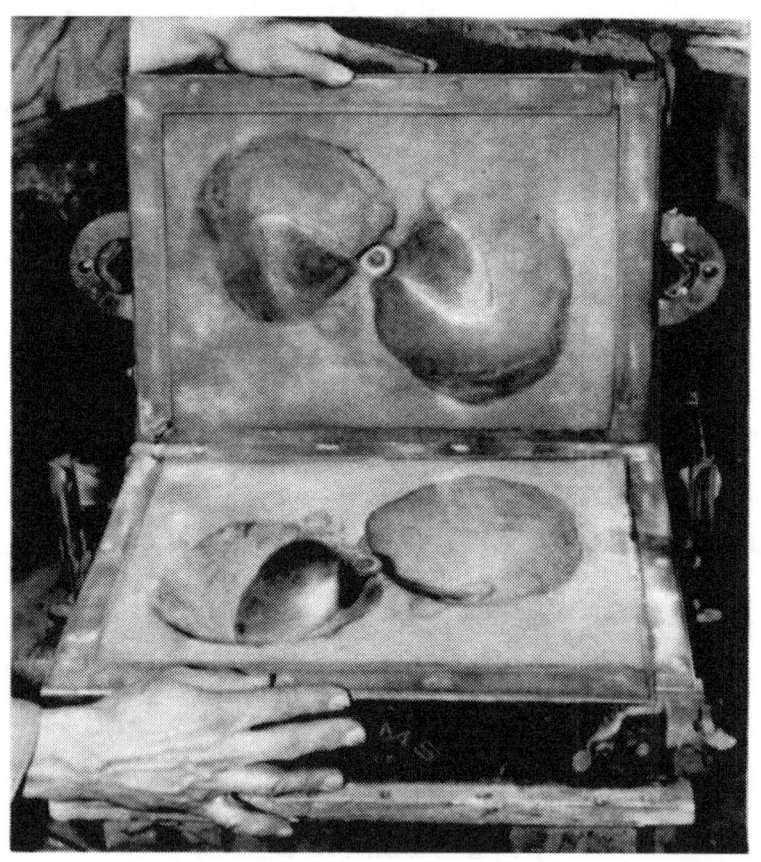

Figure 95. Drawn cope.

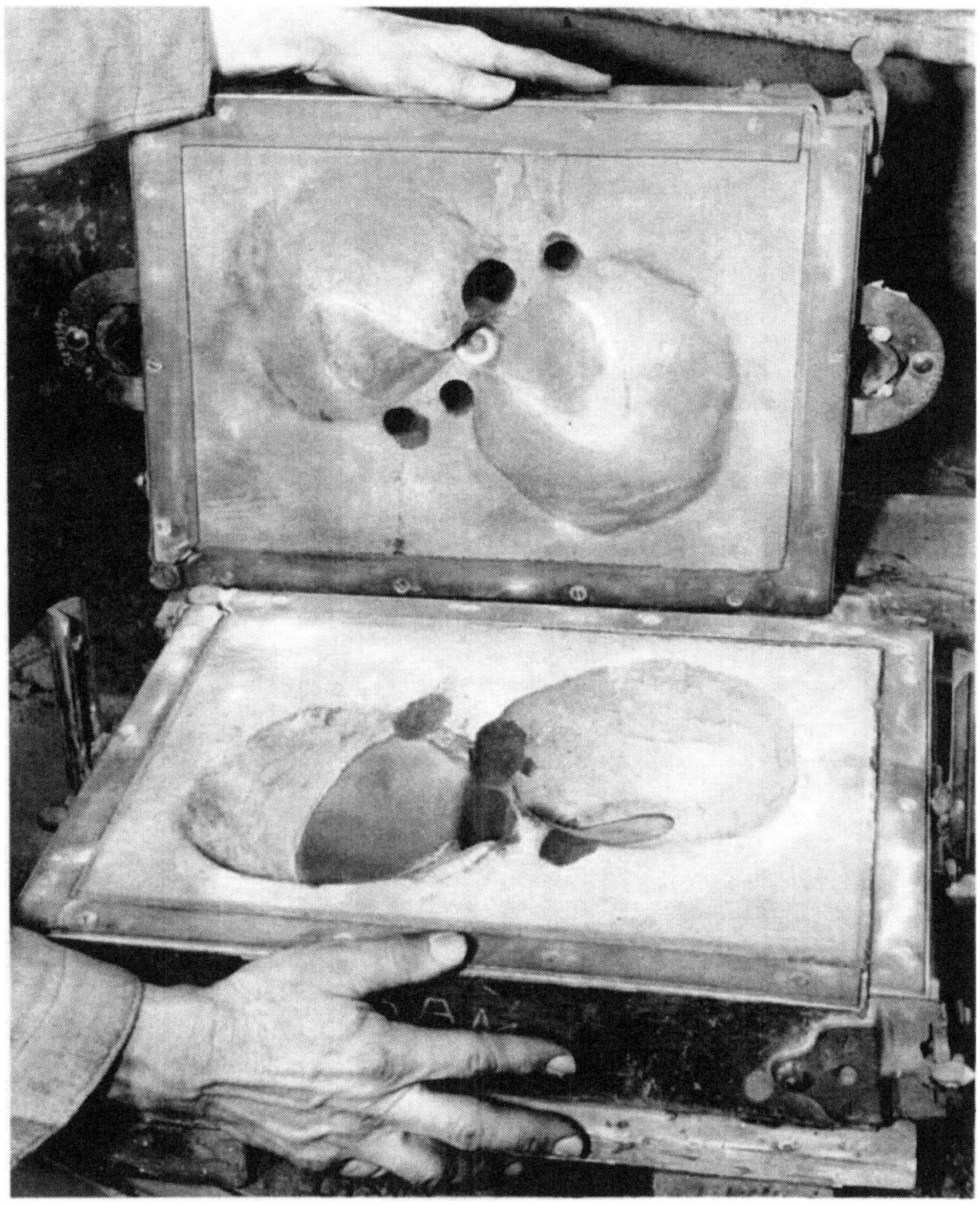

Figure 96. Mold ready for closing

Figure 97. As-cast propeller

Chapter VI
MAKING CORES

Cores are used for forming internal cavities in a casting, for forming parts of molds when the pattern is difficult to draw, or for details that are difficult to make in molding sand. The various properties required of good cores are discussed in detail in Chapter 4, "Sands for Molds and Cores." Briefly, the properties desired in a core are: (1) refractoriness, (2) some green strength, (3) high dry strength, (4) good collapsibility, (5) a minimum amount of gas generation by the core during casting, (6) good permeability, and (7) high density.

COREMAKING TOOLS AND ACCESSORIES

Tools and accessories used in the making of cores are the same as those used for making molds, with the addition of coreboxes, sweeps, core driers, and special venting rods. Cores are shaped by the use of the core boxes, by the use of sweeps, or by a combination of these methods. Sweeps are limited in their use and will not be discussed here. Core driers are special racks used to support complicated cores during baking. They are usually not used unless a large number of cores of a particular design are being made. Complicated cores can often be made as split cores, baked on flat drying plates, and then assembled by pasting.

TYPES OF CORES

BAKED SAND CORES

Core work aboard ship is concerned primarily with baked sand cores. They have the desired properties, are easy to handle, and may be made up ahead of time and stored in a dry place for future use. Baked sand cores have higher strengths than dry-sand cores. This means that complicated cores can be made most easily as baked cores.

DRY-SAND CORES

Dry-sand cores are made from green-sand mixtures to which additional amounts of binders have been added. They are dried in the air or with a torch and their strength comes from the large amount of binder. Dry-sand cores are not as strong as baked sand cores and require more internal support and careful handling. Although they can be made faster than baked sand cores, this is often offset by diadvantages of lower strength and the need for more careful handling.

INTERNAL SUPPORT

Cores are made from sand mixtures that are very weak before they are baked or dried. These mixtures often need some reinforcing. Large or complicated cores need proper arbors or reinforcing rods in the sand to permit handling of the unbaked core and to help support the baked core in the mold. When a core is made entirely of sand, the force tending to lift it is quite great when metal is poured around it, but when the core is hollowed out or filled with coke or cinders, as is often done to improve collapsibility, the force is even greater. If a core shifts, floats, deforms, or breaks, the casting is almost always defective.

Figure 98 shows a cast iron arbor used to support a core of medium size. Figure 99 shows the core rammed with the arbor in place. Arbors can also be made by tying rods or wires together, or by welding rods or strips together. Cast iron arbors are seldom used for small work; steel rods or wires are more commonly used. When placing rods or arbors, support them to avoid all twisting, bending, or breaking forces. Place the support in such a way that it does not interfere with the proper hollowing or "gutting out" of the core. Hollowing is done to improve collapsibility after the casting is poured. Care should be taken to make certain that the arbor or rod does not project through the surface of the core or even approach too close to it. All pockets or projecting parts of the core should be made with rods to make it easier to draw the core box and to give good support for the core. Figure 100 shows a typical method for supporting cores which must be suspended from the cope. Figure 101 shows lifting-hook assemblies used for handling and fastening large cores.

FACING, RAMMING, RELIEF, AND VENTING OF CORES

After obtaining the core box and selecting the proper reinforcing rods or arbor, the next operation is to put the core sand uniformly into the core box to a depth of approximately 1/2 inch or more, depending on the size of core and thickness of the casting. The sand should be free flowing and should not require hard ramming, but it is necessary that it be rammed sufficiently to develop a smooth, uniform surface. In pockets, tucking the sand in place with the fingers or suitable tools is necessary. Many core makers tend to overlook the importance of this operation and its omission is a source of continual trouble. Uniformity of ramming is a big factor affecting green and dry strength, ease of cleaning, and the quality of the casting surface.

After or during the ramming of the facing material, the reinforcements are placed. For small cores, the entire box may be filled with facing sand prior to placing the rods. In making medium and large cores, the facing material may be backed up with old molding sand, cinders or coke to support the core. This material, after drying, can easily be removed to provide space for venting and for collapsibility of the core.

One of the necessary requirements of a core is venting. In some of the simple cores, venting is easy, but in the more intricate ones, it is often difficult. A small, round core may be vented by running a vent rod through its center after ramming. A core made in halves may be vented by cutting channels through the body and core prints at the parting line before baking. When neither of these methods can be applied, a wax vent should be used. It is buried in the sand along the line or lines that the escaping gas is to follow. When the core is baked, the wax melts and disappears into the body of the core, leaving the desired vent channels. Care should be taken to avoid using too much wax, as it produces gas when heated by the molten metal. Cores made with coke cinders, gravel, or similar material in their central sections do not usually need additional venting.

The importance of good ramming, a uniformly smooth surface on a core, and of proper venting cannot be emphasized too strongly. When cores have a tendency to sag before baking or during baking, they can often be supported in a bed of loose green sand which can be brushed off the core after it is baked.

TURNING OUT AND SPRAYING

After the core box is filled and the excess sand is removed, a metal plate is placed on the box, the whole is turned over, and the box is rapped or vibrated as it is drawn away from the sand. There are several precautions to be observed in this operation. The core plate should be clean and straight and should be perforated to facilitate drying of the core. Care should be taken to avoid hard rapping of the box. This causes distorted cores.

After the core is freed from the box, all fins and irregularities must be removed. The core should be sprayed or painted with the proper wash as described in Chapter 4, "Sands for Molds and Cores."

A silica wash or spray is a good general-purpose material to smooth the surface of the core to give a smoother casting. With a little practice, washes can be applied to a core either before or after it is baked. The wash should be heavy enough to fill the openings between the sand grains at the surface of the core, but not so heavy that it will crack or flake off when it is dried.

BAKING

In the baking of oil sand cores, two things occur. First, the moisture is driven off. Following this, the temperature rises, causing drying and partial oxidation of the oil. In this way, the strength of the core is developed.

For proper baking of oil sand cores, a uniform temperature is needed. This temperature should be not over 500°F., nor under 375°F. If linseed oil cores are baked at a moderate temperature of 375°F. or 400°F., they will be quite strong. The same cores baked quickly at 500°F. will be much weaker. Baking the cores to the point where the bonding material decomposes must be avoided, or the cores will lose strength.

The size of the core must be considered in drying. The outer surface of a core will bake fast and will be the first part to develop maximum strength. If the temperature is maintained, the inside will continue to bake until it finally reaches maximum strength, but by that time the outer surface may be overbaked and low in strength. The tendency for this to happen in large cores can be partly overcome by filling the center of the core with highly porous material with a low moisture and bond content (for example, cinders or coke), by the use of well-perforated plates, and by using low baking temperatures. It is not only a matter of heating the center of the core but also of supplying it with oxygen. Thus, there is need for free circulation of air around and through the core while baking.

The most skillful and careful preparation of metal and mold can easily be canceled by poor cores. The need for proper baking cannot be overemphasized. If cores are not properly baked, the following is likely to happen to the casting:

1. Excessive stress, possible cracks, caused when the core continues to bake from the heat of the metal, thus increasing the strength of the core at the time the metal is freezing and contracting.

2. Unsoundness from core gases not baked out.

3. Entrapped dirt from eroded or spalled sand from weak cores.

When overbaked, the loss of strength of the core results in excessive breakage in handling or during casting, and cutting or eroding of the core surface.

To establish a full appreciation of the problems of drying cores, a series of 3, 5, and 8-inch cube cores should be made without rods and then baked at temperatures of 400°F., 425°F., 450°F., 475°F., and 500°F. for varying times. After being taken out of the oven and cooled, they should be cut open with a saw to determine the extent to which they are baked. Conducting this simple test will aid in determining the proper time and temperatures to use for various cores in a given oven.

Practice is necessary to accurately determine when a core is baked properly. A practical method is to observe the color of the core. When it has turned a uniform nut brown, it is usually properly baked. A lighter color indicates insufficient baking and a darker color indicates overbaking.

Mechanical venting of the core by using many vent holes and then carrying these to the prints on the core joint will facilitate baking.

CLEANING AND ASSEMBLY

At the baking temperature, cores are quite fragile. Thus, after being removed from the oven, they should be allowed to cool to below 125°F. before being taken from the core plates.

When cool, all excess materials such as fine and loose sand should be cleaned from them. Sand, gravel, or cinders used for fill-in material should be removed to provide for the collapsibility of the core and to improve venting.

Vents should be cut in such a way as to prevent metal from entering them when the casting is poured. Make sure at all times that vents are adequate to permit the full flow of core gases as they are generated. Overventing does no harm. Underventing gives bad castings.

After this, the core should be fitted together with a gage for control of size. The use of gages for assembling cores is necessary for producing quality castings.

When properly cleaned and gaged, the core sections are then assembled using a plate mixed as follows:

3 % bentonite
6 % dextrine
91 % silica flour (200 mesh or finer)
  Water to develop the correct pasty consistency.

After this, all joints are sealed with a filler mixed as follows:

3 % bentonite
3 % dextrine
94 % silica flour (200 mesh or finer)
  Water to develop the consistency of a thin putty.

The filler and paste are dried by returning the core to an oven for a short period or by local application of heat (as from a torch).

The above paste and filler developed at the Naval Research Laboratory have been found to give excellent results. A major caution to be observed in their use is to see that they are thoroughly mixed in the dry state before adding water, and again thoroughly mixed after the water is added.

STORAGE OF CORES

Storage time has an important effect on the quality of cores and upon the resulting castings. Baked or dry cores decrease in strength because they pick up moisture from the air, particularly on the surface. For this reason, it is unwise to make cores much in advance of requirements, usually not over 24 hours. If baked or dry cores must be stored, put them in a dry place.

Consideration must also be given to the storage of cores prior to baking. If cores are allowed to stand for too long a time before baking, evaporation of the surface moisture may take place and give a weak surface on the core. For thin cores of a large surface area, 10 minutes may be too much time for standing out of the oven, while for heavier cores more time may be allowed. In all cases, the time should be kept at a minimum.

MAKING A PUMP-HOUSING CORE

The following figures show the various steps in making the core for the pump-housing casting shown in Chapter 5, "Making Molds:"

The two core boxes for making the parts for the pump-housing core are shown in figure 102. A parting compound has been sprinkled on the core boxes to make it easier to turn out the core. Ramming of the sand in one of the core boxes is shown in figure 103 and striking off of the core is shown in figure 104. Placing of the reinforcing rods is shown in figure 105. Additional sand is then added, the core rammed lightly again and struck off. The reinforcing rods can also be placed when the core is partially rammed. Cutting the vents with a molding tool is shown in figure 106. Notice that one vent comes out the end of the core, while other vents radiate from the center and vent through the center of the supporting core. The idea is to give gases inside the core a free passageway out of the core. The two core halves are shown in figures 107 and 108 after they have been turned out of the core boxes. After the two core halves are baked, core paste is applied with a small rubber squeeze bulb, as shown in figure 109. Note that there is no excess of core paste. Core paste on the outside of a core can cause a defective casting because of gas formation. The assembling of the two core halves is shown in figure 110. The assembled core must then be baked for a short time to dry the paste. The completed core after it is sprayed is shown in figure 88, Chapter 5, "Making Molds."

SUMMARY

Cores should always be made with accurate, clean equipment and should have the following qualities to a degree suitable for the purpose intended:

1. Refractoriness to withstand the casting heat. This is obtained by selection of material and proper processing.

2. Strength to withstand handling and casting forces. This is obtained by the use of the proper amount of binders and by good internal structural supports.

3. Collapsibility to permit breakdown during contraction of the casting and ease of cleaning. By avoiding the use of sands bonded too strongly and by hollowing out the center or filling it with coke, cinders, gravel, or weak sand, this quality may be obtained.

4. Smooth strong surface to provide a good casting finish, internal cleanliness, and ease of cleaning. This quality is obtained by the use of an adequately bonded refractory sand, uniformly hard rammed, baked immediately after being made, and used shortly after baking.

5. Low gas content to prevent unsoundness in the casting. This quality is obtained by using the minimum of organic binding materials, baking well, and venting thoroughly. All of the above features are essential in core making and are regularly obtained only by good core practice.

Figure 98. Arbor for a medium-size core.

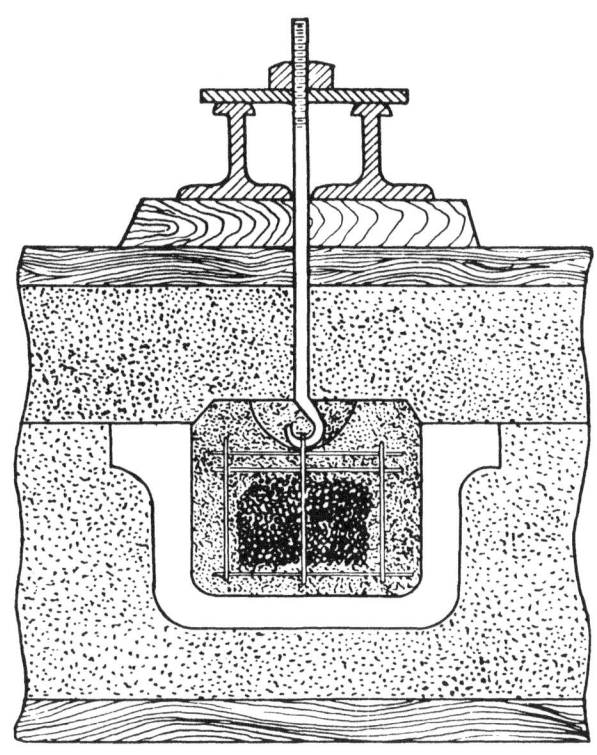

Figure 100. Section of mold showing use of lifting eye for supporting heavy core.

Figure 99. View of inside of core showing hollowing to make the core more collapsible when metal is poured around it.

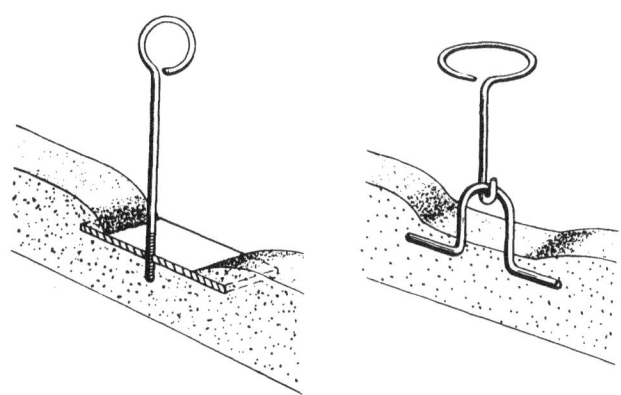

Figure 101. Typical lifting hooks for lifting cores.

- 87 -

Figure 102. Core boxes for pump housing core

Figure 103. Ramming up the core.

Figure 104. Striking off the core.

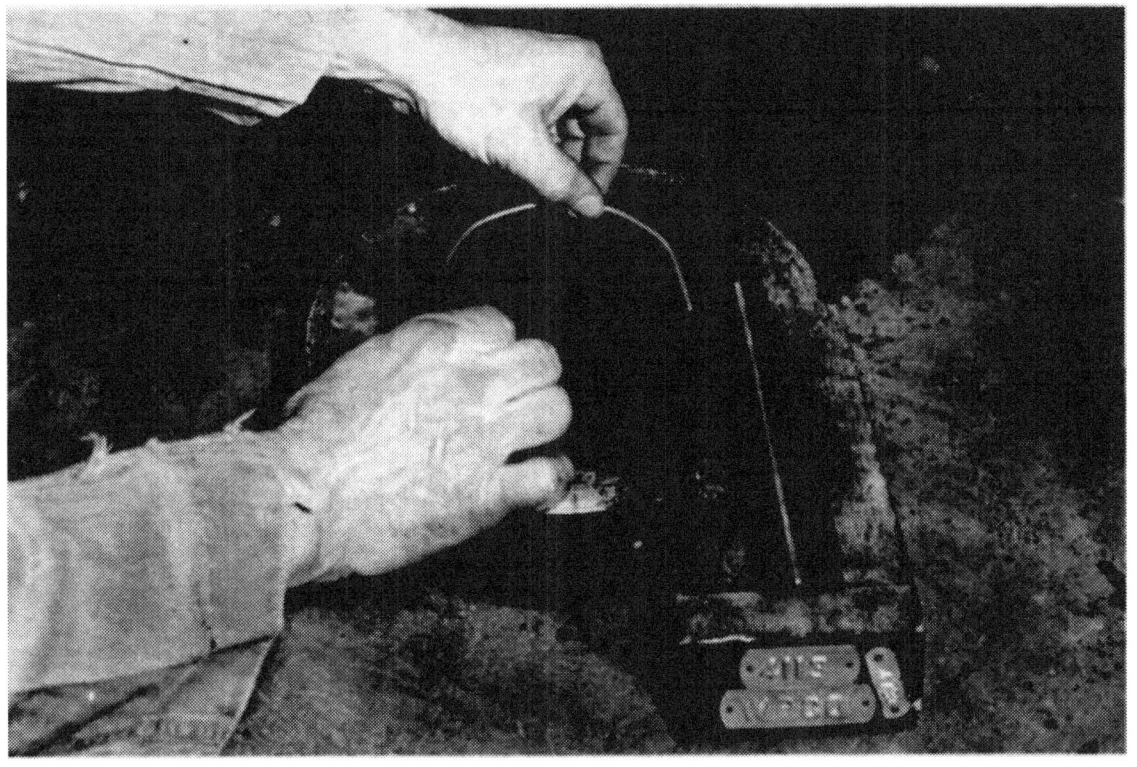

Figure 105. Placing the reinforcing rods

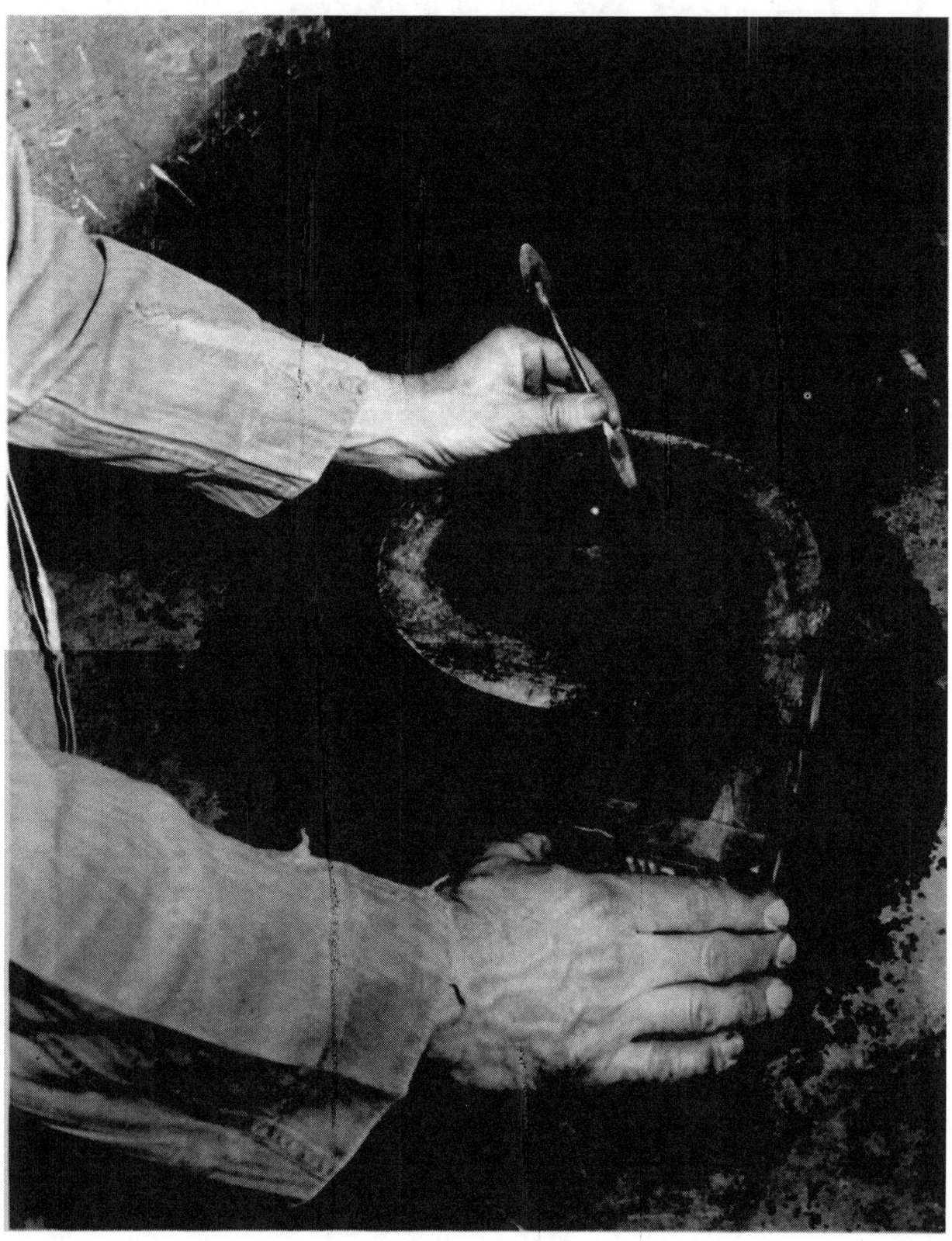

Figure 106. Cutting vents

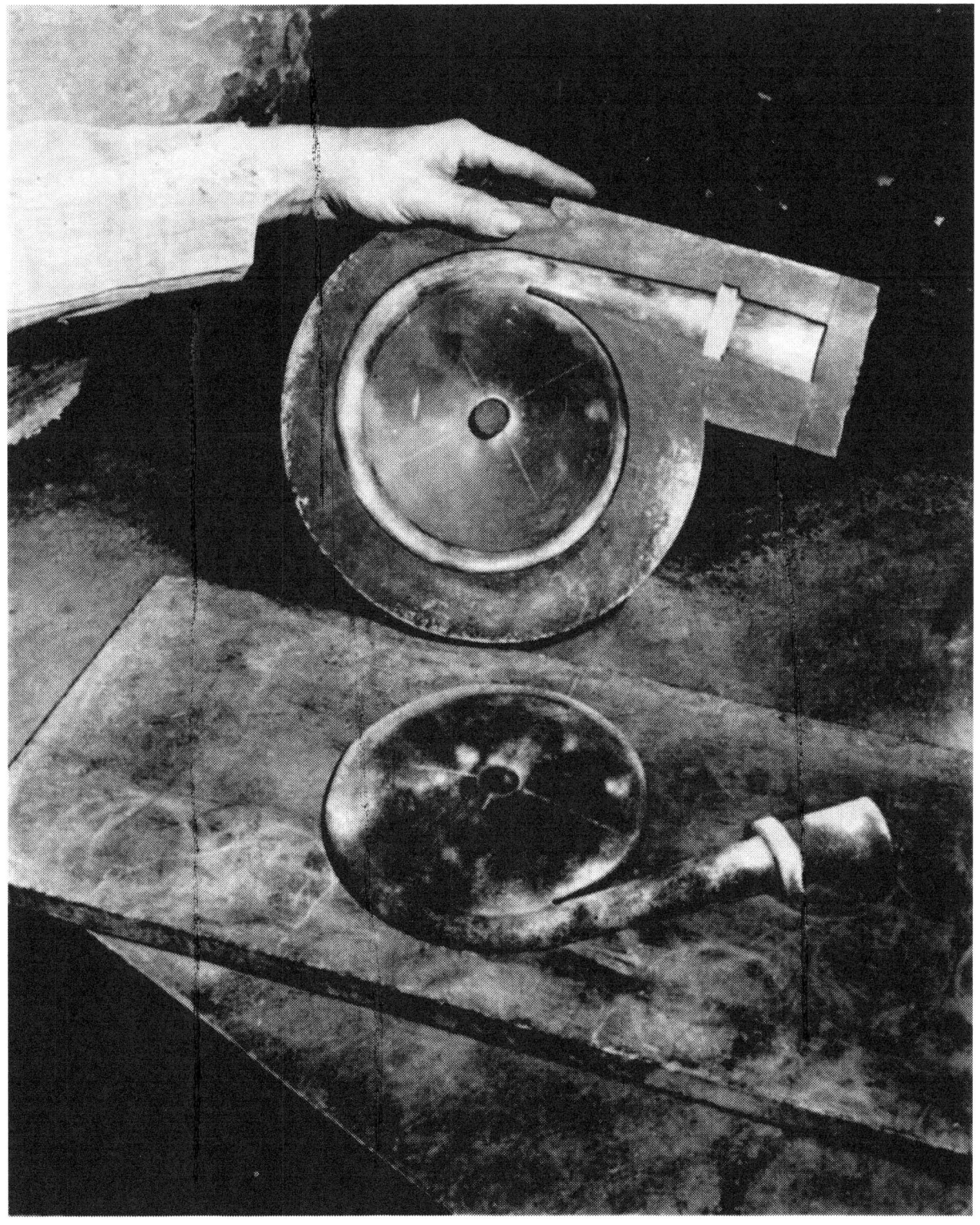

Figure 107. Drag core turned out

Figure 108. Cope core turned out

Figure 109. Applying core paste.

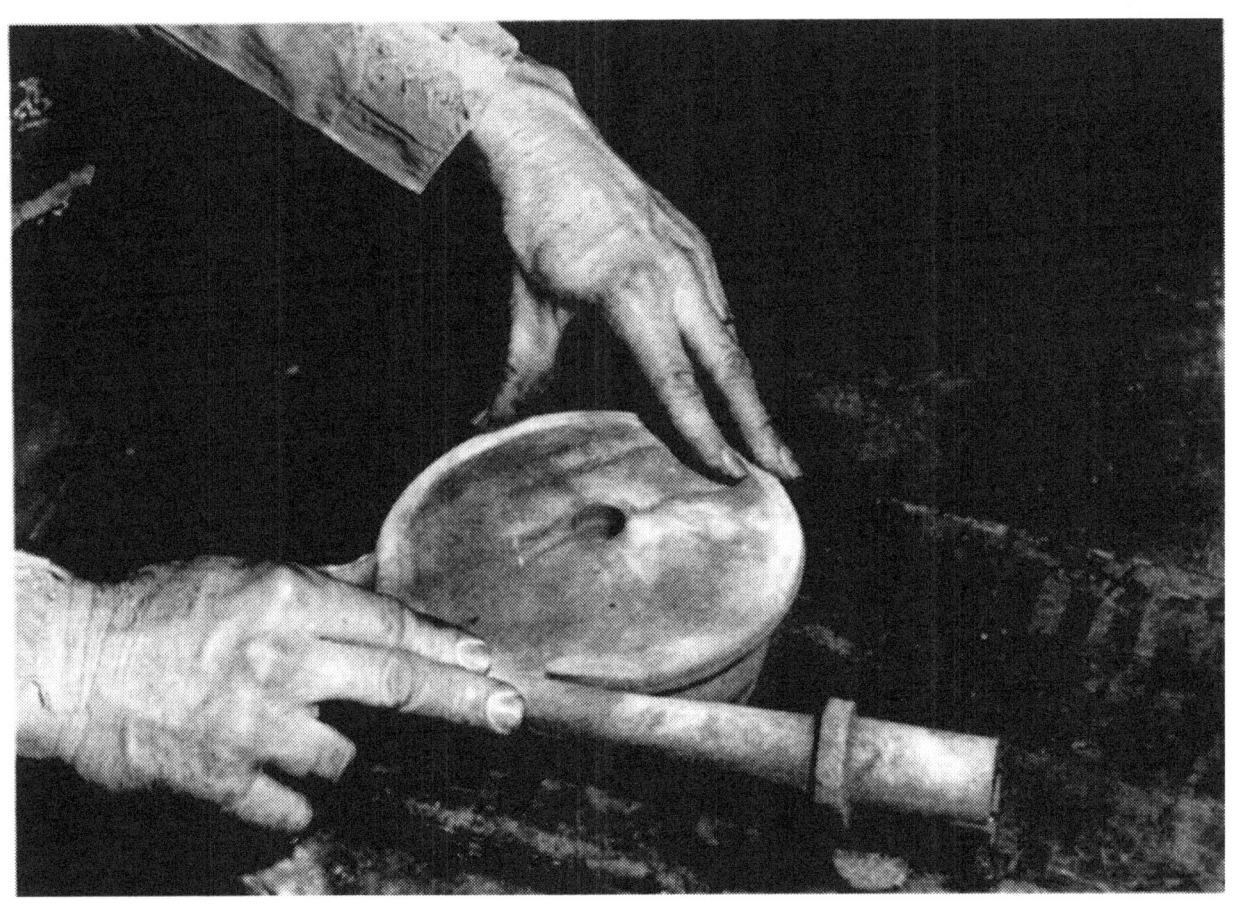

Figure 110. Assembling the two core halves

Chapter VII
GATES, RISERS AND CHILLS

GENERAL PURPOSE

Gates, risers, and chills are closely related. The function of one cannot be explained without reference to the others. This interrelationship is also carried into the casting itself. The best gating practice can be nullified by poor risering practice, and improper use of chills can cause the scrapping of well-gated and properly risered castings. The purpose of the gating system is to deliver the molten metal to the mold. The risers are used to supply liquid metal to compensate for solidification shrinkage in heavy sections; that is, they "feed" the casting. Chills are used to set up temperature gradients in a casting and permit full use to be made of directional solidification. They make one part of a casting solidify ahead of another. The proper use of gates, risers, and chills are important tools of the foundryman in producing a good casting.

GATING SYSTEM

A gating system should be able to do the following:

1. Permit complete filling of the mold cavity.

2. Introduce the molten metal into the mold with as little turbulence as possible so as to minimize gas pickup and prevent damage to the mold.

3. Regulate the rate at which the molten metal enters the mold cavity.

4. Establish the best possible temperature gradients within the casting so that directional solidification can be fully utilized, and prevent casting defects due to poor thermal gradients.

To achieve these aims, steps must be taken to control the following:

1. The type of ladle and ladle equipment.

2. The size, type, and location of sprue and runner.

3. The size, number, and location of gates entering the mold cavity.

4. The rate of pouring.

5. The position of the mold during casting.

6. The temperature and fluidity of the metal.

The various parts of a simple gating system are shown in figure 111.

GENERAL RULES OF GATING

The following general rules are given as a guide in making good gating systems:

1. <u>Use Round Sprues</u>. (a) Round gates or the closest approach to round gates are preferred. (b) A circular cross section has the minimum surface exposed for cooling and offers the lowest resistance to flow.

2. <u>Taper the Sprue</u>. The sprue should be tapered with the smaller end toward the casting. This makes is possible to keep the downgate full of metal when pouring. <u>Never</u> locate a tapered sprue so that metal is poured into the smaller end.

3. <u>Streamline the Gating System</u>. Gating systems having sudden changes in direction cause slower filling of the mold cavity, are easily eroded, and cause turbulence in the liquid metal with resulting gas pickup. Streamlining of the gating system eliminates or minimizes these problems. Avoid right-angle turns.

(a) Round sprues are preferred for sprue diameters of 3/4 inch or less. Larger sprues should be square or rectangular. However, a 3/4 inch diameter sprue is about the maximum size that can be kept full of metal while hand pouring.

(b) Wide flat gates and runners are preferred for light metal alloys.

4. <u>Use Patterns for the Gates</u>. The gating system should be formed as part of the pattern whenever possible. In the case of many loose patterns used aboard repair ships, gating patterns should be used instead of cutting the gates by hand. The use of patterns for the gates permits the sand to be rammed harder and reduces sand erosion or washing. Hand-cut gates expose loosened sand which is easily eroded by the flowing metal.

5. <u>Maintain Proper Gating Ratio</u>. There is a definite relationship between the cross-sectional areas of the sprue, runners, and ingates, to produce the best filling conditions for the mold. The rate of filling the mold should not exceed the ability of the sprue to keep the entire gating system full of liquid metal at all times. The cross section of the runner should be reduced in size as each gate is passed. An

example of such a gating system is shown in figure 112. This keeps the runner full throughout its entire length and promotes uniform flow through all of the gates. If this procedure is not followed in a multiple-ingate system, all of the metal will have a tendency to flow through the ingates farthest from the sprue.

An example of the use of gating ratio can be made with figure 112. Aluminum was used to make this flat plate casting, and one of the gating ratios that has proven successful for this type of casting is a 1:3:3 ratio. The first number refers to the cross-sectional area of the sprue base, the second number refers to the total cross section of all the runners from that sprue, and the third number refers to the total cross-sectional area of the ingates. In other words, the area of the sprue base is 1/3 that of the total area of the runners, and the total cross-sectional area of the runners equals the total cross-sectional area of the ingates.

The size of the ingate for this plate casting was selected to be 1/4 inch thick and 1-1/2 inches wide. The individual ingate then has an area of 3/8 square inch. There are four ingates, so the total ingate area is 4 x 3/8 square inch, or 1.5 square inches. The total runner area is then also 1.5 square inches, as determined by the gating ratio. Since there are two runners, each runner must have a cross-sectional area of 0.75 square inches. In figure 112, this is shown by the runner dimensions of 3/4 inch thick by 1 inch wide. To complete the gating system, the sprue base must have a cross-sectional area equal to 1/3 that of the runners. This is equal to 1/2 square inch. A sprue with a base diameter of 4/5 inch will satisfy this requirement.

6. Maintain Small Ingate Contact. The area of contact between the ingate and the casting should be kept as small as possible (unless gating through side risers as described later).

7. Utilize Natural Channels. Ingates should be located so that the incoming flow of metal takes place along natural channels in the mold and does not strike directly on mold surfaces or cores. The continuous flow of metal against a mold or core surface quickly burns out the binder and washes the loose sand into the casting.

8. Use Multiple Ingates. Unless a casting is small and of simple design, several ingates should be used to distribute the metal to the mold, fill it more rapidly, and reduce the danger of hot spots.

9. Avoid Excessive Ingate Choke. The ingate should not be choked at the mold so that it causes the metal to enter the mold at such a high speed that a shower effect is produced. Besides excessive turbulence and oxidation of the metal, the mold may not be able to withstand this eroding force. Choking of the ingate to assist in gate removal is a proper procedure if a number of ingates are used to allow an adequate amount of metal to enter the mold without jet action.

The recommended nomenclature for various types of gating is shown in figure 113. Additional information on gating systems for light metals is given in Chapter 15, "Aluminum-Base Alloys."

TYPES OF GATES

There are three general classifications for gates which are commonly used. They are: (1) bottom gates, (2) top gates, and (3) parting gates.

Bottom Gates. Bottom gates are most generally used because they keep mold and core erosion to a minimum. In spite of this, they have the very decided disadvantage of causing unfavorable temperature gradients in the casting, which make proper feeding particularly difficult and often impossible. Figure 114 shows the undesirable temperature gradients present in the bottom-gate and top-risered casting of figure 115, the latter showing the types of defect obtained with this method.

When using bottom gates, as the metal rises in the mold, it heats the mold with which it comes in contact. This produces relatively cold metal in the riser with considerably hotter metal next to the gate. In other words, there is hot metal and hot mold near the gate and cold metal in a cold mold near the riser. Such conditions are opposite to those desired for directional solidification in a casting. The risers should contain the hottest metal in the hottest part of the mold, and the coldest mold parts should be at points farthest removed from the risers.

Thus, bottom gating produces an unfavorable temperature gradient in the metal, the top of the casting is the coolest and the bottom is the hottest at the time the mold is filled. The amount of this temperature difference is related to the pouring rate, the rate of rise of the metal in the mold, and to the heat conductivity of the mold. A slow pouring rate will produce a temperature gradient more unfavorable than a fast pouring rate. When pouring slowly through a bottom gate, the metal has a greater opportunity to give up its heat to the lower portions of the mold than it would have if the mold were filled rapidly. The difference in temperature gradients due to slow and fast pouring is also shown in figure 114.

Unfavorable temperature gradients resulting from bottom gating are corrected to a slight extent by pouring through the riser as soon as the metal level becomes high enough. It is very difficult to perform this operation correctly.

Two types of bottom gates are shown in figure 116. The horn gate is also a bottom gate, but has the disadvantage of producing a fountain effect within the mold, causing mold erosion and entrapping air. In general, horn gates should not be used unless they are of the reverse type, as shown in figure 117. This type of horn gate has the large end of the horn at the mold cavity, as shown in figure 118. When using a horn gate, it is best to gate into a riser, as shown in figures 117 and 120, rather than directly into the casting.

Besides allowing easy flow of the metal into the mold, thereby reducing the erosion of the mold and core surfaces by the molten metal, bottom gating also results in quiet, smooth flow, thus reducing the danger of entrapped air.

The advantages of bottom gating without its disadvantages can be obtained if the casting is gated through a side riser as shown in figure 119. This type of gating produces the best conditions for directional solidification with a minimum of turbulence in the metal. In figure 119, it will be noticed that the wheel casting is gated through two side risers to permit a rapid pouring and filling of the mold. The molten metal will flow by two paths and meet approximately equally distant from the two risers. This will permit directional solidification to take place toward the hot risers. The riser at the hub of the wheel is necessary to feed the heavy section at the hub.

The bottom gate is often constructed with a well as part of the gating system, as in figure 121. The well acts as a cushion for the metal dropping down the sprue and prevents the erosion of sand, which is particularly apt to occur at the point of sudden change in the direction of flow. A special core (known as a splash core) may be used at the base of the sprue to minimize erosion of the sand.

Top Gates. Top gating of a casting is limited by the ability of the mold to withstand erosion, because the molten metal is usually poured through an open-top riser, such as shown in figure 122. Contrary to the characteristics of bottom gating, top gating has the advantage of producing favorable temperature gradients, but the disadvantage of excessive mold erosion. This method of gating is usually used for castings of simple design which are poured in gray iron. Top gating is not used with nonferrous alloys which form large amounts of dross when agitated.

For some heavy metal castings, the metal will be poured through a shower or pencil gate, as shown in figure 123. Pencil gates permit the metal to fall in a number of small streams and help to reduce erosion of the mold.

Parting Gates. Parting-line gates are used most frequently because they are the easiest for the molder to construct, particularly in jobbing work. In addition, it is usually possible to gate directly into a riser.

The main disadvantage of parting gates is that the molten metal drops in the mold to fill the drag part of the casting. Such a drop often causes erosion or washing of the mold. In nonferrous metals, dross formation is aggravated and air is often trapped to produce inferior castings. A typical parting gate is shown in figure 124. In this gating system, the sprue was used as a riser. A shrinkage defect formed in the indicated area because of improper feeding.

Gating through side risers should be used wherever possible. If this procedure is not used, hot spots will cause shrinkage defects. Gating directly into the casting produces hot spots, because all of the metal enters the casting through the gates and the sand near the gates becomes very hot and retards cooling of the metal. Unless risers are provided for feeding these portions of the casting with molten metal, cavities or shrinkage defects will be formed. Figure 125 shows gating into the riser with a parting gate.

Whirl gates, such as shown in figure 126, are sometimes used with heavy metals and parting gates. The purpose of these gates is to collect dross, slag, eroded sand, and to trap it, allowing only clean metal to enter the casting.

Step Gating. There is a fourth type of gating which is sometimes used. It is described here for information purposes only and its use is not recommended. The theory behind the step gate is that as the metal rises in the mold, each gate will feed the casting in succession. This would then put the hot metal in the riser where it is desired. Recent studies have shown that step gates do not work this way. To get proper step-gate feeding, a complicated step-gating design must be used. The use of step gates for castings normally made aboard repair ships is not recommended. A simple step gate is shown in figure 127.

USEFUL PROCEDURES

There are two ways in which the advantages of bottom gating can be obtained without serious disadvantages. They are: (1) mold manipulation, and (2) gating into blind risers.

Mold Manipulation. Mold manipulation makes it possible to keep mold erosion to a minimum during pouring, and by altering the position of the mold, to obtain temperature gradients even more favorable than those obtained by top pouring. The mold is tilted with the ingate end lowest. After pouring is finished, the mold is turned through an angle of 30°, 100°, or 180°, depending on the design of the part. For mechanical reasons, 100° and 180° manipulations are limited to small and medium castings of suitable design, but 30° manipulations are common for both large and small castings. A 30° partial reversal for a bottom-gated casting is shown in figure 128.

The gating system shown in figure 128 was devised to insure the flow of metal through the bottom ingate (horn gate) until the metal reached the bottom of the riser. After this, the balance of the mold is automatically filled through the upper gate and riser. This insures heating of the riser cavity and the proper conditions of hot metal and hot mold at the riser and cold mold and cold metal at the farthest point from the riser. This type of gate has the disadvantage that it is more difficult to mold and requires the use of a core. A pouring angle of 10° or 15° is found satisfactory for proper bottom gating. This enables the molten metal to travel forward as an unbroken stream, instead of fanning out over the entire mold cavity. This mold is then reversed through 30° to 40° after pouring to produce better feeding from the riser.

Total Reversal. The most favorable temperature gradient in both metal and mold may be obtained by the "total reversal" method as shown in figure 129. In this case, the feed head is molded on the bottom, with only small vents on the top of the mold, and the sprue enters the riser at the lowest point to prevent draining after reversal. After the casting is poured, the vents and the sprue are immediately sealed with wet sand and the mold reversed through an angle of 180° to bring the risers directly above the casting. The 180° reversal is used in the casting of what are commonly called "billets." There may be sufficient demand aboard ship for billets for the machine shop to warrant making a special rig to assist in reversing the mold.

Gating Into Blind Risers. By gating into blind risers attached to the lowest part of the casting, it is possible to take advantage of the bottom-gating system and not suffer from the formation of shrinkage cavities. In order to make a blind riser function well in such cases, it is best to have the gate enter directly into it. The proper use of blind risers is discussed later in this chapter.

POURING CUPS AND BASINS

Pouring cups make it easier to pour the molds. There are a few general principles which must be considered when designing a pouring cup. The inside diameter of the cup at the top should be about 2.5 to 3 times that of the sprue diameter. The inside walls should be at a steep angle, so that the cup is easy to make. Cups for small sprues usually require shoulders as shown in figure 130, so that the cup will have sufficient depth. The hole in the bottom of the cup should exactly match the top of the sprue.

When designing a pouring cup, it should not be too small or it will be impossible to pour metal into it fast enough at the start to completely fill the sprue in time to prevent dirt or slag from flowing down the sprue. A shallow cup is difficult to fill without splashing and is more difficult to keep filled during pouring.

Pouring cups can be made out of backing sand with extra bonding material added so the cups will bake hard in an oven. The inside surface of the cup should be coated with silica wash to make it more resistant to erosion.

In the pouring of steel, it is necessary to use larger sprues and larger pouring cups than for cast iron, bronze, or aluminum. These latter metals are much more fluid than steel. The cups shown in figure 130 are adequate for steel and may be reduced in size for the other metals.

Performed pouring cups are much better than a simple depression scooped out of the top of the cope at the sprue. The disadvantage of a hand-cut pouring aid is that the sand is loosened and sharp corners are present so that sand is readily eroded by the flowing metal and carried into the casting. A pouring cup should be used whenever possible.

A pouring basin serves two additional purposes as compared with a pouring cup. It not only makes it easier to pour the mold, but it also regulates the flow of metal into the mold and aids in trapping and separating slag and dross from the metal before it enters the sprue. A simple pouring basin is shown in figure 131. To make a pouring basin work properly, a plug should be used with it. The plug can be made from core sand or a graphite rod. It should be long enough to extend well above the pouring basin. It is good practice to have a wire or thin metal rod fastened to the plug to make it easier to pull the plug from the basin. Refer to Chapter 9, "Pouring Castings," for the proper use of a pouring basin.

RISERS

The principal reason for using risers is to furnish liquid metal to compensate for solidification shrinkage in the casting. In addition to this main function, a riser has other reasons for its use. It eliminates the hydraulic-ram effect (similar to water "pound" when a valve is closed suddenly), shown when the mold is full, flows off cold metal, and vents the mold.

Just at the time that a mold is completely filled with metal, there can be a sudden and large increase in pressure in the mold because of the motion of the flowing metal. This added pressure may be enough to cause a run-out of the casting or may produce a deformed casting. A riser permits the metal to flow continuously into it instead of coming to a sudden stop. This reduces the pressure or hydraulic-ram effect which produces these defects. An open riser permits the man pouring the mold to see how rapidly the mold is filling and provides him with a means to regulate the flow of metal.

When a casting must be poured rapidly, the permeability of the sand is not capable of permitting air and gases to escape quickly enough. In such a case, a riser provides an easy exit for the gases.

GENERAL RULES OF RISERING

The most important function of a riser is that of a reservoir of heat and molten metal. To be effective, it must be the last portion of the casting to solidify. There are four primary requirements which a satisfactory riser should meet:

1. The volume of the riser should be large enough to compensate for the metal contraction within the area of the casting it is designed to feed.

2. Enough fluid metal must be in the riser to penetrate to the last cavity within its feeding area.

3. The contact area of the riser with the casting must fully cover the area to be fed, or be designed so that all the needed feed metal in the riser will pass into the casting. See figure 143.

4. The riser should be effective in establishing a pronounced temperature gradient within the casting, so that the casting will solidify directionally toward the riser.

Accordingly, the shape, size, and location of the riser must be effectively controlled.

Riser Shape. The rate of solidification of a metal varies directly with the ratio of surface area to volume. In other words, for a given weight of metal, the shape which has the smallest surface area will take the longest time to solidify. The ratio of surface area to volume is obtained by dividing the surface area by the volume. In table 20 are listed some of the solidification times for various shapes of steel castings having the same weight.

TABLE 20. COMPARATIVE TIME FOR SOLIDIFICATION OF VARIOUS STEEL SHAPES

Form and Size of Riser	Volume, cu inch	Weight, lb	Area, sq inch	Amount Solidified in 1 Minute, lb/cu inch	Time to Completely Solidify, minutes	$\frac{A}{V}$
Sphere: 6-inch diameter	113	32	100	42.7	7.2	0.884
Cylinder: 4-1/4 inches by 8 inches	113	32	120	51.2	4.7	1.062
Square: 3-5/8 inches by 3-5/8 inches by 8-5/8 inches	113	32	135	57.5	3.6	1.194
Plate: 2-1/4 inches by 6-1/4 inches by 8 inches	113	32	160	68.4	2.7	1.416
Plate: 1-25/64 inches by 10-5/32 inches by 8 inches	113	32	220	93.8	1.5	1.947

When the ratio of surface area to volume is plotted against the solidification time (as in figure 132), a smooth, curved line is produced. The sphere which has the lowest ratio of surface area to volume and the longest solidification time would be the ideal shape for a riser. Because of molding difficulties it is impossible to use the sphere as a riser. Therefore, the next best shape, that of a cylinder, is often used. Blind risers make the closest approach to the spherical riser because they use a cylindrical body with a spherical dome.

Molten metal in the corners of square or rectangular risers solidifies rapidly because of the large amount of surface area to which the metal is exposed. Figure 133 is a sketch showing that square risers are only as effective as an inscribed circular riser would be. The metal in the corners of square or rectangular risers is wasted.

There may be times when risers must be elliptical, square, or irregularly shaped where they join the casting, but they should be constructed in such a manner that they are cylindrical above the neck of the riser.

Riser Size. Practical foundry experience has shown that the most effective height of a riser is 1-1/2 times its diameter in order to produce maximum feeding for the minimum amount of metal used. Any riser higher than this is wasteful of metal and may be actually harmful to casting soundness. A riser having incorrect height and one with recommended height are shown in figure 134.

The problem of determining the correct riser diameter for feeding a given section is somewhat more difficult. A safe approximation is to assume that the riser has the same volume as the section it is to feed. As an example, a flat plate 1 x 4 x 8 inches has a volume of 32 cubic inches and a surface area of 88 square inches. The volume to area ratio (V/A) is 0.364. The riser necessary to feed this section will also have to have a volume of 32 cubic inches. Since the riser height is set at 1.5 times its diameter, the formula for the volume of a cylindrical riser is $V = 3\pi r^3$. The solution of $32 = 3\pi r^3$ results in a radius of 1.5 inches. The riser then has a 3-inch diameter and 4.5-inch height. The volume to area ratio of the riser is 0.566, as compared to the casting V/A ratio of 0.364. As a result, the riser will solidify after the casting and should feed properly. These figures were computed for a side riser. The riser size may be changed slightly, depending on the experience gained with various castings.

Records of successful risering arrangements are useful in determining the size of the risers that feed various shaped sections correctly. Records that are used as a reference for determining riser size and location reduce the time necessary in making the mold and producing a good casting

Good and bad risering practice with respect to size and shape are shown in figures 135 and 136. Figure 135 shows a cylinder casting in which risers of excessive height and square shape were used. This casting was also overgated. The yield on this casting was 43 percent. That is, only 43 percent of the metal poured was in the casting. The same casting, with proper gating and risering, as in figure 136, had a 77 percent yield. Notice that the risers are round and that the height is approximately 1-1/2 times the diameter.

Riser Location. Heavy sections of a casting have a large amount of solidification shrinkage which must be compensated for from an outside source. Heavy sections, therefore, are the locations for risers. Figure 137 shows a cast wheel with top risers at the rim and spike junctions and at the hub. An important point to remember in the risering of a casting is that the hottest metal must be in the riser if it is to be effective. A risering arrangement that resulted in cold metal in the riser is shown in figure 138. This system produced a leaky casting when tested under hydrostatic pressure. When the gating and risering were changed so that the last and hottest metal was in the riser, as in figure 139, a sound casting was produced.

Another factor that must be remembered in risering is that it is impossible to feed a heavy section through a thin section. The thin section will freeze before the heavy section has completely solidified and a shrink will result. The diagram in figure 140 illustrates this. The two heavy sections are fed by their respective risers. The section on the right, however, has its heaviest part (C_2) separated from the riser by a reduced section (C_1). Section C_1 will solidify and be fed from the riser C before part C_2 has solidified. As a result, a shrink defect will be found at D. A method of preventing this defect would be to use a blind riser to feed the section at D.

Many times, a heavy section is so located in a casting that it cannot be fed with an open riser. In such a case, a blind riser is effective in feeding the section. Figure 141 shows a flanged casting which has been gated with an open riser. This type of riser location has the disadvantages that the riser is hard to remove from the casting and it is not possible to gate into the riser. This same casting is shown in figure 142 with a blind riser. It will be noticed that it is gated through the riser to make the best use of directional solidification and arranged to make removal of the riser easier.

Another point which must be considered in the location of a riser is that of the contact area of the riser and the casting. It must be remembered that the contact area of the riser must be large enough to permit feeding and small enough that it is not too difficult to remove the riser.

A good method for determining the size of riser necks involves the use of inscribed circles in roughly drawn sectional layouts as shown in figure 143. The maximum circle possible is inscribed in the area to be fed, and its diameter determined. The diameter of the contact (B), should be at least 1-1/2 times the diameter of the circle inscribed in the casting (A). For contacts of rectangular shape, the minimum dimension should likewise be 1-1/2 times the diameter inscribed in the casting section. There is no need for having the contact diameter more than twice the diameter of the circle inscribed in the casting section.

TYPES OF RISERS

There are two general types of risers, the open riser and the blind riser. The open riser is open to the air while the blind riser is not cut through to the surface of the mold. A blind riser cannot be seen when the mold is closed.

Open Risers. Open risers are used widely because they are simple to mold. Their greatest use is in large flat castings which have numerous heavy sections.

Blind Risers. Blind risers are advantageous because:

1. They facilitate bottom gating into castings by feeding the hot spot at the point of entry of metal. Gating into the riser also preheats the riser cavity and promotes greater feeding efficiency as well as proper temperature gradients within the casting.

2. They can be located at any position in a mold to feed otherwise inaccessible sections.

3. They are more efficient than open risers because they can be designed to closely approach the ideal spherical shape, thus substantially reducing the amount of riser metal required for satisfactory feeding. In addition, they are completely surrounded by sand, which eliminates the chilling by radiation to the air and keeps the metal liquid longer.

4. They are easier to remove from castings than open risers because they can be more strategically positioned.

An idea of the relative efficiencies of open and blind risers may be gained from the fact that open risers do not usually deliver more than 20 percent of their volume to the casting, whereas blind risers deliver as high as 35 to 40 percent. For the same casting, blind risers can be made much smaller than open types.

Blind risers operate in fundamentally the same manner as the open type except that it is not necessary to place them above the casting in order to feed properly. In common with open risers, the molten metal they contain must be kept open to the atmosphere, in order that atmospheric pressure may bear upon it, and proper feeding of the casting result.

Before discussing the means commonly used to keep blind risers open to atmospheric pressure, an explanation of the effect of this pressure on solidifying castings is necessary.

With metals solidifying as a continuously thickening envelope or skin, and contracting in volume as they freeze, a vacuum will tend to form within the casting, if the molten metal in the casting system (casting, risers, or gates) is not acted upon by atmospheric pressure, and if properly melted metal with a low dissolved-gas content is used.

Figure 144 illustrates the difference between keeping top risers open to atmospheric pressure and not keeping them open. Figure 145 shows the same condition for blind risers.

Figure 146 shows other fairly common types of casting defects attributable to this same phenomenon, which can be explained in the following manner. When the vacuum starts to form, the atmospheric pressure of 14.7 p.s.i. may collapse the casting walls, if they are weak enough, as shown in C, figure 147. It also may penetrate at a hot spot where the solidified skin is quite thin and weak. This usually occurs around a small core or at a sharp corner of the casting as shown in figure 146a and 146b. The importance of this effect cannot be overemphasized. It must be understood by foundrymen if their efforts are to be consistently successful, since its influence is felt in many ways in the production of castings.

The most successful method of introducing atmospheric pressure into blind risers involves the use of either a small-diameter sand core, or a graphite rod, placed in the riser cavity as shown in figure 147. A sand core is permeable enough to allow atmospheric pressure to enter and act on the last molten metal in the riser, which is at the center of the riser. Metal does not solidify rapidly around the sand core because it is small and does not conduct heat very rapidly. The cores are generally made of a strongly bonded oil sand and are reinforced with small wires or rods. The sizes used are as follows:

Risers up to 3-inch diameter—3/8 inch or 1/2 inch core.
Risers from 3-inch to 6-inch diameter— 5/8 inch or 3/4 inch core.
Risers from 6-inch to 10-inch diameter— 7/8 inch or 1-inch core.

The graphite rod has its best application in risers for steel castings although the sand core is satisfactory. The graphite rod is not altogether impermeable but most of the atmospheric pressure enters along the outside of the rod. The steel in the riser absorbs the carbon of the graphite rod which lowers its freezing point by 100°F. or more, thus, keeping it molten longer. Because of this carbon absorption, and the small mass of the graphite rod, the metal does not solidify around it, thus, permitting the entracen of atmospheric pressure. The high-carbon area left in the riser makes it necessary to use care in the selection of the riser size to make sure it does not extend into the casting. This method can also be applied to cast iron. The graphite rod works as well as the sand core in all cases and has the advantage of greater structural strength. (CAUTION: When graphite rods are used in risers for steel castings and a dead-melting practice is followed in the Rocking Arc or Induction furnace, undissolved graphite rods should be picked out of the risers to avoid carbon pickup when these risers are remelted.) In general, the sizes of graphite rods used are as follows:

Riser Diameter	Graphite Rod Diameter, inch
Up to 3 inches	3/16
3 to 5 inches	1/4
5 to 8 inches	5/16
8 to 12 inches	3/8

Blind risers have an advantage in addition to those previously given; namely, they have the ability to feed sections of castings in positions higher than their point of attachment. For example, sections of steel castings as much as 30 inches higher than the riser contact have been fed by blind risers. These castings were made in a laboratory under ideal conditions, however, and such practices should not be applied in the production of emergency castings where there is no time for experimentation.

Theoretically, it should be possible for a blind riser, when properly kept open to the atmosphere, to force steel upward into a void to a height slightly greater than four feet. Actually, this cannot be done because a true vacuum never exists in even the best-made steel, some gas coming out of solution to partially fill the void. Further, to count upon a completely sound skin in every case would not be practical, there always being the danger of eroded sand, ladle slag, or local mold disturbances rupturing this skin and thus breaking the vacuum and the feeding system.

Figure 147 shows a layout for the use of a blind riser for feeding the heavy section of a casting. The sketch shows the casting actually seven inches higher than the top of the blind head. In the manufacture of average castings, such as valves, this situation seldom exists. In general, the mass of the flange or other section is great enough to require a blind head very nearly as high as, or slightly higher than, the part it is to feed.

Referring again to figure 147, the metal poured into the ingate must flow first through the riser and then into the casting. As soon as the mold is completely filled, the metal loses temperature rapidly to the sand, and a skin of solid metal quickly forms at the mold-metal interface. This initial skin formation is shown as the cross-hatched areas of the figure. As temperature drops, more and more metal solidifies.

The atmospheric pressure acts like a piston on the metal in the blind riser, forcing it into the casting to feed shrinkage. In other words, the system is functioning on the principle of a barometer. Shrinkage is constantly tending to create a partial vacuum in the casting, and atmospheric pressure, acting through the medium of the molten metal in the riser, is constantly relieving it. If solidification proceeds properly with the parts most remote from the riser freezing first (cold metal in a cold mold cavity) and progressing thence toward the riser (hot metal in a hot mold cavity), each successive amount of shrinkage is compensated by additional fluid metal forced in from the riser. The ingate into the riser, being smaller than the neck leading to the casting, freezes off first and completes this part of the closed system.

Blind risers with pencil cores to produce atmospheric pressure on the molten metal in the riser work satisfactorily with most heavy metals. Blind risers may be used with light metals also, but the pencil core is not effective. Oxide films, formed on the surface of the molten light metals, prevent the pencil cores from functioning properly.

Use of Blind and Open Risers Together. When several risers are used at different levels in the same casting, it is essential that a particular zone of feeding be assigned to each riser. Figure 148 is a sketch showing the necessity for this precaution. The blind riser failed to function. A cavity was found in the casting in the position shown. The reason for this is that both risers were initially open to the atmosphere. The net advantage of the blind riser in this respect was zero. Because of the higher position of the open riser, the metal was forced through the system to actually feed the blind riser. By the time the narrow section of metal shown at (c) had solidified and shut off the hydraulic contact between the two risers, metal had solidified beyond the end of the sand core and made it impossible for atmospheric pressure to act. The section to be fed further was choked at (c) and solidified with the shrinkage shown. It is to be noted that the sand core, had it been placed through the blind riser at a lower point, would have aided in preventing this.

The casting could have been made perfectly sound without changing the method of risering to any appreciable extent by either one of two methods. One method would be to place external metal chills cast to shape, around the neck of the casting at point (c). This would chill the metal at this point, separate the two heavy sections, and allow each riser to function independently of the other. The other method would be to apply hydrostatic pressure to the blind riser by extending it to the surface of the mold as an open riser. This would then have equalized the hydrostatic pressure in the two risers. The assignment of an independent feeding zone to each riser is a very important part of properly feeding castings.

VENTS

Vents are a necessary part of any system of gates and risers. The function of a vent is to permit gases to escape from the mold cavity fast enough to avoid developing back pressure which would oppose the inflowing metal. Vents also prevent gases from becoming trapped in the metal and forming gas cavities. Vents should be taken off all high parts of molds, such as flanges, bosses, lugs, and care should be taken to make sure that they are open to the top of the mold.

Experience has shown that round vents large enough to evacuate mold gases at a proper rate will frequently reveal a fine shrinkage cavity in the casting when they are removed. Vents of rectangular section are preferred, and they should be kept comparatively thin so that the metal which flows into them will solidify quickly.

For small and medium castings, such as will be made for most emergency work, vents made by the use of molder's lifters or by a saw blade are satisfactory. This represents the desired thickness. It is better to use many thin vents than few large ones.

PADDING

Padding is used mainly in conjunction with risers to obtain directional solidification and is discussed here rather than under casting design. It is often possible to avoid the use of chills or extra risers by padding between heavy metal sections. If weight is important, or if the mechanical functioning of the part is affected, this padding is removed when the casting is machined. Figure 149 shows some typical cases in which padding is applied to avoid the use of chills or risers.

Padding is used to encourage directional solidification in members of uniform thickness. When used for this purpose, it is a tapered section of metal with the taper increasing in the direction of the feed heads. When uniform sections are made without padding, centerline shrinkage may occur. Uniform solidification in a member generally causes centerline shrinkage, but progressive solidification along the member gives a sound casting.

Uniform solidification can be prevented to a large extent by gating and risering, but in many cases this is not sufficient. Figures 150 and 151 show some typical applications of padding to obtain soundness or freedom from shrinkage.

Centerline shrinkage actually occurs on the thermal centerline of the member, since it is the last portion to solidify. Figure 152 shows where centerline shrinkage will occur in unpadded sections. In A, because of the lower heat-extracting capacity of the core completely surrounded by metal, the centerline shrinkage will be nearer the cored surface. To a lesser extent, this applies to B. In C, the thermal centerline will coincide with the section centerline.

The use of a special core to obtain padding is shown in figure 153.

When padding is used, it should be applied if possible on surfaces where it can be removed by machining.

HOT TOP AND ANTIPIPING COMPOUNDS

Methods of keeping top risers open to the atmosphere so that maximum feeding can be obtained are of great value to the foundryman. Materials which aid the foundryman in this respect fall into two calsses: (1) insulating compounds and (2) exothermic compounds.

Insulating Compounds. As the name implies, insulating compounds are used to insulate the riser and to reduce the heat lost by radiation to the air. They are usually spread on top of the riser after the pouring has been completed. Any of the commonly known insulating materials can be used. Examples of suitable materials are asbestos, sawdust, blacking, talc, and even dry sand. In unusual cases requiring small risers, insulating sleeves may be made from some of these materials and rammed into the mold to make the riser cavity. This procedure, along with insulating material on top of the riser, provides complete insulation of the riser. Care should be used to prevent excessive contamination of molding sands with these materials.

A comparison between an insulated riser and one not insulated is shown in figures 154 and 155. By good insulation of the top of a riser, it can be made smaller and still feed well. An important factor which must not be overlooked in the use of insulating compounds is that of absorbed moisture. Before using insulating compounds, they should be dried, especially if they are molded and rammed up in a mold.

Exothermic Compounds. Exothermic compounds are usually mixtures of aluminum with a metal oxide. When ignited by the molten metal in a riser, they burn and produce aluminum oxide, metal, and a large amount of heat. The reaction between aluminum and iron oxide in the thermite reaction for steel results in a temperature of 4,500°F.

The principal function of this type of compound is to supply heat to keep the riser molten longer. In the use of these compounds, a careful check should be made of their analysis to prevent any harmful elements from being picked up on later remelting. Because of the heat supplied by these compounds, shorter risers are required. A high riser will not permit the molten metal to feed properly and the desired effect will be lost. A comparison between an ordinary riser, an insulated riser, and one using an exothermic compound is shown in figure 156.

Carbonaceous materials may be used to produce better feeding in steel castings but the function is different. Some of the carbon in the material becomes dissolved by the steel in the riser. This lowers the melting point of the steel in the riser and it remains fluid for a longer period of time. There is a disadvantage to this type of riser compound; it produces an area high in carbon in the casting immediately below the riser, and this will produce variable carbon content when the risers are remelted.

CHILLS

When a heavy section of a casting is remote from a source of feed metal, and it is difficult to mold a riser in place, or hard to remove it in cleaning the casting, internal or external chills can be used to good advantage. Chills are metal shapes used to speed up the solidification in heavy sections, thus permitting the shrinkage that takes place to be fed through adjoining sections.

Chills are of two basic types, internal and external. Internal chills are cast into the casting and become a part of it. External chills are rammed up in the mold to form part of the mold surface and can usually be recovered for re-use.

INTERNAL CHILLS

It is difficult to set rules for the use of internal chills, their successful use being affected by many variables, such as chill composition, metal analysis, location, metal temperature, rate of pouring, chill surface, type of mold. In general, because of the many variables attendant to their use, they should be applied only in exceptional cases.

When used, their composition should be basically similar to the metal being cast; i.e., low-carbon-steel chills for steel and cast iron, copper chills for brass and bronze, and aluminum chills for aluminum.

Internal chills should always be very clean and dry. If they are not, gas will form as the molten metal surrounds them. This gas formation is the largest factor in the unsatisfactory behavior of chills. Oxide films, grease or oil, paint, mold washes, and moisture are all harmful. In green sand molds, the chills should not be placed until just before closing, and the mold should be poured immediately. If the mold cannot be poured immediately, it should be disassembled and the chills removed and kept dry. Internal chills should not be allowed to remain in molds during oven drying, since the fumes and moisture given off in drying will affect their surface adversely. Holes for receiving the chills should be made in the green mold, however, before it is dried.

The size of the internal chill is very important because its effect may prevent feeding if it is too large, and fail to accomplish anything if it is too small. The chill must fuse perfectly into the casting if soundness is to be obtained. Internal chills that are too large sometimes cause cracks in the cast metal.

The chill shape is very important. Figure 157 shows several types used in practice. By eliminating flat, horizontal surfaces and using surfaces which are streamlined, any gases formed are better able to rise and avoid becoming enveloped by the metal.

The location of internal chills is important. When used in bosses (a popular use for such chills), their location and size should be such that they will be completely removed in machining. Their location with respect to the metal flow in the mold is important. Internal chills placed directly in front of an ingate are quite likely to be melted and thus have no value, often being quite harmful. It is desirable to have some metal flow past a chill to wash away gases and aid in proper fusion. The amount of metal which will pass the chill must be considered in determining the size to use. Internal chills should not be used in sections which must be pressure tight or which must withstand radiographic inspection and magnetic powder testing. Their use, even in the hands of an expert, is not always completely successful.

EXTERNAL CHILLS

The use of external chills is favored whenever it is necessary to increase the rate of solidification in any part of a casting. These chills may be cast to shape in either iron, steel, bronze, or copper, or they may be formed of plates, bars, or rods. Figures 158, 159, 160, 161, and 162 show some typical applications of such chills.

General rules to be followed in applying external chills can be summarized as follows:

1. Their surfaces should be clean and accurately fit the area to be chilled.

2. The ends and sides of large, massive chills should be tapered. Too drastic cooling at the edges of chills may form casting stresses resulting in cracks (See figure 163).

3. Chills should be large enough so that they don't fuse to the casting.

4. They should not be so large that they cause cracking of the casting or interfere with feeding.

5. The area of contact between chill and casting should be controlled (See figure 164). Note that in this figure, the chills on the left have a larger contact area with the casting than the chills on the right. As a result, the amount of metal solidified by the chills on the left is greater than the amount of metal solidified on the right. This can be verified by comparing the thicknesses of metal shown in the figure. This shows that contact area between the chill and the metal is very important in determining the effectiveness of a chill.

Rules 3 and 4 may usually be met by using a chill equal in thickness to that of the casting section being chilled. An increase in thickness over this will not appreciably increase the rate of solidification.

If the surface of a long chill is rough, the normal contraction of the cooling metal may be restrained and a crack produced in the casting.

A common use for external chills is in corners or parts which are inclined to crack in the mold, due to contraction stresses. The use of brackets in preventing defects of this sort was previously discussed in Chapter 2, "Designing a Casting." Many times, brackets are of little help and chills must be used. By placing a chill in contact with such areas, the metal is more rapidly cooled to give increased strength at the time when the stresses would normally cause hot tearing of the casting.

In the discussion of casting design in chapter 2, it was mentioned that member junctions of L, T, V, X, and Y design which were inaccessible for feeding would be discussed further in this section. When designing such sections, fillets, must be kept to a minimum to avoid excessive increases in section thickness. In this regard, these sections differ from those which can be fed. With small fillets and the tendency to unsoundness in the center of the section, cracks are quite likely to be formed. External chills definitely reduce such defects (See figure 165). Figure 166 shows the preferred location of external fillet chills. This method also prevents cracking at the ends of the chills.

External chills applied to cast iron increase its solidification contraction and cause the iron in the chilled areas to become very hard. For these reasons, their use should be avoided in cast iron except in such cases where either of the above results is desired.

SUMMARY

It is difficult to consider or select gating, risering, and the use of chills separately because the three factors are interdependent. All three have a definite influence on each other and must be considered together if a sound casting is to be made.

A good gating system must supply clean metal to the mold cavity at a temperature and rate which will produce a casting free from defects. Risers must be capable of supplying hot molten metal to the casting to compensate for solidification shrinkage without causing any defects in the casting. Chills should be used only when necessary and then to assist in establishing proper temperature gradients within the casting.

In this chapter, no attempt has been made to discuss any problems relating to specific metals. Information on these metals will be found in following chapters dealing with the specific metals.

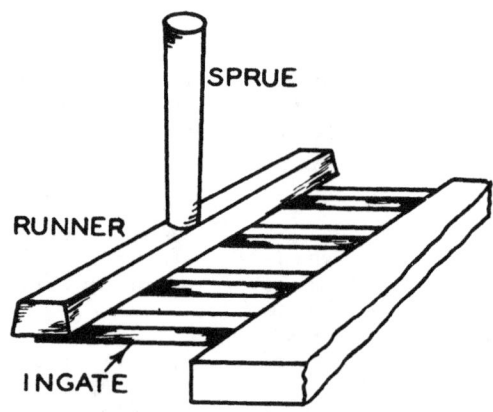

Figure 111. Parts of a simple gating system.

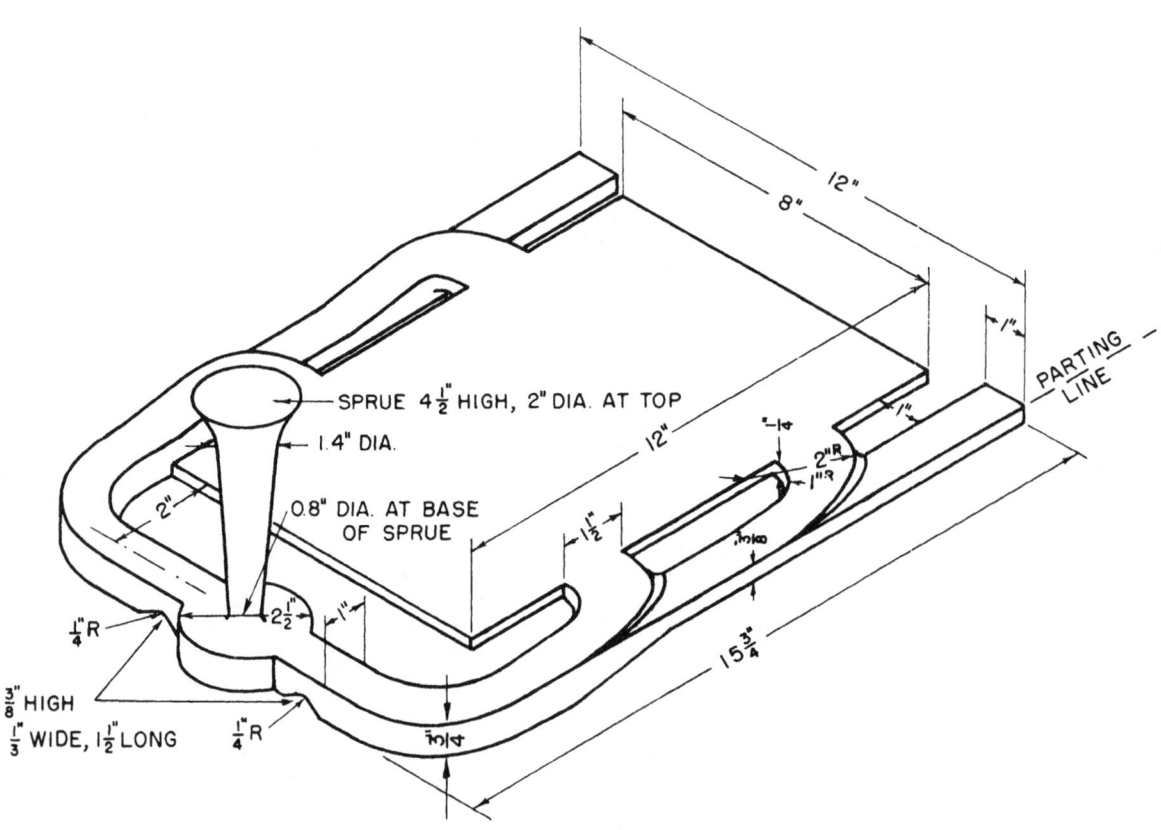

Figure 112. Illustration of gating ratio.

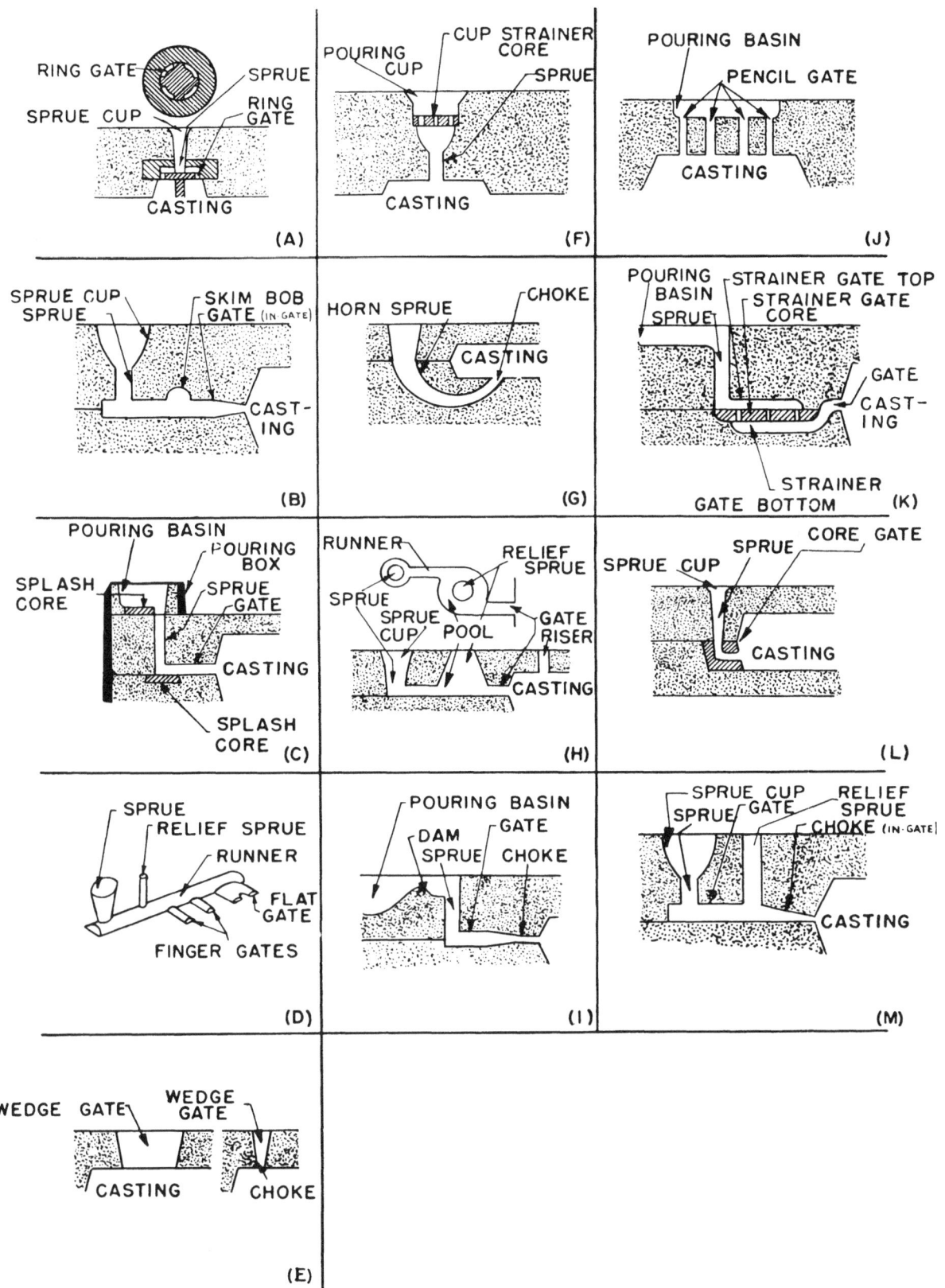

Figure 113. Gating nomenclature.

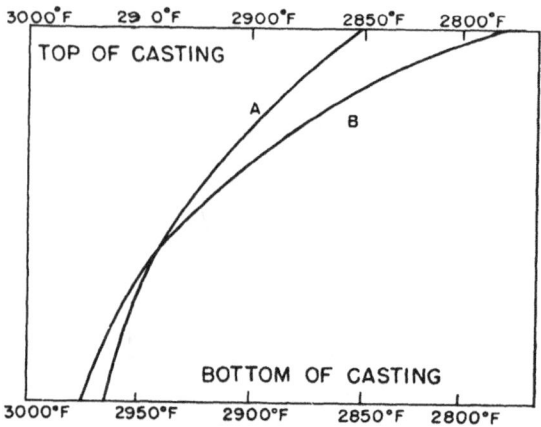

A - FAST POURING
B - SLOW POURING

Figure 114. Unfavorable temperature gradients in bottom gated casting.

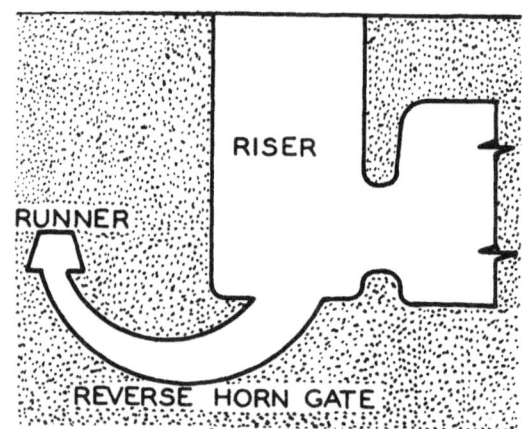

Figure 117. Reverse horn gate.

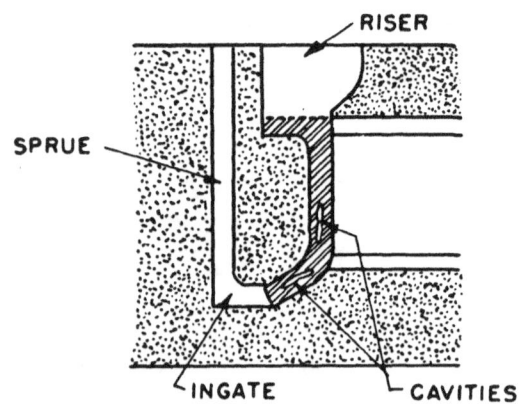

Figure 115. Defect due to bottom gating.

Figure 118. Reverse horn gate.

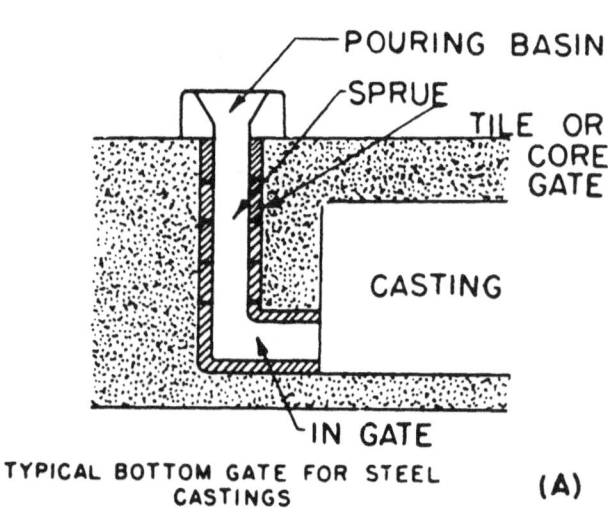

TYPICAL BOTTOM GATE FOR STEEL CASTINGS (A)

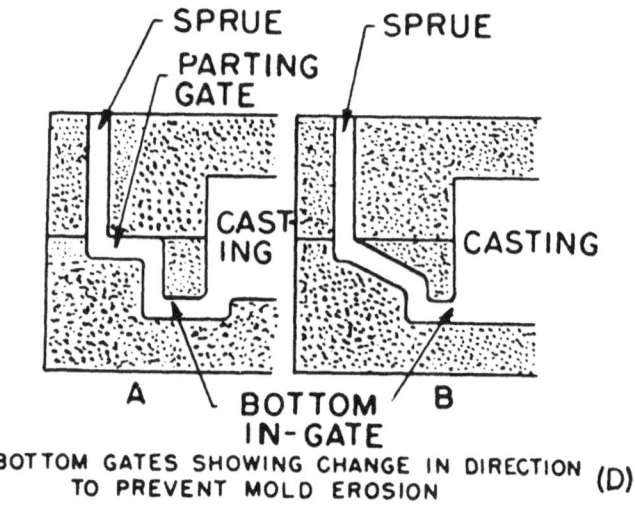

BOTTOM GATES SHOWING CHANGE IN DIRECTION TO PREVENT MOLD EROSION (D)

Figure 116. Bottom gate.

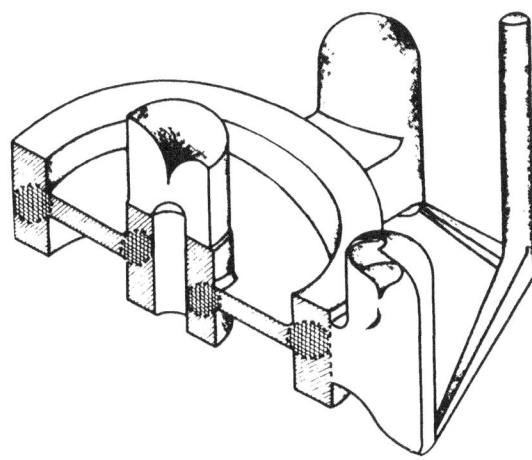

Figure 119. Bottom gating through side risers.

Figure 120. Bottom gating through riser with horn gate.

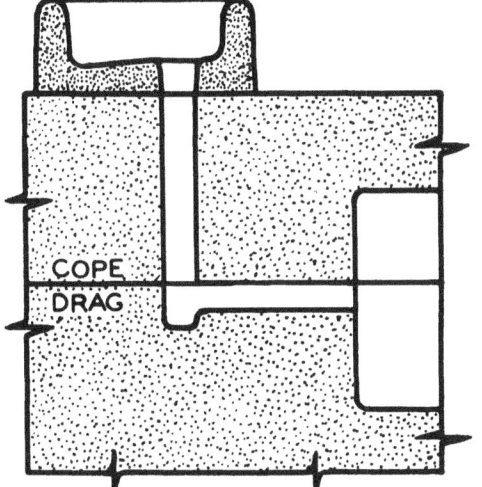

Figure 121. Sprue with well at base.

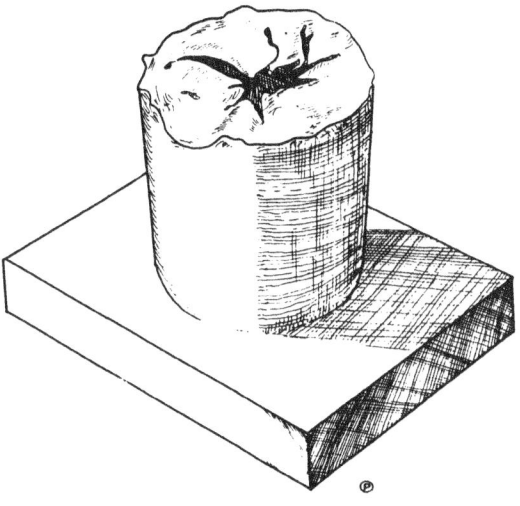

Figure 122. Simple top gating.

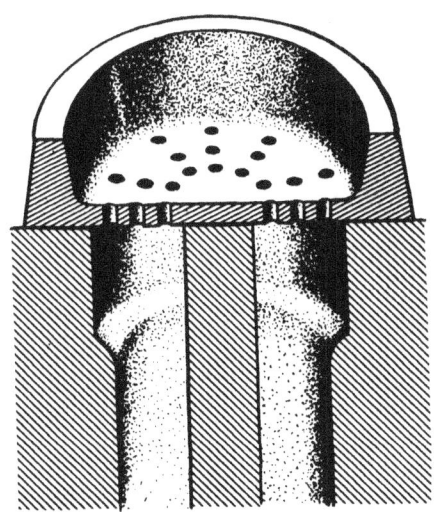

Figure 123. Pencil gate.

Figure 124. Typical parting gate.

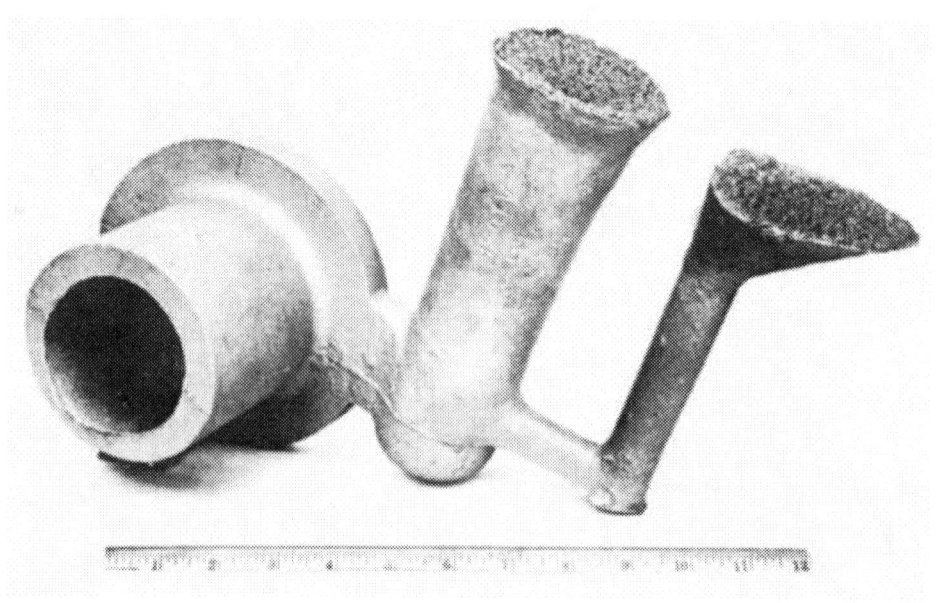

Figure 125. Parting gate through the riser.

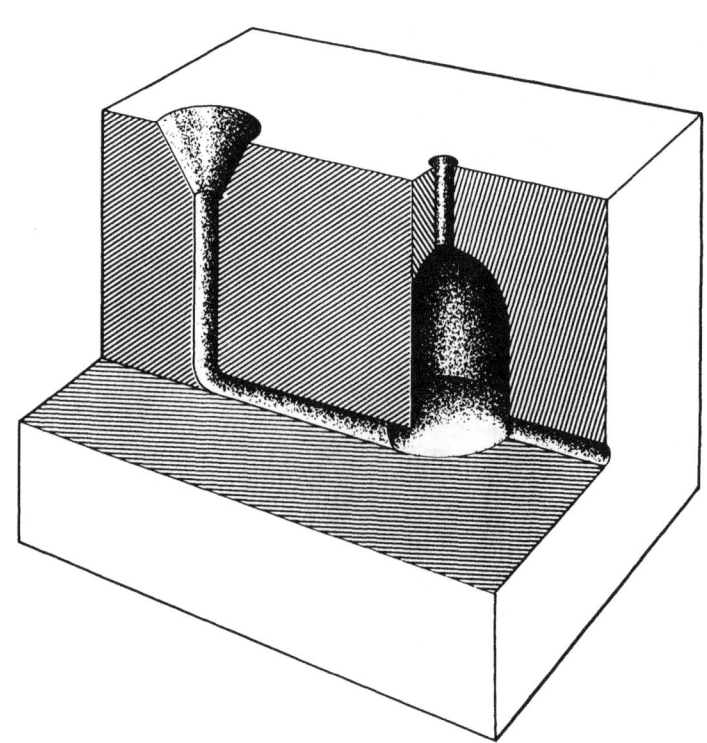

Figure 126. Whirl gate.

- 110 -

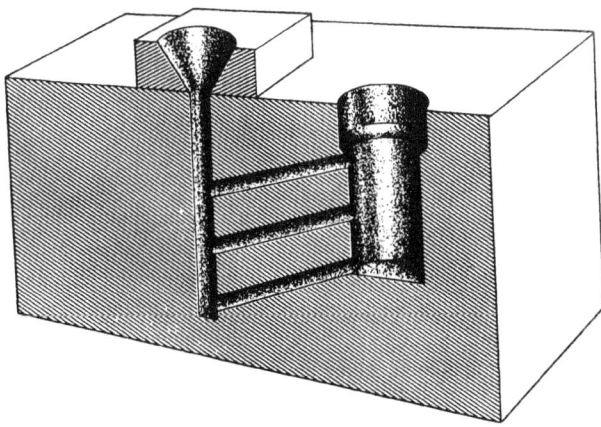

Figure 127. Simple step gate. (Not Recommended)

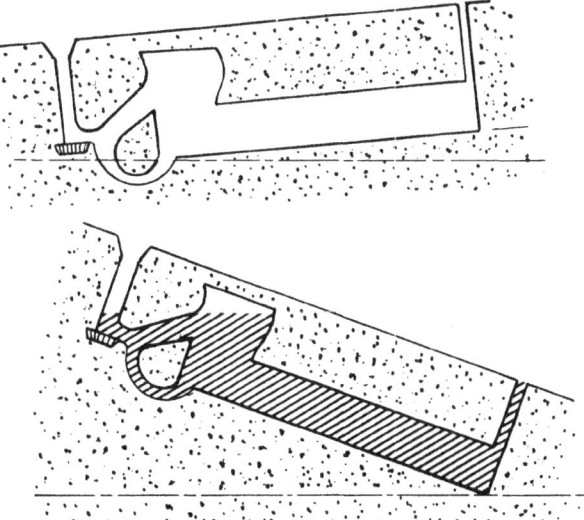

Figure 128. Thirty-degree mold manipulation.

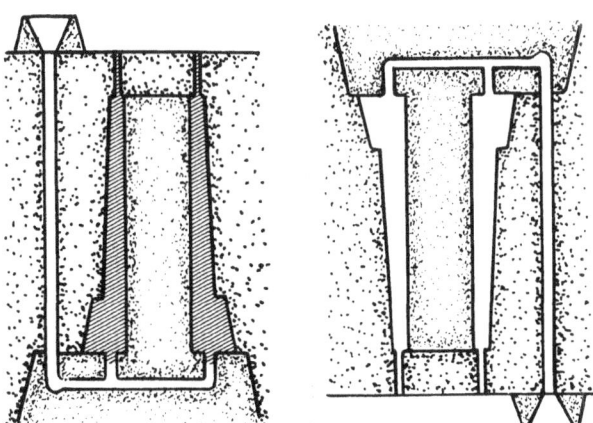

Figure 129. Complete mold reversal.

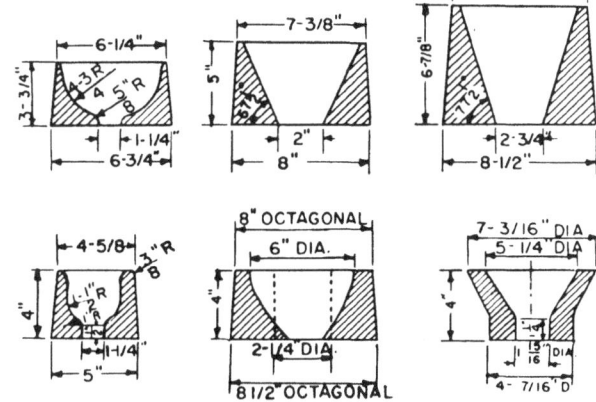

Figure 130. Pouring cups.

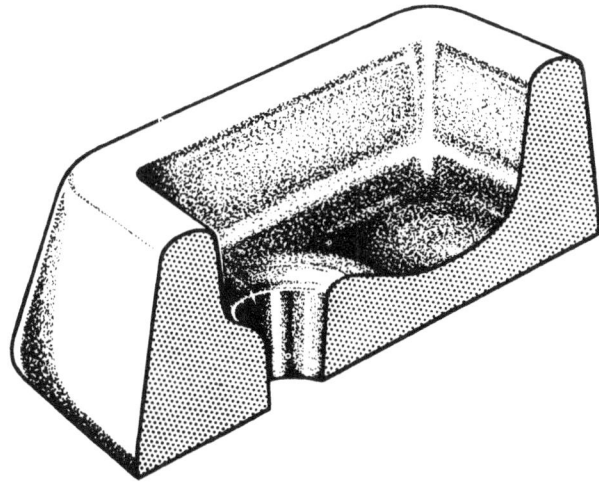

Figure 131. Pouring basin.

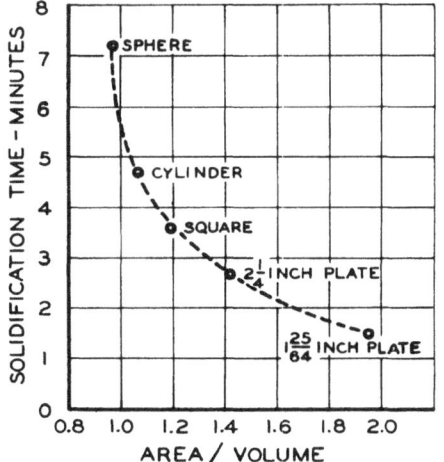

Figure 132. Solidification time vs A/V ratio.

- 111 -

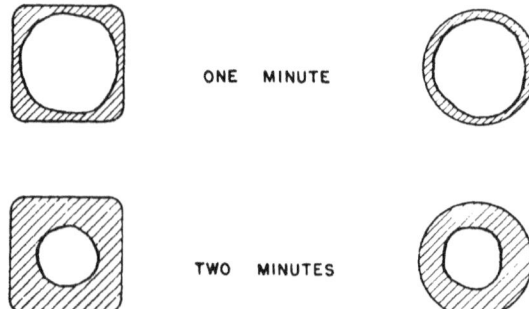

Figure 133. Effectiveness of square and roung risers.

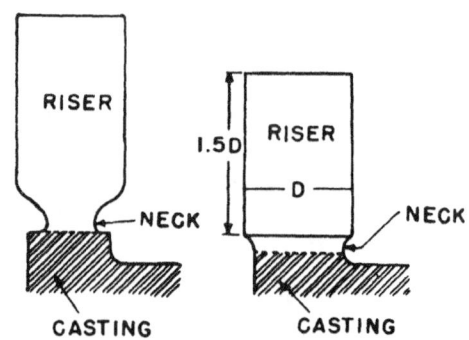

Figure 134. Proper and improper riser height.

Figure 135. Poor riser size and shape.

Figure 136. Proper riser size and shape.

Figure 137. Riser location at heavy sections.

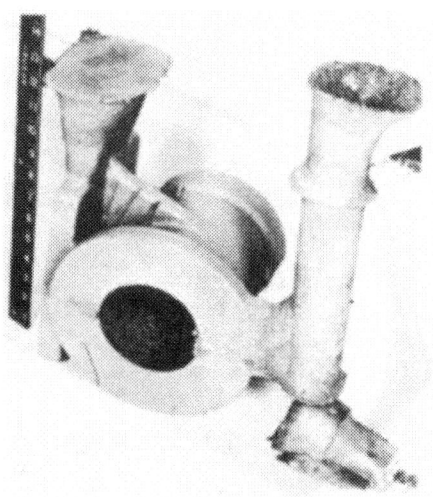

Figure 138. Cold metal riser. (Not Recommended)

Figure 139. Hot metal riser.

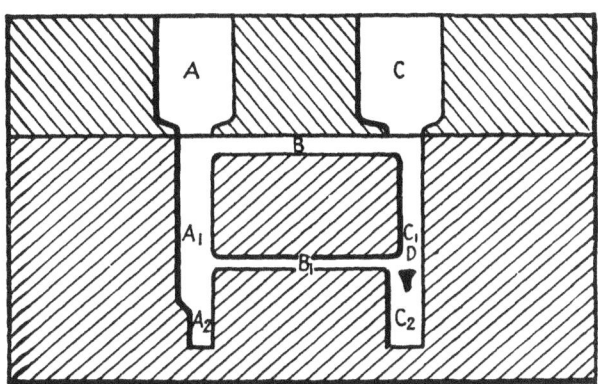

Figure 140. Feeding through a thin section.

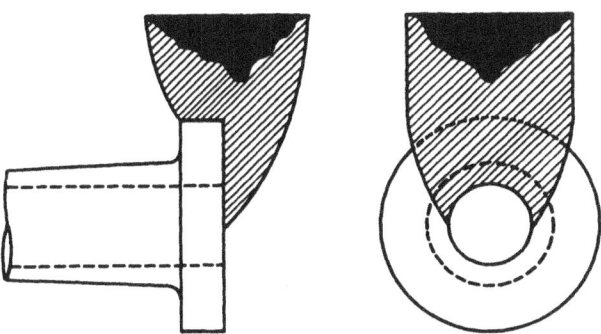

Figure 141. Flanged casting with open riser.

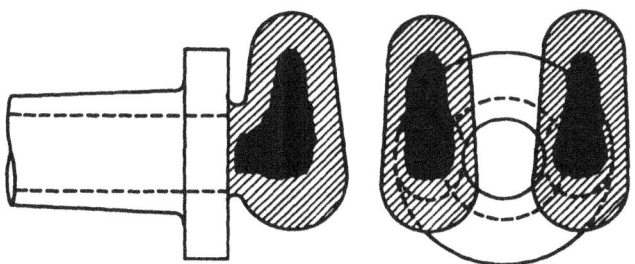

Figure 142. Flanged casting with bind riser.

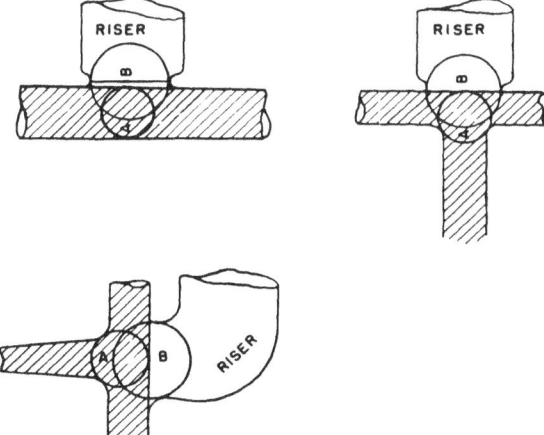

Figure 143. Inscribed circle method for riser contact.

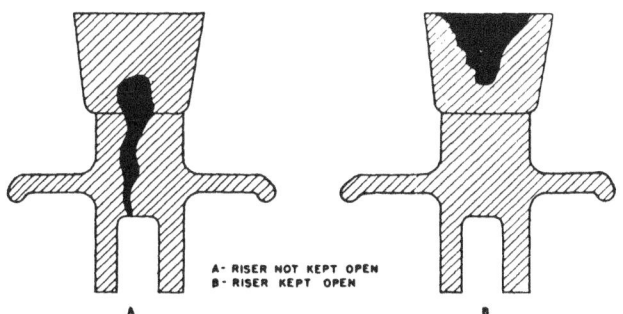

Figure 144. Effect of keeping top risers open.

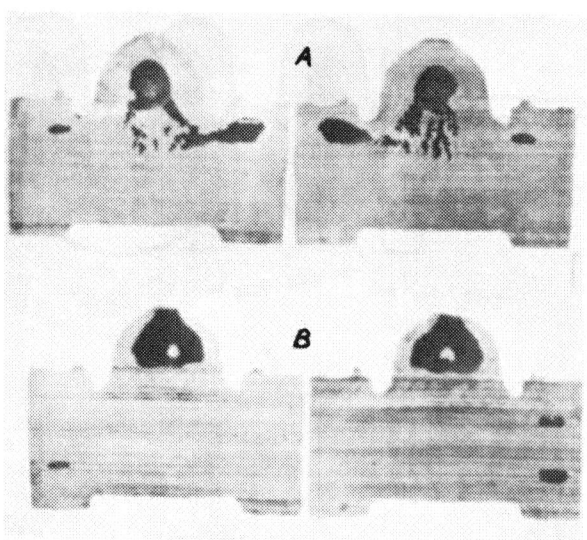

Figure 145. Effect of keeping blind risers open.

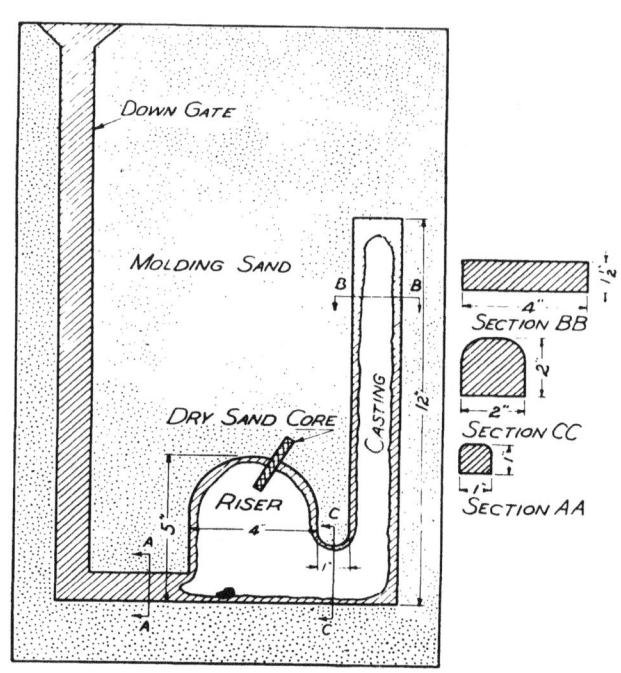

Figure 147. Blind riser principle.

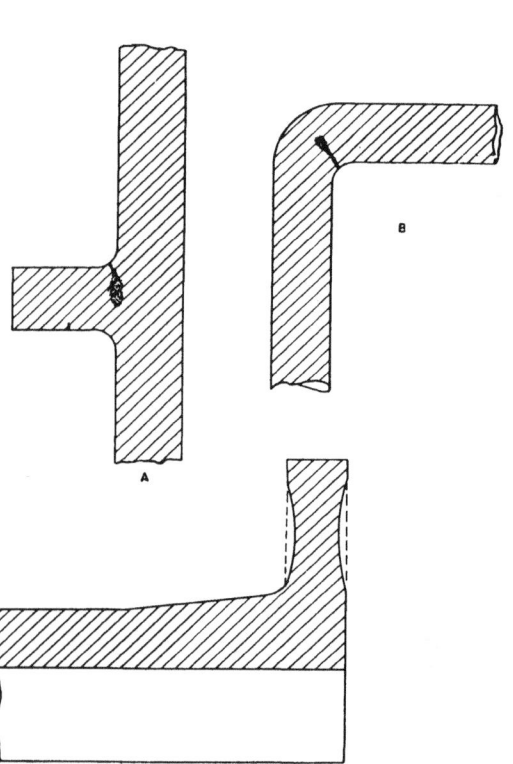

Figure 146. Casting defects attributable to shrinkage voids and atmospheric pressure.

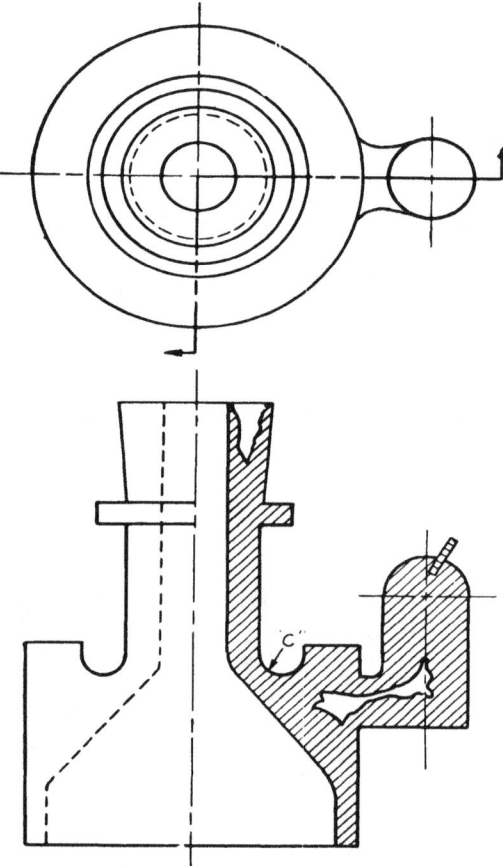

Figure 148. Individual zone feeding for multiple risers.

- 114 -

 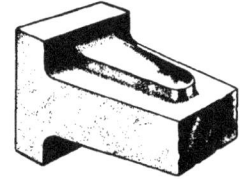

Figure 149. Padding to avoid the use of chills or risers.

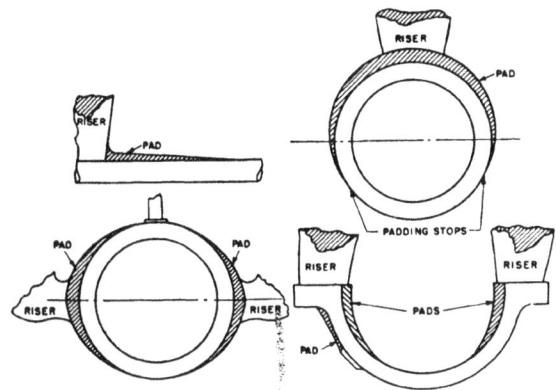

Figure 150. Padding to prevent centerline shrinkage.

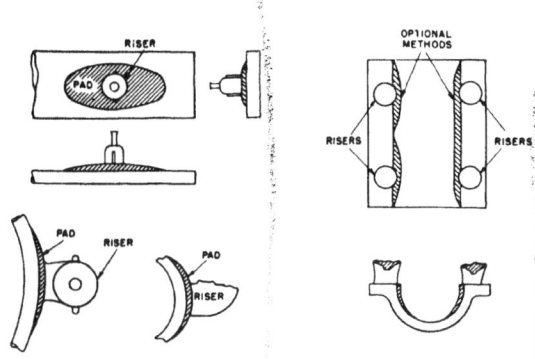

Figure 151. Typical padding of sections.

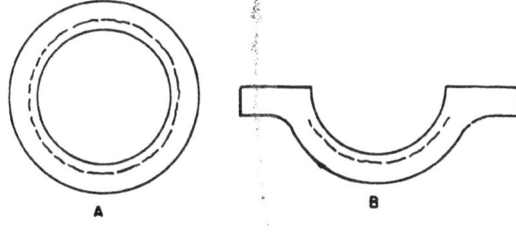

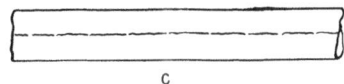

Figure 152. Shrinkage on the thermal centerlines of unpadded sections.

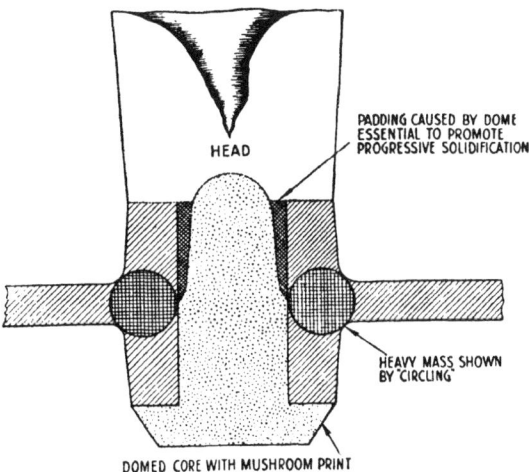

Figure 153. Use of a core to make a padded section.

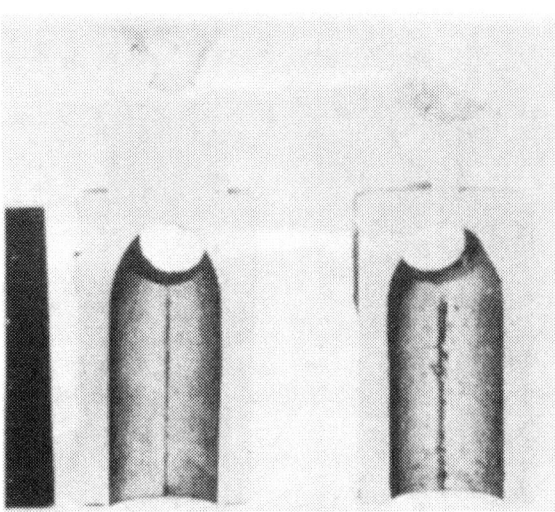

Figure 154. Effect of insulated risers.

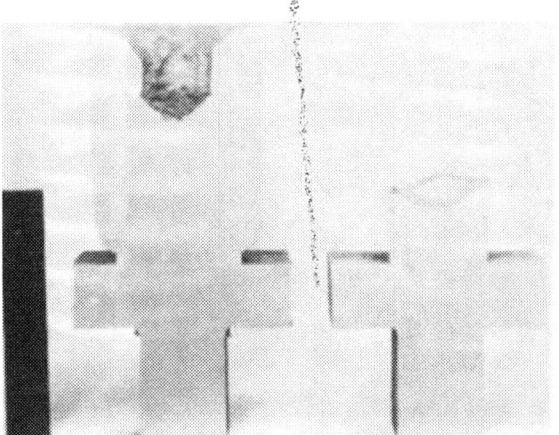

Figure 155. Reduction in riser size due to insulating.

- 115 -

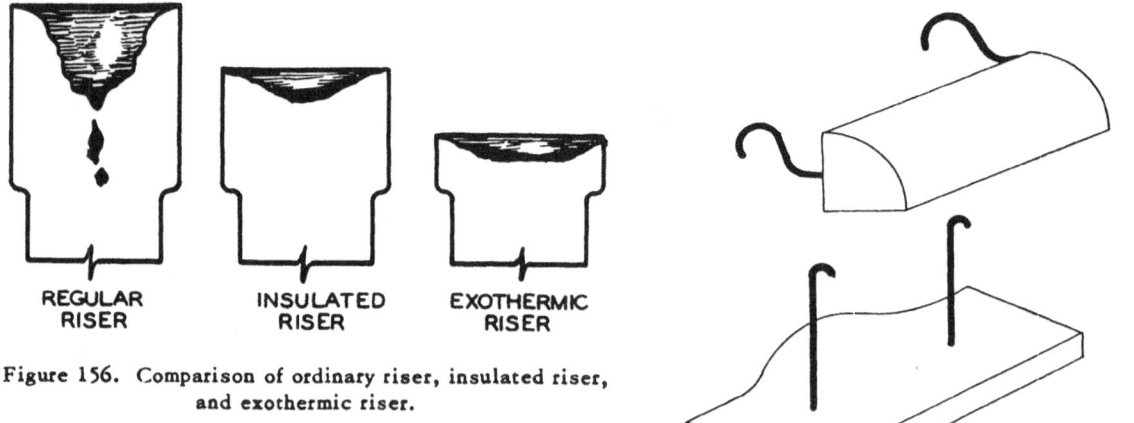

Figure 156. Comparison of ordinary riser, insulated riser, and exothermic riser.

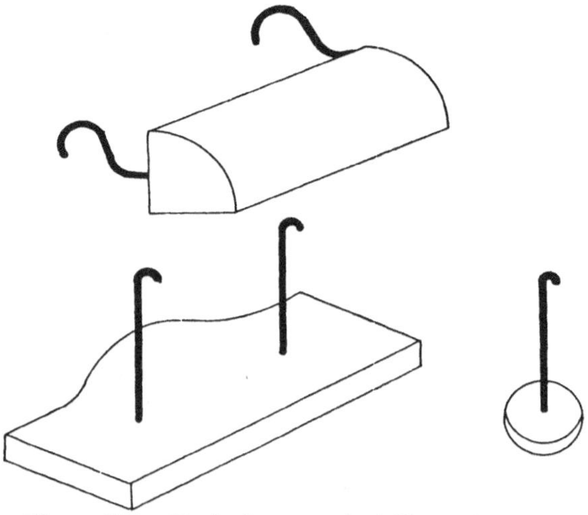

Figure 158. Typical external chills with wires welded on or in to hold chill in place.

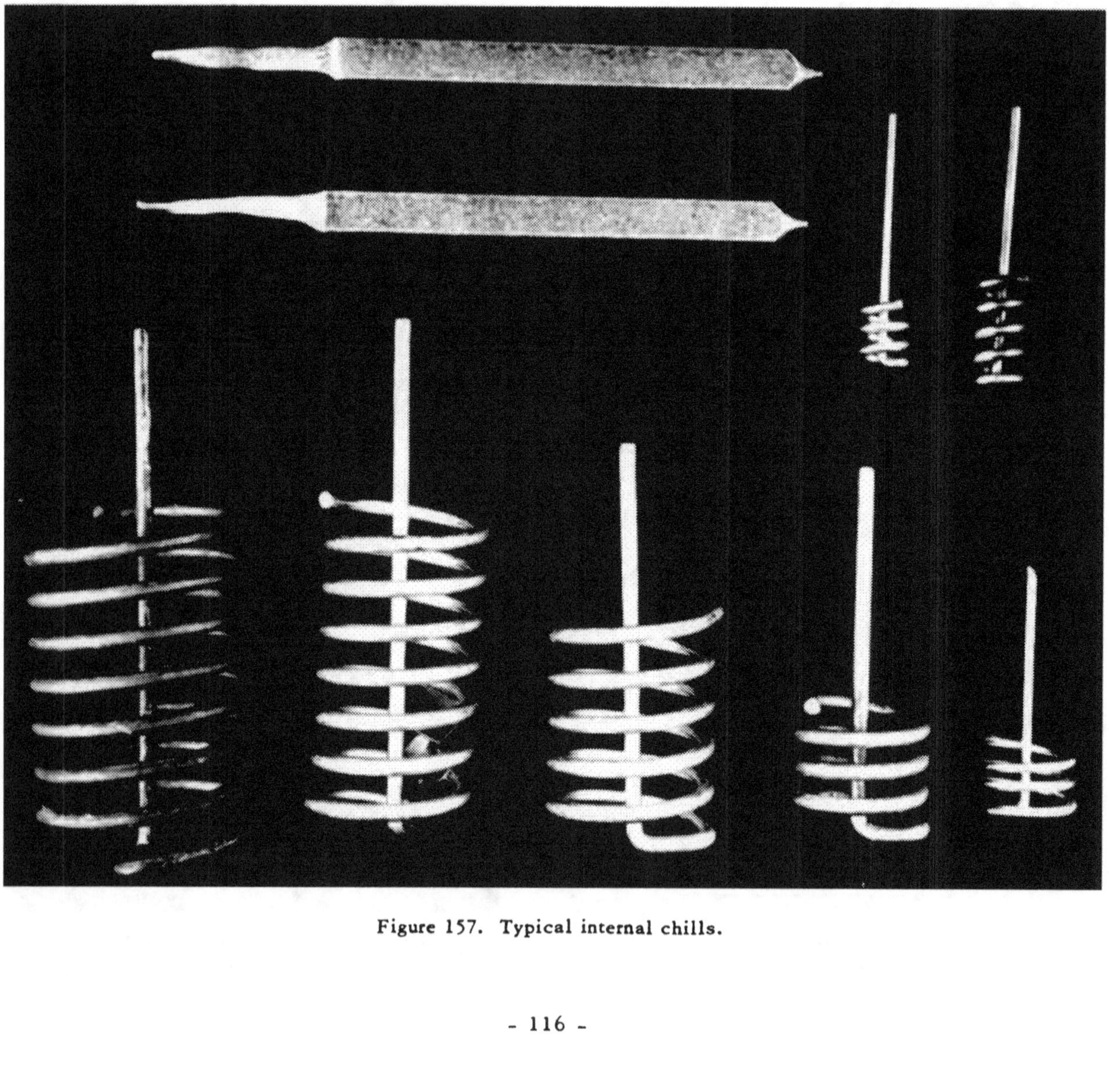

Figure 157. Typical internal chills.

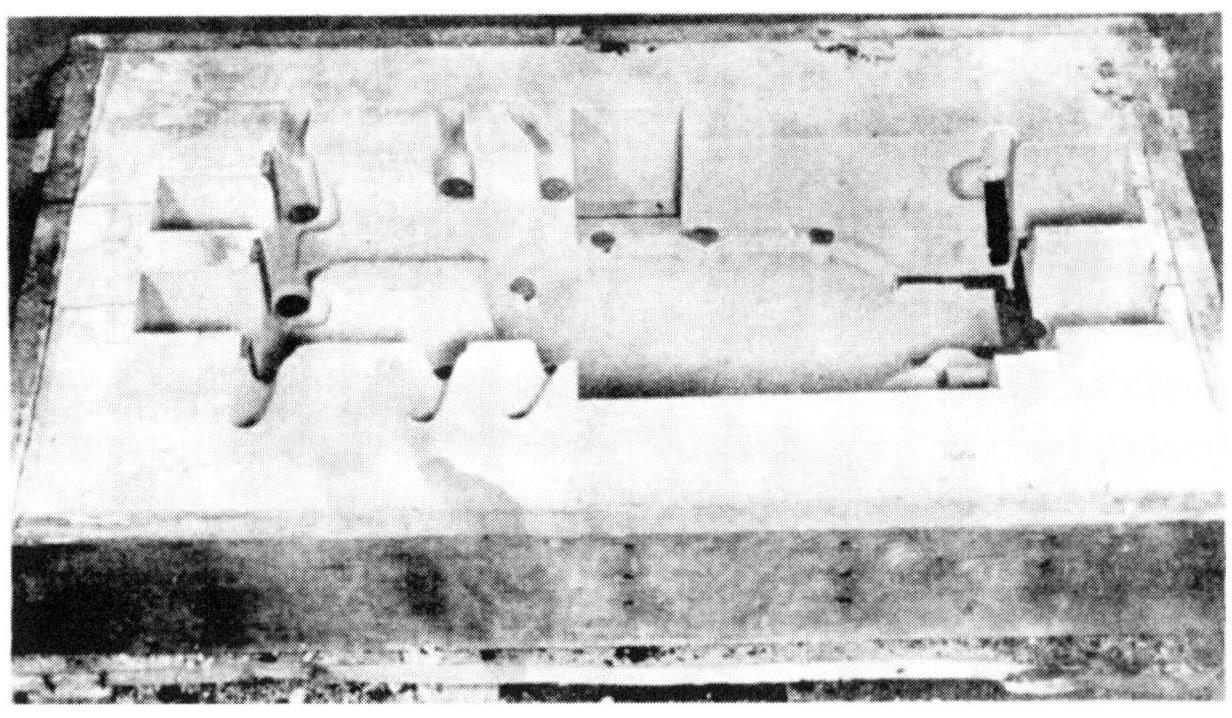

Figure 159. Use of external chills in a mold for an aluminum casting.

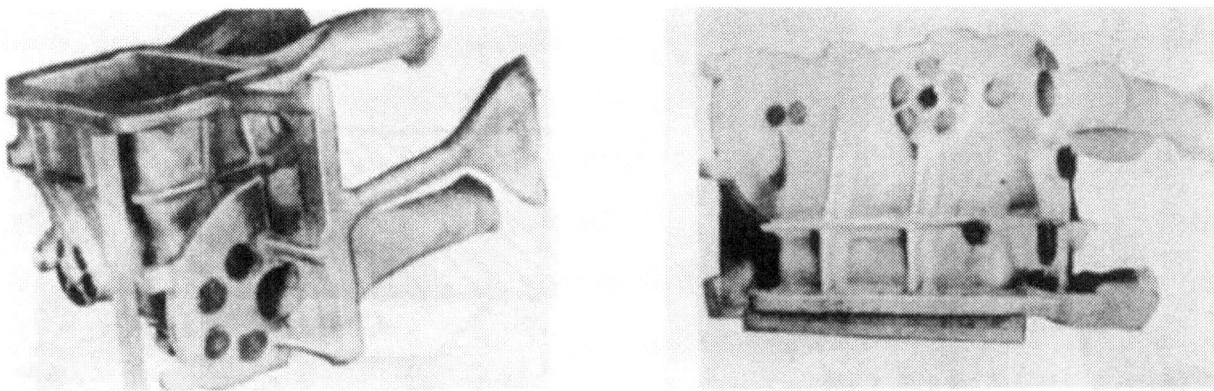

Figure 160. Use of external chills on a bronze casting.

Figure 161. As-cast aluminum casting showing location of external chills.

Figure 162. Gear blank mold showing location of external chills.

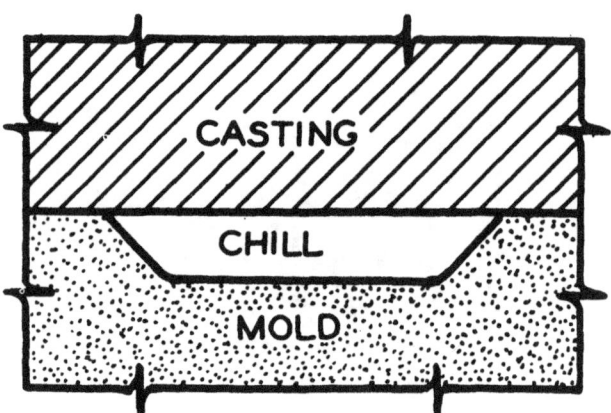

Figure 163. Principle of tapering edges of external chill.

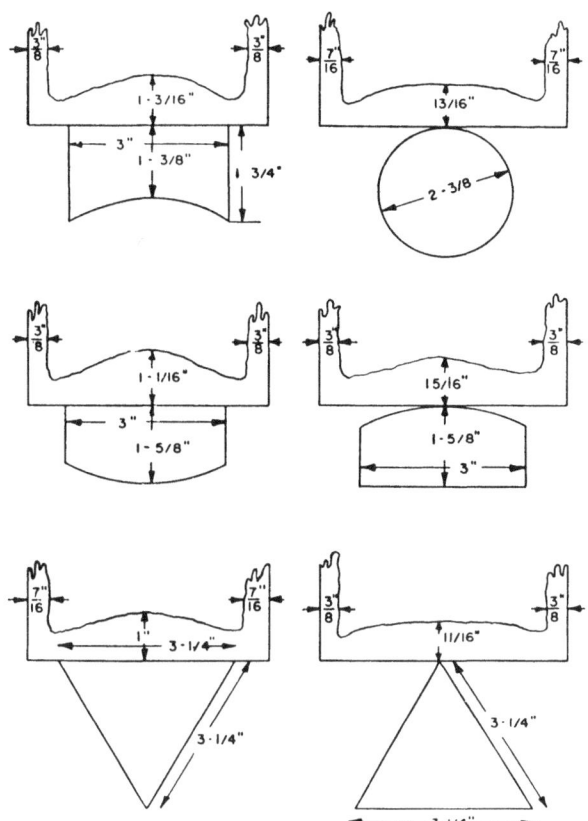

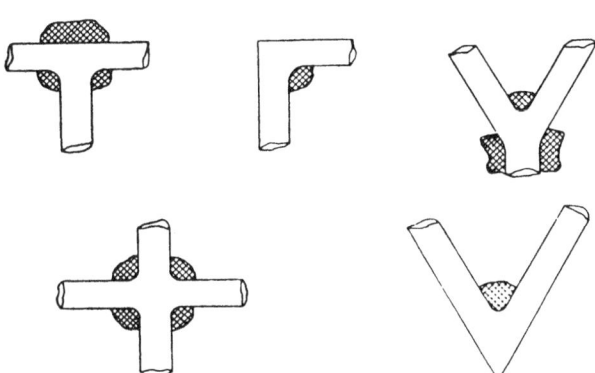

Figure 165. Typical application of external chills to unfed L, T, V, X, and W junctions.

Figure 164. Effect of chill mass and area of contact.

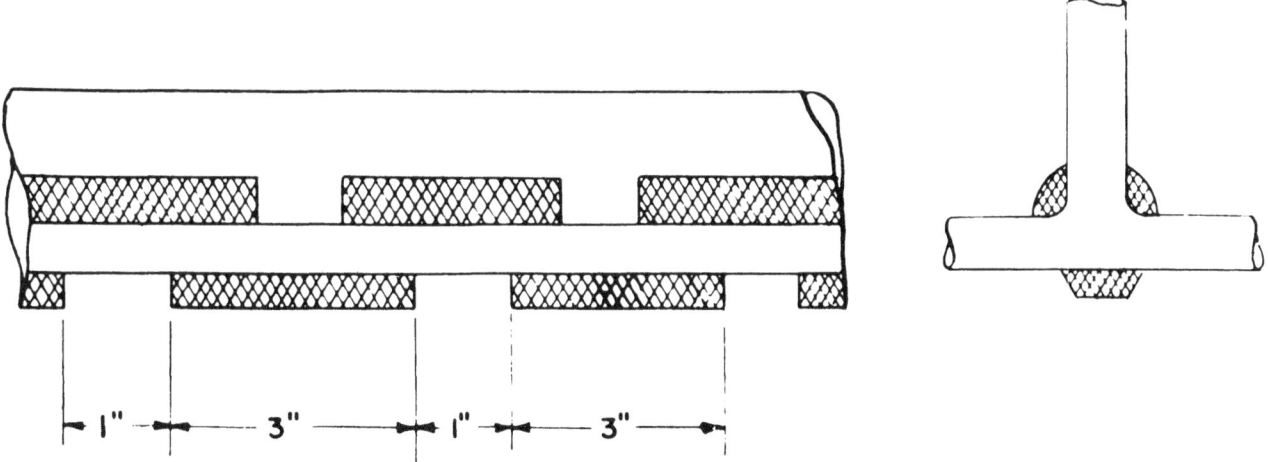

Figure 166. Preferred method of applying external chills by staggering.

- 119 -

Chapter VIII
DESCRIPTION AND OPERATION OF MELTING FURNACES

OIL-FIRED CRUCIBLE FURNACE

Oil-fired crucible furnaces are satisfactory for melting aluminum, brass, bronze, other nonferrous metals and occasionally, cast iron, but will not get hot enough to melt steel.

CONSTRUCTION

Oil-fired crucible furnaces are of two basic kinds: the stationary type and the tilting type. The stationary type requires the crucible to be lifted in and out of the furnace. When a stationary furnace is recessed into the foundry floor or deck, it is known as a pit-type furnace. A pit-type furnace is shown in figure 167. In a tilting furnace, a crucible with a special lip is used, as shown in figure 168, and the molten metal is poured from the melting crucible by tilting the furnace. A typical tilting furnace is shown in figure 169.

A cross section of a stationary furnace is shown in figure 170. The furnace consists of four principal parts: (1) shell, (2) lining, (3) base block or pedestal, and (4) combustion unit. The shell is heavy-gage steel. The lining is usually a preformed, highly refractory unit which is cemented into place. The base block is used to support the crucible. The combustion unit is usually of the premixing type, which mixes the fuel oil and air for proper combustion. A tilting furnace has about the same general construction, except for the addition of a tilting device.

LINING THE FURNACE

The best linings are preformed and fired shapes of highly refractory material, such as high-alumina clay or silicon carbide bricks set into the furnace and cemented into place with a refractory cement. When the preformed linings are unavailable, a lining may be made by ramming in a suitable refractory.

A rammed lining can be made in an emergency from a stiff mixture of crushed firebrick, sand or gravel, 15 percent fire clay, and water. Thorough mixing is important; the mixture works better if it is made up a day ahead of time, rammed into a solid slab and then cut off with a shovel as it is needed for ramming. The crushed brick ("grog") or silica should be refractory, and of a suitable size. When using grog, successive layers should be rammed into place until the lining has been built up to the required thickness. Rammed around a form (as described below) is best. If no form is available and successive layers have to be rammed against the furnace shell, one must be sure that each layer is roughened before the next is applied. Otherwise, the layers may separate later. If delays occur and the lining is allowed to dry out between layers, it should be thoroughly dampened before ramming is resumed. The finished lining must be dried slowly and completely before it is used.

There are also available heavy-duty plastic refractories which can be used to line furnaces of this type. Prepared mixtures of this type are preferred over those mentioned above. Water is added to the refractory mix to make it workable and then the lining is rammed in place around a form. Ramming mixtures containing the proper amount of moisture are also available. Care should be taken to center the form properly after the bottom has been rammed. The sides should be rammed into place by gradually building up the refractory around the form and roughening each layer before the next is rammed in. Heavy grease or aluminum foil can be used to cover the form to keep it from absorbing water from the refractory mix and to make it easier to draw the form. The proper thickness of lining can be obtained from the instructions supplied by the manufacturer of the unit. Always dry a new lining slowly and completely.

The proper thickness of lining should be maintained at all times by patching. When patching the lining, the patch should be undercut into the lining so that it is keyed into place as shown in figure 171. After the cavity has been made, it should be thoroughly dampened before the patch is rammed into place. The refractory mix should be made with the smallest amount of water possible. This will tend to minimize the drying shrinkage in the patch. The patch must be dried before the furnace is used. Proper maintenance of the furnace lining is necessary because a poor lining affects the combustion of the fuel in the furnace. Turbulence in the burning gases causes poor heating and melting takes longer.

Attention must be paid to the proper location of the burners with respect to the base block and the crucible. The burners should be directed into the chamber so that the center line of the burners is level with the top of the base block and so that the flame is directed between the furnace wall and the crucible. This is shown in figure 172. To obtain the best melting results from a furnace of this type, the size of crucible suggested by the manufacturer should not be changed. With this practice, the volume around the crucible available for combustion will always be the same and more consistent furnace operation will result. It is better to melt a small charge in a large crucible than to melt the same charge in a smaller crucible that will cause a change in the operating characteristics of the furnace and possibly increase the time for melting.

Base blocks for crucible furnaces should be made of the same or similar materials as the crucible. The block must be refractory enough that it will not soften or slump at the operating temperatures of the furnace. If the block softens, it will stick to the bottom of the crucible and make the removal of the crucible difficult. If the block slumps, it will cause the crucible to tilt and possibly spill molten metal in the furnace, or in the case of the long-lip crucible, cause it to crack. Silicon carbide base blocks are preferred.

Along with lining upkeep, the care of crucibles must be taken into consideration for good furnace operation. Crucibles should be stored in a warm dry area. If proper storage facilities are not available, crucibles should be dried at a temperature of 300°F. for 8 hours before use. A gas-fired or oil-fired core oven should never be used to dry crucibles. Both of these fuels have moisture as a product of combustion. This makes the proper drying of crucibles difficult. If the precaution of carefully drying crucibles is not observed, any moisture retained in the crucible will cause cracking or spilling when the crucible is used. Soft-burned (unvitrified) clay-bonded crucibles should be annealed slowly before use for melting. The crucibles should be placed upside down in a cold or nearly cold furnace. NEW CRUCIBLES SHOULD NEVER BE PLACED IN A HOT FURNACE. The fuel is turned on with only enough air blast to bring the temperature up slowly until the crucible reaches a cherry-red color. It may then be carefully turned right side up, charged, and used in a melt down. Crucibles are very fragile, either cold or hot, and should be handled carefully at all times. Annealing is not necessary with vitrified hard-burned, clay-bonded crucibles and should be avoided with graphite crucibles.

Further important information on the proper care of clay-graphite crucibles is given on page 128 under "Electric Induction Furnace." These precautions should be used for crucibles of _all_ types.

CHARGING THE FURNACE

Tongs and shanks used for handling crucibles should fit properly so as to avoid damage to the crucible. Proper fit for a pair of tongs is shown in figure 173. Notice that the tongs clear the top of the crucible and that the blades bear evenly against the sides of the crucible. Whenever a crucible is handled with tongs, they should be gently lowered, NOT DROPPED, onto the crucible and centered vertically before gripping and raising the crucible.

When charging the crucibles, the remelt (such as gates, risers, and sprues) should be charged on the bottom and ingot material on top, providing that there is room enough for the ingot without exposing it to the flame. Do not overfill the crucible so that any solid metal sticks out of the furnace where it can pick up gas from the flame. When charging both scrap and ingots, care should be taken to prevent any of the charge material from becoming wedged in the crucible. Wedged material will expand when heated and crack the crucible. If ingot material cannot be charged with the scrap, it should be added after the first charge has started to melt. Ingot material should be preheated before charging so as not to chill the molten metal. Ingots or heavy pieces of metal should not be dropped or thrown into crucibles.

Whenever possible, a separate crucible should be kept for each type of metal melted. This reduces the possibility of contamination in successive heats.

Where it is impossible or impractical to keep separate crucibles for the various metals, the same crucible may be used only if a wash heat is made between the required heats. A wash heat consists of melting a scrap charge of the same composition as the desired heat. This wash is used to remove the undesirable metal that has penetrated the crucible and to prevent contamination of the following heat.

FLAME ADJUSTMENT AND FURNACE ATMOSPHERE

Fuel oil is usually supplied to the furnace at a pressure of 20 to 30 pounds per square inch. The air required for combustion is supplied by a blower at 16 ounces per square inch average pressure. All the air necessary for combustion should come from the blower and be thoroughly mixed with the fuel before entering the furnace. The introduction of secondary air around the burner nozzle is to be avoided because it results in erratic furnace operation. A proportional-mixer oil burner maintains a constant mixture of oil and air regardless of the volume of fuel being supplied to the furnace. This type of burner makes furnace control easier.

If manually operated fuel and air valves are used, extreme care should be taken to control the combustion of fuel. If too much air is used to make an oxidizing atmosphere, the excess air will consume the carbon or graphite in the crucibles. If too much oil is used and a strongly reducing atmosphere is made, an excess of moist gases will be produced and will attack the crucibles and cause gas pickup in the melt. Careful control of an oil-fired melting furnace is an absolute necessity for good furnace operation for obtaining good service from the crucibles and for melting metal low in gas content.

The atmosphere in the furnace should be slightly oxidizing, that is, there should be just a little more air than is required to completely burn all of the fuel. When the furnace is operating under the correct slightly oxidizing atmosphere, a slight green tinge will appear around the outer fringe of the flame. A reducing atmosphere is usually indicated by a smoky, yellow flame. A quick test can be made by passing a freshly broken piece of virgin zinc through the open flame. If it turns black, the atmosphere is highly reducing. If it turns straw-yellow to light gray, the atmosphere is slightly reducing, and if it does not change color, the atmosphere is oxidizing. Another quick test is to throw a small block of wood into the furnace. If it burns with a flame, the atmosphere is oxidizing. If it chars slowly, the atmosphere is reducing and more air is needed at the burner.

MELTING AND TEMPERATURE CONTROL

Melting in an oil-fired crucible furnace should be as rapid as possible. The shorter the time the metal is held in the furnace, the less opportunity there is for excessive oxidation of the charge and absorption of gas by the molten metal. As soon as the charge is melted down, the temperature should be determined with an immersion pyrometer and repeated checks made until the desired temperature is reached. The flame should be cut back when measuring the temperature of prevent any damage to the thermocouple due to localized heating. It is generally necessary to shut off the fuel and air before the desired temperature is reached. The temperature of the metal in the crucible will continue to rise because of the heat retained in the furnace. The point where the fuel should be cut off will depend on the type and amount of charge metal and the operating characteristics of the furnace. The rise in temperature after the fuel and air has been shut off may vary from a few degrees to several hundred degrees. Considerable skill on the part of the melter is necessary to prevent over heating in this type of melting.

TAPPING

The tapping temperature should not be more than 100°F. above the desired pouring temperature. If the melting crucible is used for the pouring operation, a properly fitted chank should be used. The crucible should be removed from the furnace with well-fitting tongs. Any pieces of slag or refractory which may stick to the crucible should be removed and the crucible set in a bed of dry sand. Any pieces of slag or refractory which are not removed will cause damage to the crucible.

The molten metal may also be tapped into a well-dried, preheated ladle for pouring. Any small additions of alloys may be added to the metal during tapping and should be placed in the bottom of the pouring ladle after about an inch of metal has been tapped into it. The alloy additions or material to compensate for melting losses should be added in small pieces so that they melt easily and do not produce hard spots in the castings.

After the heat has been poured, all slag and remaining metal should be drained and scraped from the crucible. Metal should never be permitted to solidify in a crucible. On reheating, the solidified button will expand and crack the crucible. If crucibles are not to be used immediately again, they should be cooled slowly, away from any draft.

ELECTRIC INDIRECT-ARC FURNACE

PRINCIPLE OF OPERATION

This furnace gets its name from the fact that the electric arc does not come in direct contact with the metal being melted. Any metal melted in the electric indirect-arc furnace receives heat from two sources. Radiant heat is obtained from the electric arc. A secondary source of heat comes from the refractory lining, which is heated by the arc. The rocking action of this type of furnace exposes more lining to the heat of the arc, which in turn permits the lining to deliver more heat to the metal. Rocking permits the lining to heat the metal and the metal to cool the lining. The melt-down time would be longer if the furnace were not rocked, and the lining would be damaged. Another advantage of the rocking furnace is the mixing action given to the molten metal. This produces a more uniform heat.

CONSTRUCTION

The construction of the furnace itself is fairly simple. It consists of a barrel-shaped steel shell with a heavy-duty refractory lining. The electrodes for supplying the necessary energy for melting are centered on the axis of the barrel. A charging door is provided on the front of the barrel. There is a special device supplied to rock the furnace automatically while it is in operation. An electric indirect-arc furnace installation is shown in figure 174.

ELECTRICAL SYSTEM

Power is supplied to the furnace from a transformer that "steps-down" the 440-volt current delivered by the generator to 100 volts. The reactance control, located on the transformer, serves to stabilize the arc. Electrode regulation and power in-put to the furnace is controlled on the power panel. A knife switch on the power panel is used to isolate the panel from the source of power or to permit power to pass from the source of power to the panel. The amount of current passing through the carbon electrodes is controlled by the load-adjusting rheostat. For more detailed information on the electrical system, refer to the manufacturer's literature.

LINING THE FURNACE

A preformed lining of mullite or sillimanite refractory is generally furnished for the electric indirect-arc furnace. The lining should be cemented into place with a highly refractory cement of the same composition as the lining; that is, either a mullite or a sillimanite cement. The life of a new refractory lining and the quality of the metal produced depend largely on the initial drying-out period. The following instructions should be followed in preparing a new lining for the first heat.

PROCEDURE FOR DRYING A NEW LINING

1. Place the lined barrel on the furnace rollers.

2. Fasten the automatic and manual electrode brackets.

3. Insert and clamp two new carbon electrodes. Make certain that they are properly aligned by introducing shims between the brackets and shell end plates. The electrodes should now appear as one continuous unit.

4. Wrap one turn of heavy wrapping paper around the electrodes and push the port sleeves in place.

5. Tamp the port sleeves in place with Alundum cement. Recheck the alignment of the electrodes.

6. Remove the paper and operate the electrodes manually over their full travel. No binding or sticking should occur.

7. Cut out the silicon carbide refractory door brick to conform to the contour of the furnace door and ram the spout and the space around the door with mullite or sillimanite refractory furnished with unit. The purpose of cutting away the door-brick refractory is to permit air to get into the barrel during the melting cycle.

8. Build a charcoal fire in the barrel. Allow it to remain for at least twelve hours. Clean out most of the charcoal at the end of that time.

9. Insert the stationary carbon electrode approximately 1-1/2 inches past the center of the barrel. Adjust the automatically controlled electrode so that it is approximately 1/2 inch from the stationary electrode. This practice should be followed on every heat, as it insures that the arc will be approximately centered during the melting cycle, thus, preventing damage to the end walls of the furnace lining.

10. Circulate the cooling water. This practice should be adhered to at all times prior to starting the furnace in order to prevent damage to the electrode clamps or any other jacketed parts. The temperature of the outlet water should never exceed 200°F.

11. Set the "Rocking Center" on the index mark.

12. Rotate the "Constant Rocking Period" knob until the "Range Pointer" is on the "off" position, which is full normal rock.

13. Close the d.c. electrode-motor switch. The closed circuit is indicated by the light on the d.c. contactor panel.

14. Place the arc-circuit toggle switch in the "on" position. Throw the circuit breaker in by means of the remote-control switch. This is indicated by the lights on the meter panel.

15. Push in the button marked "hand" on the regulator panel.

16. Advance the operating or automatically controlled electrode until it makes contact with the stationary electrode and strikes the arc. Withdraw the electrode rapidly until the kilowatt meter shows the desired rate of input. This should be less than the rated input of the furnace unit for the drying-out period. In general, a shortening of the arc increases the input rate and a lengthening of the arc by withdrawing the operating electrode decreases the input rate. Place the furnace on automatic control by pushing the button marked "automatic" on the regulator panel and make the necessary adjustments with the load-adjusting rheostat.

NOTE: Difficulty may be experienced in obtaining the desired, steady input on a green or cold lining when the furnace is started. If on automatic control, the unit tends to stabilize itself. However, hand adjustment of the operating electrode can be made until the input does not fluctuate excessively.

17. Place the rocking motor contactor box switch in the "on" position.

18. Place the "Automatic Rock" switch in the "on" position.

19. The drying-out period should be as slow as practical. The following schedule should not be exceeded. Apply the heat intermittently at:

    (a) 6 kilowatt hours once each 1/2 hour for 2 hours.

    (b) 9 kilowatt hours once each 1/2 hour for the next 2 hours.

    Additional heat should now be applied continuously until the lining temperature reaches approximately 2700°F. (white heat). Turn the remote-control switch, arc-circuit toggle switch, and the d.c. electrode-motor circuit switch off and allow the lining to cool off to a dull red before removing the charging door. Loosen the electrode clamps and run the electrode back and forth in the port sleeve to remove any foreign matter and to prevent binding. If necessary, blow out the port sleeve with a compressed-air blast. The lining is now ready for its first heat, which preferably should be made the next day.

20. The next day, patch any cracks in the lining with a mortar cement refractory of the type used in cementing the preformed shapes. A general assembly view of the subject melting unit is shown in figure 175. The accessory equipment (gear) furnished with the unit for the preparation of the furnace for use is shown in figure 176.

Lining Repair. A carefully maintained lining is essential for the production of good quality metal. If slag or dross are allowed to accumulate in a lining, or a patch is placed over a slag area, a "choked" arc will result. Under such conditions, a poor quality melt will be produced. Therefore, it is necessary to remove all slag or dross from the lining before patching. Patching should always be deep and keyed in place. The patch material should be the same grade as the original lining. This is usually a mullite or sillimanite refractory. Patch the area around the furnace door and make up the spout daily. Port sleeves should be patched with Alundum cement daily and reamed to insure a tight electrode fit, without binding.

Shell Replacement. To change furnace shells, remove the electrode brackets and lift the shell off the furnace rollers. In replacing another shell on the furnace base, make certain the shell cam does not injure the limit switch. Keeping the charging door in top-center position will minimize this possibility. Replace the electrode brackets and check the alignment of the electrodes.

Check the "limit" and "overtravel" switches to ascertain that they are in proper adjustment. A mishap can be created by the use of the push-button with the "Automatic Rock" switch in the "on" position in checking the operation of the "overtravel" switch.

CHARGING THE FURNACE

Preheating. Preheat the lining with the furnace at full normal rock before charging the first heat. Note the starting instructions and precautions listed under "Procedure for Drying a New Lining." NEVER CHARGE METAL INTO A COLD FURNACE. The lining should preferably be preheated to the tapping temperature of the alloy to be melted. The approximate kilowatt-hour input necessary to preheat linings of various size units are listed below:

Furnace Type	Kwhr Required to Preheat
LF	60
LFC	80
LFY	115
AA	175
CC	275

CAUTION: Do not overheat the lining, as it will decrease the life of the refractory and introduce operating difficulties.

Charging. The ideal position for charging is with the furnace door in the top-center position. In any case, it should be within 45° of the top center position. Before charging, slide the electrodes back until they are flush with the furnace wall and will not be damaged during the charging period. The charging position should be varied from time to time to prevent excessive wear on one section of the lining.

First, charge foundry returns (gates and risers) which have been thoroughly cleaned of any excess sand. Excessive sand will cause a slag blanket to form on the surface of the molten metal during the melting cycle. This condition should be avoided, as it insulates the bath from the heat generated by the arc and will make it difficult to reach or determine the desired tapping temperature. Any unusually heavy pieces (such as large risers) should now be charged to the rear of the barrel. If borings of any kind are used in the charge, they should be added at this time. They will filter down through the foundry returns and give a more compact charge free from direct contact with the arc. Ingots are added to the charge last. Pile most of the charge toward the rear of the furnace so that a larger angle of rock may be obtained with greater electrode safety. A properly charged furnace with electrodes in position for striking the arc is shown in figure 177. Charging should be accomplished as quickly as possible to prevent excessive loss of heat from the lining. (It is poor practice for an inexperienced operator to exceed the rated capacity of his furnace.) At this point, the furnace door is closed and clamped securely.

ROCKING MECHANISM

Adjust the rocking mechanism as follows:

1. Set the "Rocking Center" on the index mark. This synchronizes the furnace shell with the angle of rock.

2. Push in the button marked "Hand" on the regulator panel.

3. Set the "Range Pointer" and "Selector" on settings that will give the greatest angle of rock and permit the barrel to reach full normal rock as quickly as possible, and compatible with electrode safety. The "Selector" provides a means of changing the rate of the rocking angle. The no. 1 setting is the fastest and the No. 6, the slowest. The "Constant Rocking Period" knob provides a means of delaying the angle of rock. A "Nameplate Table" printed on the rocking controller assembly lists the time required for the barrel to reach full normal rock from various settings and charging positions. One distinct advantage of the rocking feature is the absorption of heat by the charge from the refractory lining. Thus, it is obvious that reaching full normal rock as quickly as possible will result in rapid melting, more uniform melt, lower power consumption, increase in refractory life, and a decrease melting loss. Actual settings of the "Selector" switch and "Constant Rocking Period" knob can best be determined through experience and will depend upon the physical position of the charge in the barrel. For the first heat, place the "Selector" in the No. 6 position and the "Constant Rocking Period" knob on the 20° rock setting. Delay the increasing angle of rock. Observe the heat closely and increase the rock manually as the heat progresses and the charge settles. After a few heats, the operator should be able to determine the most efficient and safe setting for any charge. Make certain that the "Selector" is placed securely in the notch intended.

ELECTRODE CONTROL

Center the carbon electrodes as follows. Insert the stationary electrode approximately 1-1/2 inches past the center of the barrel. Adjust the automatically controlled electrode so that it is approximately 1/2 inch from the stationary electrode. This practice should always be followed on every heat as it insures that the arc will be approximately centered during the melting cycle, thus, preventing damage to the end walls of the lining. Make certain that the electrodes are not shorted.

Strike the arc manually. Advance the operating or automatically controlled electrode until it makes contact with the stationary electrode and strikes the arc. Withdraw the electrode rapidly until the kilowatt meter shows the desired rate of input. In general, a shortening of the arc increases the input rate and a lengthening of the arc, by withdrawing the operating electrode, decreases the input rate. Place the furnace on automatic control by pushing the button marked "automatic" on the regulator panel and make the necessary adjustments with the load-adjusting rheostat. Mark the setting for the normal operating input on the load-adjusting rheostat dial for future reference. Place the electrodes on automatic control by pushing in the push-button marked "Automatic" on the regulator panel.

Put the "Automatic Rock" switch in the "on" position. Observe the operating characteristics of the arc. It should be sharp and clear. The input should be steady and not fluctuating. A cloudy smoke arc from which small graphite particles are emitted is not a healthy operating condition for this unit and will result in casting losses due to porosity and low physical properties. This condition may be caused by low voltage across the electrode clamps or spongy electrodes. It may be corrected by increasing the voltage, decreasing the transformer reactance, or inserting good electrodes.

ELECTRODE PRECAUTIONS

Various steps that should be followed to provide good electrode operation are listed below.

1. Make certain that there is good contact between the clamp and the electrode.

2. Do not grip the electrode joints within the clamp as it may result in a broken nipple or a reduced area of electrical contact.

3. In joining two electrodes, insert the threaded carbon stud in the socket of one and screw in the other, finishing with a snapping action to insure good electrical contact. Do not use excessive force or the nipple will break.

4. Electrodes should be stored in a warm dry place to avoid sponginess. Sponginess will cause poor operating characteristics that will produce a poor-quality melt.

5. If an electrode breaks during the melting cycle, shut the furnace down and remove the broken section from the hearth immediately. Insert a new electrode.

6. When the nipple approaches the arcing end of the electrode, it should be broken off because it might slip off and fall into the hearth of the furnace and cause carbon pickup by the metal.

MELTING AND TEMPERATURE CONTROL

During the melting cycle, the charge should be observed periodically to check on the rate of meltdown and to make sure that it is melting satisfactorily. The angle of rock should be increased as quickly as conditions will permit. The initial melting will take place under the arc where a molten pool of metal will collect. As the angle of rock increases and melting proceeds, the pool will wash over the rest of the charge and make the meltdown faster. When the entire charge is molten, the furnace should be at full rock. Once the charge becomes completely molten, a close check should be kept on the temperature with either an immersion thermocouple or an optical pyrometer, depending on the metal being melted.

During the time the heat is being superheated to tapping temperature, it is important that the bath be kept clear of any slag or dross which may have formed. Any such substance not only prevents proper heating of the metal bath but also makes the determination of temperature difficult.

Various techniques used during meltdown for determining temperatures are described in the chapters pertaining to the particular metals.

TAPPING

Just before tapping, the furnace should be operated at reduced input, just sufficient to maintain the temperature of the bath during the tapping period. Place the "Automatic Rock" switch in the "off" position. The furnace is now operated by the portable push-button station to carry it through the pouring stage.

Tapping should not be delayed once the proper temperature has been reached. If a slight delay is unavoidable, shut the arc off and set the "Automatic Rock" switch at "off." The temperature of the bath will not fall or rise appreciably during the first few minutes. If a longer delay is necessary, the furnace should be operated intermittently at reduced input and at full rock in order to maintain the desired temperature. The molten metal should always be tapped into a <u>dried</u> and <u>preheated</u> ladle.

After the furnace has been completely drained of molten metal, remove the door. With push-button control, roll the furnace barrel over until the charging-door opening is completely underneath. Allow the slag or dross to drain out by rocking the barrel back and forth or by applying the arc for a few minutes.

The barrel should then be returned to its original charging position and any slag removed from the spout. The spout may then be repaired by hot patching, if necessary. The electrodes are then run in and out to prevent binding by slag or metallic particles. The port sleeves should be blown or cleaned out. The furnace is then ready for another heat.

ELECTRIC RESISTOR FURNACE

PRINCIPLE OF OPERATION

The operating characteristics of the resistor furnace are different from those of the indirect-arc furnace. The indirect-arc furnace melts with a heat which is produced by the electric arc between electrodes. The resistor furnace produces the heat for melting by using continuous graphite resistors and no arc. The electric current passing through the resistors causes them to become heated to temperature sufficient to melt the charge. The principal is the same as that of an electric toaster.

CONSTRUCTION

The construction of the resistor furnace proper (including the rocking mechanism is the same as that of the electric indirect-arc furnace. The electrodes and electrical system have been modified to use resistor heating elements. The furnace uses two pairs of 1-3/4 inch-diameter resistors, which meet in the center of the furnace. They are seated and locked by spring tension from the electrode brackets. The use of male and female electrode sections permits easier withdrawal of the electrodes for charging or replacement. The resistors thread into a 4-inch-diameter terminal and form a reduced section, which in operation forms two continuous graphite bars through the melting chamber. A view of the furnace showing the resistors through the center of the barrel is shown in figure 178. The electrode-bracket assembly is shown in figure 179.

ELECTRICAL SYSTEM

The current for the resistor furnaces on repair ships is supplied by a 440/36 volt, 150-kva, 3-phase to 2-phase Scott-connected transformer. The input voltage is varied from 440 to 184 volts in 11 steps by means of solenoid-operated tap switches. This results in a variable secondary voltage from 36 volts to 10 volts at the furnace terminals. Numerals on a disk visible through a circular window in the transformer housing indicate the nominal open-circuit voltage for each position. Taps can be changed under load. Push buttons on the control panel permit raising or lowering of the voltage and current on each phase so that the power input can be controlled at almost any level. During melting, the power input to the furnace is generally held at 150 kw.

LINING THE FURNACE

Lining of the resistor furnace and its repair are the same as for the electric indirect-arc furnace. For a description of the lining and its repair, see the section, "Lining the Furnace for the Electric Indirect-arc Furnace," page 124.

CHARGING THE FURNACE

The same precautions in charging are necessary with the resistor furnace as with the indirect-arc furnace (see page 123). The electrodes must be withdrawn until they are flush with the furnace end walls to prevent any damage during charging.

ELECTRODE CONTROL

Manipulation or control of the electrodes is not necessary during the melting operation. After the furnace is charged, the electrodes are run into the center of the furnace until they meet and form two continuous resistors. Once this is done, there is no further control necessary other than to make certain that there is always strong enough spring tension at the electrode brackets to maintain the resistors in position with good electrical contact. This avoids arcing at the joints.

MELTING AND TEMPERATURE CONTROL

Melting practice and temperature control are the same as for the electric indirect-arc furnace. (see page 123.)

ELECTRIC INDUCTION FURNACE

PRINCIPLE OF OPERATION

In the operation of the furnace, a high-frequency electric current is passed through the primary coil (figure 180), inducing a secondary current in the charge, heating it by resistance to the desired temperature. The charge may consist either of a single lump of metal or a quantity of loose pieces such as ordinary scrap. Even comparatively fine turnings can be melted successfully if a moderate amount of heavier scrap is used. While high-frequency heating is effective over a wide range, about 1,000 cycles have been found to be the most practical for a 1-ton furnace. Smaller furnaces will require higher frequencies.

The heat is developed in the outer part of the charge and is quickly carried to the center by conduction, which is rapid through solid metals. After the charge starts to melt and a pool is formed in the bottom of the furnace, a stirring effect occurs. This not only carries heat to the center of the charge but accelerates melting by washing molten metal against the unmelted solid metal. It also mixes the charge thoroughly, thus assuring uniformity. The flow lines in the molten bath, indicated in figure 181 show there are no "dead spots" and that every part of the bath is moved. The vigor of the stirring can be controlled by varying the power input.

FURNACE CONSTRUCTION

The high-frequency induction furnace is essentially an air transformer in which the primary is a coil of water-cooled copper tubing and the secondary is the mass of metal to be melted. The essential parts of the furnace are shown in figure 182. The outer shell, "S," is made of asbestos lumber (transite is one brand of asbestos lumber) and carries the trunnions, "T" on which the furnace pivots in tapping or pouring. (Most of the shell has been cut away to show the section of the furnace.) The coil, "C," consisting of a helix of water-cooled copper tubing, is lined with a layer, "L," of refractory material, which forms a protective coating against metal leaks. This layer is continued above and below the coil against the asbestos support "R," the firebrick top, and base, "F." The coil lining and the firebrick bottom provide a cavity into which the refractory lining of the furnace is built. This lining may take the form of a thin-wall crucible or supporting shell, "M," packed into the cavity with granular refractory, "G," or it may be in the form of a sintered lining, which holds the charge or bath of metal, "B," and is molded at the top on one side to form the pouring spout, "D."

Due to the peculiarities in construction of the furnace, in which the primary coil is fairly close to the metal bath, the selection of suitable refractories is an important consideration. A typical electric induction furnace with a tilting mechanism is shown in figure 183. The power-control panel for this furnace is shown in figure 184.

ELECTRICAL SYSTEM

The high-frequency induction furnace receives its power from a high-frequency motor-generator set. Power to the furnace is regulated by a control panel. The power put to effective use by the melting unit is very low, so capacitors are used to correct the low power factor and to permit the generator to produce full power. To make full power available during all the melting stages, tap switches are used to regulate the effective voltage on the furnace.

LINING THE FURNACE

There are two materials used in the lining of these furnaces in addition to preformed and fired crucibles. These materials (Norpatch and Normagal) are furnished by the furnace manufacturer. Norpatch is a refractory cement, while Normagal is a granular refractory material.

Lining the Furnace Coil. The furnaces are shipped with a 3/8 inch lining of Norpatch installed on the inside of the coil. This lining prevents the dry furnace-lining materials from leaking through the coil turns and protects the coil from serious damage by the molten metal in case of a lining failure. Each time a new crucible or lining is placed in a furnace, the coil lining should be examined for cracks. If small cracks are present, they should be filled in with a brush coat of Norpatch mixed with water to a consistency of heavy paint. If large cracks or holes are present, they should be patched with the Norpatch moistened with water to a thicker consistency, rammed in place, and smoothed with a wet brush or trowel.

When it becomes necessary to install a new coil lining, all the old Norpatch should be removed from the coil. A new batch of Norpatch is then mixed with water, making certain all lumps are well pulverized. This is applied to the inside of the coil with the hand or a trowel to a thickness of 3/8 inch. The Norpatch is pressed firmly between the coil so that it squeezes through to the outside, which provides a suitable anchorage.

After the entire inside surface has been covered, it is good practice to scrape off the high spots with a straight edge and fill in the low spots. This coating should be reasonably smooth and uniform. A 1,000-watt strip heater or the equivalent placed in the furnace permits the lining to dry slowly. Drying should take about 30 hours, and when complete, the cracks should be filled in as explained above.

Crucibles. The temperatures involved in the melting of nonferrous metals are not sufficient to sinter a monolithic lining. Therefore, preformed clay-graphite crucibles are used. In installing crucibles, the first step is to turn on full water pressure and then check for leaks in the coil and hose connections. Moisture will cause short circuits (possibly explosions) if it comes in contact with molten metal.

A good grade of dry silica sand (or the material "Norsand" furnished by the furnace manufacturer) is tamped solidly in the bottom of the furnace to a depth of 3 or 4 inches. The thickness of the sand layer is adjusted so that when the crucible is placed on this lining, the top of the crucible is 1-1/2 to 2 inches below the top of the furnace. The space at the top is left for top patching. The crucible is centered in the coil and held in place by three wooden wedges between the coil and the crucible.

The sand is then placed around the crucible to a depth of 4 or 5 inches and tamped solidly in place to prevent pockets and to insure good contact between the crucible and the coil lining. This procedure is continued until the sand level is near the furnace top before the wedges are removed.

Norpatch cement, mixed with water to a ramming consistency, is then trowelled in place between the top of the crucible and the furnace shell. The pouring lip is also formed at this time. The cement seal and lip are slowly but thoroughly dried and the furnace is ready for operation.

Care of Clay-Graphite Crucibles. Proper care of clay-graphite crucibles will materially increase their life. Before use, they should be stored in a warm, dry place. A rack should be provided so that they will not be placed on a wet or damp floor and they should be stored bottom side up to prevent accumulation of moisture. Before using, crucibles should be heated to about 300°F. for approximately 8 hours to be sure all moisture is removed. If the preheating is not done, a sudden application of heat will crack them. After being rammed in place, it is good practice to heat crucibles (1) by means of a large electric bulb placed in the bottom of the furnace, (2) by use of an ignited charcoal pot in the crucible, or (3) by charging with a few pieces of carbon of sufficient volume to provide radiated heat when the furnace switch is on <u>low</u>.

After each heat, all excess metal clinging to the sides and bottom of the crucible should be removed. Care should be taken not to exert too much pressure, as the crucible tends to become brittle at high temperature. It is then checked for cracks of sufficient size to allow metal to leak through. Small hairline cracks will not give trouble as they will be sealed by the next heat when the pot expands.

At the end of the day's run, or particularly if there is a long period between heats, the crucible should be covered to allow it to cool slowly, as fast cooling tends to develop cracks. Water should be allowed to run through the cooling system until the crucible has reached room temperature.

Crucibles which have worn down to 5/8-inch wall thickness are unsafe for further use, and should be replaced. Relining of the furnace requires that the old crucible and lining be removed—sometimes, by hammering and chiseling. Care should be taken to preserve the refractory cement next to the coil. If it is damaged, it should be repaired before relining.

Rammed Linings. The use of monolithic linings for ferrous metals (iron and steel) is generally favored because of their long life and because steel cannot be melted in a clay-graphite crucible. Also, the steel will absorb carbon from the crucible. Cast iron, however, can be safely melted in a clay-graphite crucible.

Because of the difficulty of installing monolithic linings, the allowance list for each base and shipboard foundry includes magnesia crucibles in addition to materials for installing the monolithic lining. Magnesia crucibles are relatively fragile and must be handled with great care before installing and during use. The same general practice given for the clay-graphite crucibles should be carefully followed in installing them.

Although considerable skill and experience are necessary to properly install a monolithic lining for melting ferrous metals, if the instructions given below are carefully followed, a fair measure of success should be attained after the first two or three linings. Also, the experience thus gained should be sufficient to correct all faults which become apparent.

The first step consists of firmly ramming 5-1/2 inches of Normagal in the bottom of the furnace. This is most effectively done by the use of a rammer formed by welding a long handle to a steel disk 5 or 6 inches in diameter.

Next, the asbestos form, supplied by the manufacturer, is prepared. This is done by drilling four evenly spaced holes about 1/8 inch diameter in the disk about 1/2 inch from the edge. Directly opposite, four similar holes are drilled in the sleeve about 1 inch from the bottom. The disk is then fastened to the sleeve by the use of soft iron wire.

This form is next placed on the Normagal bottom and accurately centered. A weight of several hundred pounds is placed on the inside to prevent its shifting while ramming the sides. If the form is not centered at the top, it is because the bottom is not level, and proper adjustment should be made. About 3 inches of Normagal is then placed around the form and rammed uniformly and hard with a suitable hand rammer, followed, if possible, by an electric or air rammer. This practice is continued until the Normagal is within 5 or 6 inches of the top of the furnace. It is then sealed and the spout formed with Norpatch cement, trowelled firmly in place and thoroughly dried.

An alternate and popular method involves the use of a steel form as shown in figure 185. When the charge becomes molten, the steel shell melts, the asbestos fluxes to form a slag, and the lining material is properly sintered. The first heat in a monolithic lining must be steel, as cast iron does not properly sinter the lining. The quality of a monolithic lining depends largely upon proper ramming of the lining material and the importance of this cannot be overemphasized.

The thickness of the lining must be carefully controlled and maintained during use, since it has a major bearing on the power consumption and melting speed of the furnace. Thin linings give better "coupling" (more induced power in the charge) than thick ones. A perfect coupling would be obtained if the charge were the same diameter as the coil. Naturally, this is not possible, so a satisfactory compromise must be obtained.

After each heat, the furnace should be drained completely of metal and slag, and all holes should be carefully patched. (No attempt should be made to patch over steel or slag). In general, all patching below the metal line should be done with Normagal mixed with sodium silicate, and patching above the metal line should be done with Norpatch mixed with water.

To reline the furnace, the old lining is removed as in the case of clay-graphite crucibles by hammering and chiseling, care being taken to preserve the refractory cement on the coil. The unsintered Normagal should be saved, since it can be used again in relining. If the refractory cement on the coil is intact, the furnace is relined in the manner previously described.

CHARGING THE FURNACE

The charge is preferably made up of carefully selected scrap and alloys of an aggregate composition to produce as nearly as possible the composition desired in the finished metal. Final additions are made to deoxidize the metal or to adjust composition.

The heavy scrap is often charged first, and as much of the charge as possible is packed into the furnace. The durrent is turned on, and as soon as a pool of molten metal has formed in the bottom, the charge sinks and additional scrap is then introduced, until the entire charge has been added. The charge should always be made in such manner that the scrap is free to slide down into the batch. If the pieces of the charge bridge over during melting and do not fall readily into the molten pool, the scrap must be carefully moved to relieve this condition. Severe poking of the charge must be avoided at all times, however, because of danger of damaging the furnace lining. Bridging is not serious if carefully handled, but if allowed to go uncorrected, overheating of the small pool of metal may damage the lining seriously and will have a deleterious effect on the composition of the metal.

The compactness of the charge in the furnace has an important influence on the speed of melting. The best charge is a cylindrical piece of metal slightly smaller in diameter than the furnace lining. This will draw very close to the full current capacity of the equipment. Two or three large pieces with considerable space between them will not draw maximum current, since the air cannot be heated by induction. The charge should not be so tightly packed that upon heating and expanding it cracks the crucible or lining.

SINTERING THE MONOLITHIC LINING AND MAKING THE FIRST STEEL HEAT

After the furnace has been lined, a suitable charge of low-carbon scrap or Armco iron, consisting of pieces weighing between 1/4 pound and 2 pounds, should be selected and placed in the asbestos liner in such a manner that the charge will be reasonably compact. This charge should be brought to the top of the asbestos liner. Long or irregular shaped pieces should not be used for this initial charge, as such pieces may become bridged and prevent the solid metal from coming into contact with the bath as the melting progresses. Before the power is applied, it should be ascertained that water is flowing through all coils which are to be energized.

The many precautions to be observed in the first heat in a new lining are: (1) the charge should be brought to the top of the Normagal lining when molten, and (2) the metal should be held at 3,000°F. for 15 to 20 minutes to permit proper sintering of the lining. Neither of these practices should be followed in later production and are only necessary initially to secure a well-sintered lining.

MELTING AND TEMPERATURE CONTROL

As soon as the charge is completely melted and refining or superheating operations finished, further necessary additions of alloys or deoxidizers are made. The furnace is then tilted to pour the metal over the lip. If the entire heat is poured into a large receiving ladle, the power is turned off before tilting. If, however, the metal is taken out in small quantities in hand ladles, power may be kept on while pouring. This maintains the temperature of the bath and facilitates slag separation by keeping it stirred to the back of the bath. When the heat is poured, the furnace is scraped clean of adhering slag and metal and is then ready for the next charge.

Because of the convexity of the bath surface resulting from the induced current, it is difficult to keep a slag blanket on the metal and usually, no attempt is made to do so. Should a slag blanket be desired, it is necessary to control the degree of convexity of the bath, which can be done by controlling the rate of power input to the melt.

It is important that dissimilar metals not be melted in the same lining or crucible. When melting cast iron or steel, the lining absorbs iron. Brass or bronze subsequently melted in the same lining will become contaminated with iron. The reverse will also be true; cast iron or steel can become contaminated with copper, tin, or zinc. If it ever becomes necessary to melt dissimilar metals in the same furnace, a wash heat similar in composition to the next heat planned can be used to cleanse the crucible. It is always better practice to have separate furnaces or linings for use in such emergencies.

The methods of melting nonferrous alloys vary considerably for different compositions. Questions often arise regarding which metal to melt first, the temperatures at which the additions of other metals are made, the use of slags and fluxes, deoxidation practice and pouring temperatures. Frequently, several alternative procedures are available for melting the same metal. For melting specific metals, refer to the later chapters dealing with each metal.

TEMPERATURE CONTROL

Temperature control in the induction furnace is measured by an immersion or optical pyrometer, depending on the metal being melted. The power should be reduced (or better, shut off) while taking a temperature reading with an immersion pyrometer to prevent an incorrect reading. Good melting records with proper temperature readings can prove useful in the event there is a failure in pyrometer equipment. If temperatures and power input are recorded during various melting operations, the power input, along with the time at a given power input, can prove useful in making a close approximation of the temperature of the heat.

SUMMARY

The operation of assigned melting furnaces is relatively simple and even inexperienced personnel can use them to melt metal. Whether this metal will make a good casting is another matter. For a strong, sound casting free of defects, it is necessary that the metal be melted under proper conditions and handled properly. It is relatively easy to see whether a metal is hot enough to pour into a particular mold, but it is hard to see when the metal has been damaged by improper melting methods. This shows up when the casting is shaken out of the mold.

Oil-fired crucible furnaces are convenient for melting nonferrous metals (brass, bronze, or aluminum) but usually are not satisfactory for iron or steel. The quality of the metal can be damaged severely by improper furnace atmosphere, melting so that the flame hits some of the metal, or by using crucibles which have not been stored and prepared properly. CRUCIBLES AND FURNACE LININGS ARE AN ALL-IMPORTANT ITEM IN ANY FURNACE. They must be thoroughly dried before they are used for melting. If not, the damage they will cause to the metal will be more far-reaching than possible damage to the lining.

Electric furnaces are fast, clean and convenient. They also permit high temperatures to be reached.

Figure 167. Pit-type crucible furnace.

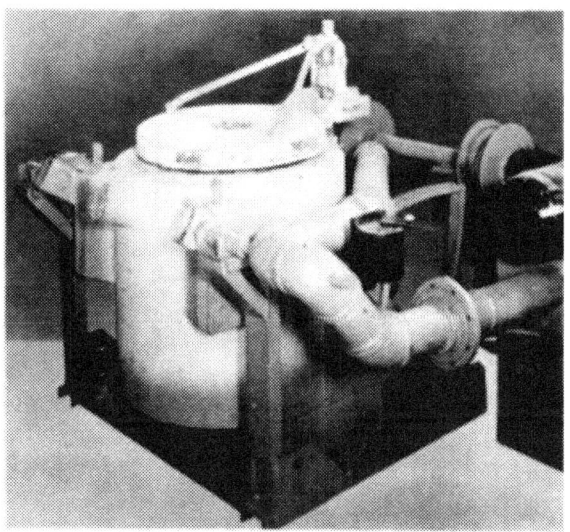

Figure 169. Tilting crucible furnace.

Figure 168. Crucible for tilting crucible furnace.

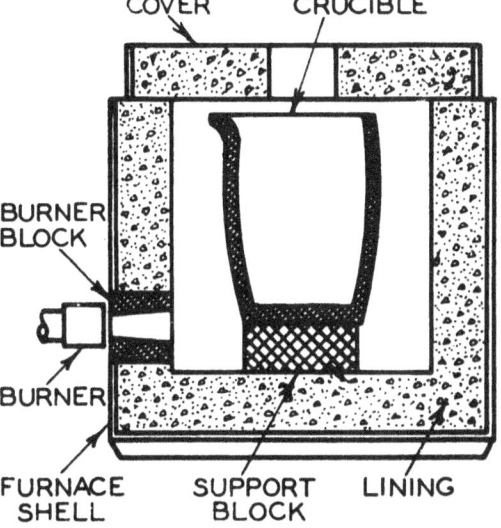

Figure 170. Cross-section of a stationary crucible furnace.

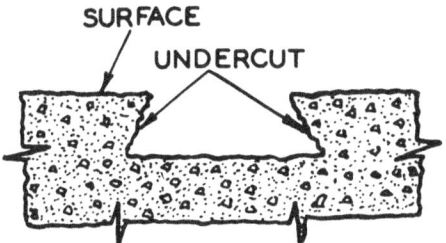

Figure 171. Undercutting a refractory patch.

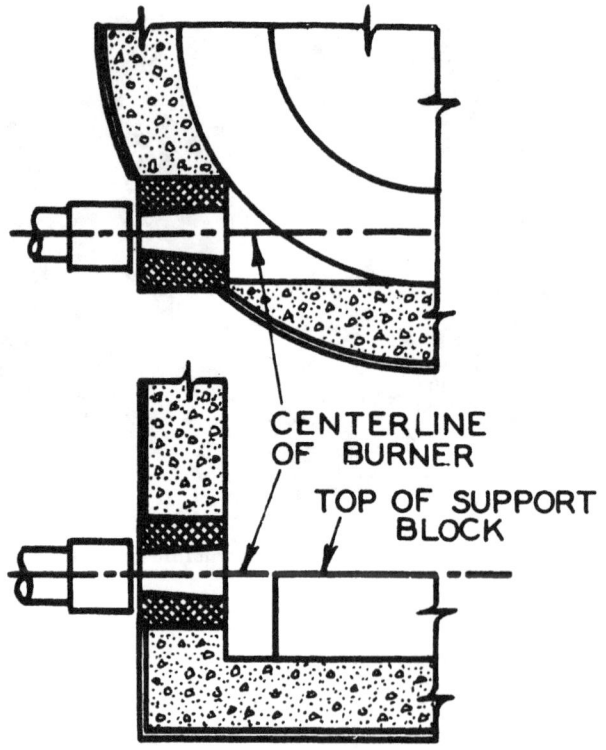

Figure 172. Proper burner location.

Figure 174. Electric indirect-arc furnace.

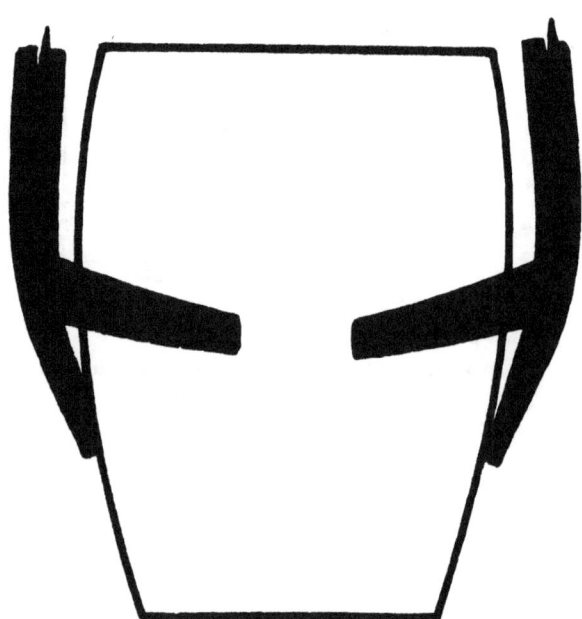

Figure 173. Proper fit for crucible tongs.

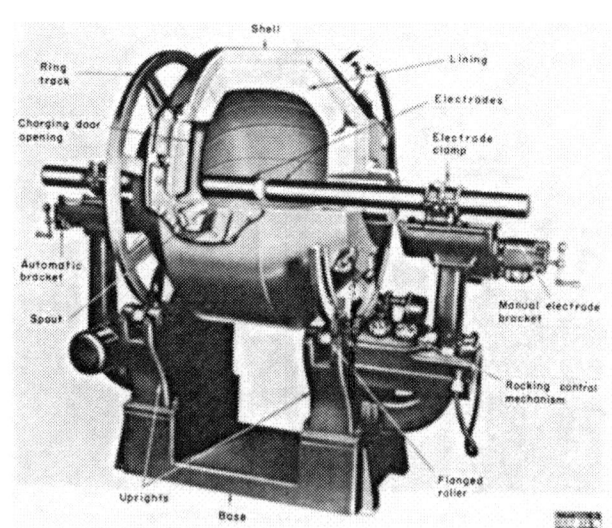

Figure 175. General assembly view of electric indirect-arc furnace.

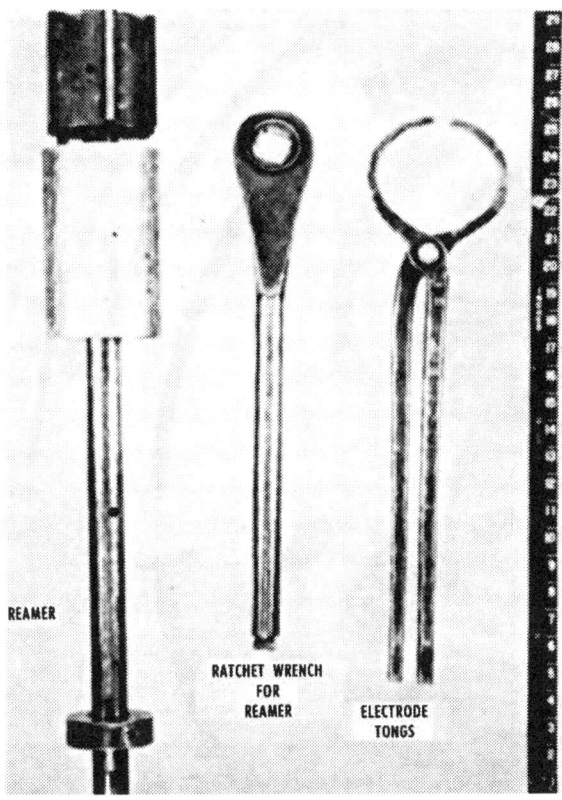

Figure 176. Accessory equipment for electric indirect-arc furnace.

Figure 177. Properly charged electric indirect-arc furnace.

Figure 178. Electric resistor furnace.

Figure 179. Electrode-bracket assembly for electric resistor furnace.

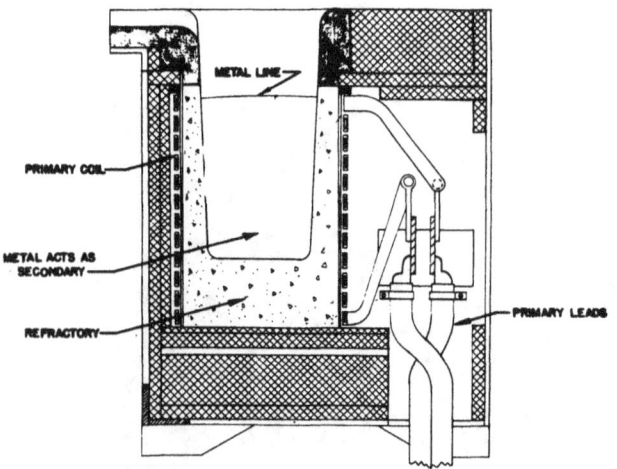

Figure 180. Cross section of electric induction furnace.

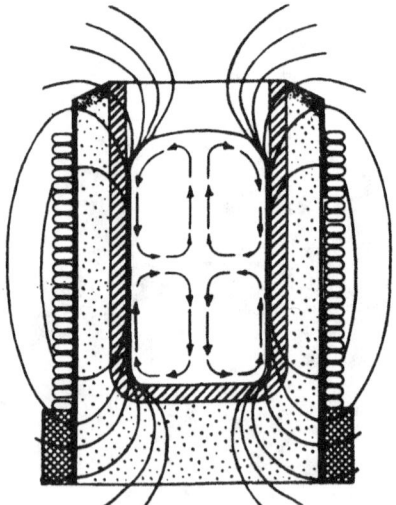

Figure 181. Flow lines in an induction furnace melt.

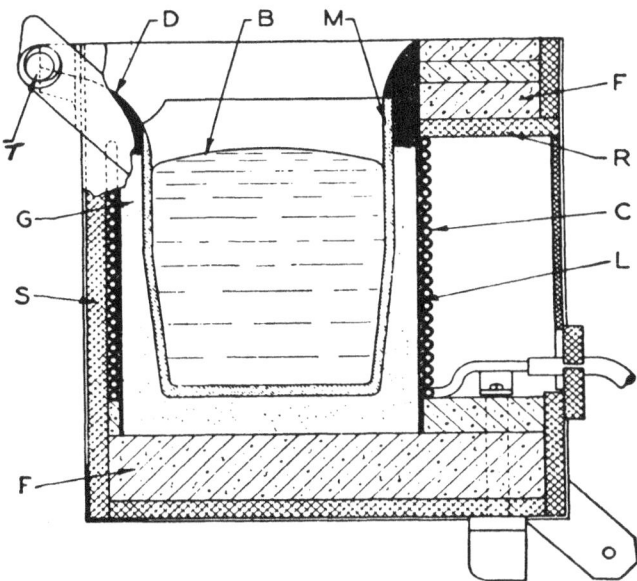

Figure 182. Essential parts of an induction furnace.
B - Metal bath
C - Coil
D - Pouring spout
F - Firebrick
G - Refractory
L - Protective refractory or insulation
M - Crucible
R - Asbestos support
S - Outer shell
T - Trunnions

Figure 183. Typical electric induction furnace.

Figure 184. Induction-furnace control panel.

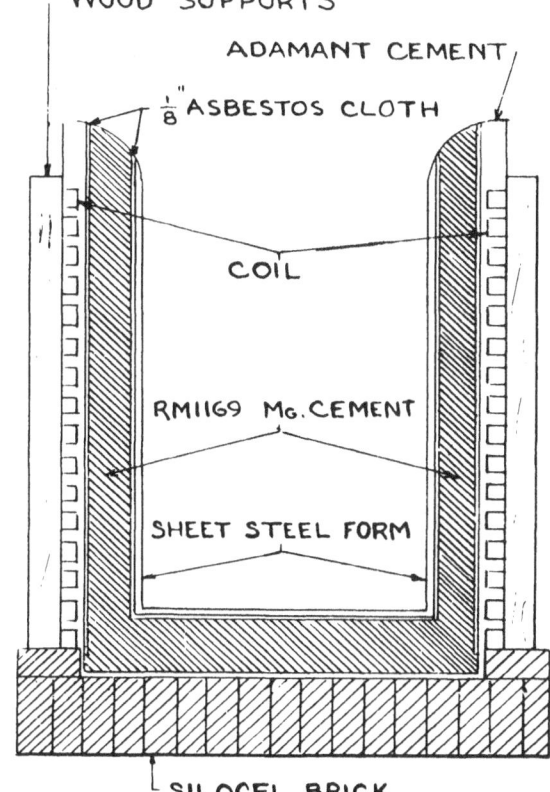

Figure 185. Method of lining induction furnace using a steel form.

Chapter IX
POURING CASTINGS

Pouring is a critical operation in the production of a casting and one which should be carefully conducted. The ladle equipment should be designed for high structural strength and, in the case of geared ladles, for foolproof mechanical operation. Because of the high temperatures involved, a reasonable factor of safety should be used when lining a ladle or when deciding whether an old lining is fit for use.

To avoid accidents, frequent and careful inspection should be made of all parts of the metal-handling equipment (bowls, bails, trunnions, etc.) to detect badly scaled or cracked areas. Defective equipment causes bad castings. When geared ladles are used, any indication of trouble should be immediately checked by carefully examining all gears for excessive tooth wear or broken teeth. Any ladle in a doubtful condition should not be used.

TYPES OF LADLES

Ladles used aboard repair ships are of two basic types. The lip-pouring ladle is shown in figure 186 and the teapot, or bottom-pouring ladle, is shown in figure 187. Crucibles, such as shown in figure 188, are a type of lip-pouring ladle. Teapot crucibles are shown in figure 189.

Ladle bowls are usually of welded or pressed steel construction. Trunnions on the larger ladles are often attached by both riveting and welding for maximum safety.

LADLE LININGS

Ladle linings have an important bearing on the cleanliness of castings produced. If not sufficiently refractory, linings will melt and form slag, which is difficult to keep out of the casting. Because of the high temperature involved, this is most apt to occur in pouring steel. Slagging of the ladle lining is less with gray iron and bronze, and negligible with aluminum.

If the lining material has insufficient dry strength, it will crumble around the upper part of the ladle. The upper part of the lining is not in contact with the molten metal and does not develop strength by fusing or fluxing. When the ladle is tilted to a pouring position, parts of the crumbled rim will fall into the stream of metal entering the mold or into open risers.

A lining mixture which will have the desired properties can be obtained by thoroughly mixing the following materials in a sand muller.

Used steel backing sand (All-purpose backing sand - chapter IV)	81.5 percent
Silica flour	15.0 percent
Bentonite	2.0 percent
Dextrine	1.5 percent
Water	5.0 percent of the dry ingredients

A mixture of silica sand or ganister, fireclay, and water can also be used when available. This mixture contains 85 to 95 percent of silica sand or ganister; the rest is fireclay. The exact percentage of silica sand or ganister and fireclay are determined by how workable a mix is desired. More clay gives a more workable and sticky mixture but increases the amount of shrinkage when the lining is dried. With all types of refractory mixes, only enough water should be added to make the mixture workable. An excess of water, although making the mixture easier to handle, causes more shrinkage and cracks in the lining when it is dried. For low shrinkage and fewer cracks, use small amounts of water and clay.

Before ramming a lining in the ladle, arrangements must be made for venting the lining during drying. This is done by drilling 3/16 inch or 1/4 inch holes through the sides and bottom of the ladle shell on 3 to 4-inch centers. If this is not done, drying will take too long. Numerous injuries to personnel have resulted from the use of improperly dried ladles. When moisture is pocketed under molten metal, a large volume of water vapor is rapidly formed and the metal is blown out of the ladle with explosive force. In addition to this, even slight traces of moisture in ladle linings will cause porosity and casting unsoundness. The most practical way to determine when a ladle is dry is to apply heat until steam flows from the vent holes, and then continue to apply heat until this flow stops completely.

With a properly vented ladle shell, the lining is then rammed in place. It is best to use a wood or metal core to form the inside of the ladle. The form can be made with taper and allowance for lining thickness. After the bottom of the ladle is rammed into place, the form is centered with wedges and the sides of the lining rammed. A harder and more dense lining can be made and the water kept to a minimum when a form is used. Also, the job is a lot easier than trying to ram a lining against vertical walls. When ramming a lining in layers, be sure to roughen the top of each layer before ramming in the next layer. After the lining has been rammed, the form is rapped lightly to loosen it and then drawn from the ladle. To make drawing of the form easier, and to keep the form from absorbing water from the mix, it should be covered with a thin layer of grease or with aluminum foil. If aluminum foil is used, it is peeled from the lining after the core has been removed. Care should be exercised so that aluminum foil does not fold and cause a deep crack in the lining. The ramming of the lining must be very hard and uniform. If the lining shows a tendency to crack into layers when the form is withdrawn, each layer was not roughened enough before the next was rammed in. Such a ladle can be dangerous to use.

The thickness of the lining varies with the metal to be handled and with the size of the ladle. For example, a ladle for the pouring of steel requires a heavier lining than one for cast iron, bronze, or aluminum because steel, at the high temperatures required, attacks the lining material much more rapidly than any of the other metals.

A lining for a ladle to hold and pour 75 pounds of steel will have a thickness of about 1 inch on the bottom and 1 inch on the sides at the bottom and will taper to about 3/4 inch at the top of the ladle. This thickness is also satisfactory for any of the other metals. For the lower-melting-point metals, the main consideration in determining the thickness of lining is proper insulation in order to prevent chilling of the molten metal and to avoid damage to the ladle shells as a result of overheating.

Drying of a new or patched lining is an operation that can cause a lot of trouble if it is not done properly. A new lining or patch should be heated gently at first to get rid of most of the water without blowing a hole in the lining or cracking it because of steam pressure. Aboard ship, this can best be done by first drying the ladle in a core oven and then completing the drying with a torch. The torch should be positioned, with respect to the ladle, so as to insure complete combustion of the gas and delivery of maximum heat to the lining. If a new lining is heated too fast at the start, the water travels back to the shell and makes that part of the lining weak and soggy. After the lining or patch is thoroughly dried slowly, the temperature can be safely increased to as high as obtainable. It is desirable to maintain a new lining at red heat for several hours before using it. If a slowly heated new lining cracks, the ramming mixture contained too much clay or water.

ANYONE WHO POURS METAL FROM A LADLE WITH A DAMP LINING OR DAMP PATCH CAN EXPECT TO FIND UNPLANNED HOLES IN HIS CASTING.

Pouring lips of ladles are a frequent source of trouble because they are often patched and then proper drying of the patch is overlooked. A wet patch on a pouring lip will put gas in the metal and cause blowholes in the castings.

Patching of a ladle is best done when the lining is cold. All adhering slag and metal must be removed in the area to be patched. If possible, undercut the old lining to help hold the patch. Brush loose dirt from the old lining and wet the lining thoroughly. Patch large holes with the same mixture used for lining. Patch small holes and cracks with a mixture of four parts clean sand and one part fireclay. DRY A PATCH THE SAME AS A NEW LINING.

In figure 190 are shown the various steps in the lining of a teapot ladle. Part (a) is a cutaway view of the ladle shell. The bottom is rammed, the forms set in place, and refractory rammed on the side. Many times, a heavy-duty refractory brick is placed in the bottom of any ladle to take the force of the molten metal stream when the ladle is filled. This reduces erosion of the ladle bottom. Part (b) shows the sides partly rammed after the forms are set. The completed lining with forms still in place is shown in (c). Order of withdrawing the forms is shown in (d).

Ladles for steel are commonly used only once per lining because of the fluxing action of the metal at the high temperatures. In an emergency, if great care is used in skimming slag, ladles may be used twice for steel, but it is not good practice. This does not hold true for the other metals, however, and the ladles may be used for many heats. Care should be taken to remove all metal and slag after each use, but it is impossible to remove all of the debris. Therefore, ladles should be used for only one metal. A separate ladle should be used for each metal or the metal will become contaminated and unfit for use.

POURING THE MOLD

The placing of weights and clamps on a mold is only a minor operation in the making of a casting but one that will produce defects if not properly done. Weights are used to prevent the force of the molten metal from lifting the cope as it fills the mold, thereby producing a swelled casting or a runout. The position of the weight on the mold should be determined and the weight placed on the mold gently, without any movement of the weight across the top of the cope.

Any such movement can cause the cope to break or force sand into the gating system or open risers.

Clamps serve the same purpose as weights and are used to clamp the cope and drag together when the casting is poured in the flask. In placing the clamps, a wood wedge is usually used to tighten the clamp down on the flask. Before placing the clamp, the area next to the clamp should be cleaned of excess sand to prevent mold damage when the clamp is set. The wedge should be placed between the clamp and the top edge of the flask, the clamp brought up snug, and then tightened by driving the wedge. Care must be taken to hit the wedge and not the clamp or flask.

Although clamps and weights are used for the same purpose, clamps are much more dependable. It is too easy to underestimate the lifting power of the metal and to use too few weights. On the other hand, too many weights can crush a mold.

With the ladle thoroughly dried, preheated to a red heat, and securely in the bail, molten metal from the furnace is tapped into it. Filling the ladle to its brim is unwise from the standpoint of safety and for the production of good castings. Filling the ladle to its brim should be done only when absolutely necessary, and then extreme caution should be exercised in handling the ladle and pouring.

If the ladle is filled to about 3/4 of its capacity, metal will not flow over the lip until the ladle is inclined to an angle of approximately 60° from the horizontal. This permits a good control of the stream, making it possible to keep the ladle quite low and thus, keep the height of fall of the metal low. This lessens mold erosion, entrapment of air, formation of oxides, and metal spills.

Figure 191 shows the proper method of pouring, while figure 192 shows poor pouring technique. In the good pouring technique, notice that the lip of the ladle is as close as possible to the mold.

Slag on the metal should be skimmed carefully prior to and during pouring. If a steel or iron slag is too fluid to be skimmed properly, dry silica sand should be spread across the surface of the molten metal to thicken the slag. Dry metal rods or special metal skimmers should be used for skimming or stirring metal. Wood skimmers or stirrers should never be used because the wood contains moisture which often produces unsoundness in castings.

SPEED OF POURING

The pouring basin should be filled quickly to prevent nonmetallics and slag from entering the mold cavity and must be kept full. In order to do this, the ladle stream must be controlled carefully. Once pouring has started, it must continue without interruption until the mold is filled. One allowable exception is to stop pouring through the sprue when the metal has filled 1/3 of a top riser. The riser is then filled last with hot metal to improve feeding. This exception applies only to top risers. With side risers, the mold might not be full when metal is seen in the riser.

The use of a pouring basin and plug to get more uniform pouring is shown in figure 193. Part (a) shows the basin ready to receive the metal. In (b) the basin is partially filled. When the basin is properly filled, the plug is withdrawn as in (c). The use of a pouring basin permits better control of the metal entering the gating system. Another variation of this method is to put a thin sheet of the metal being poured over the sprue opening. It will melt out when the basin is filled with hot metal. Keep the basin full of metal at all times.

When pouring a metal that forms dross (especially aluminum, aluminum bronze, or magnesium), every effort must be made to avoid turbulent entry of the metal into the mold. It is particularly important in such cases that the lip of the ladle be as close to the pouring basin as possible. The sprue must be filled quickly and kept full so that the tendency for dross and entrapped air to enter the mold will be at a minimum. Here again the pouring basin and plug can be used to advantage. The use of skim gates or perforated cores placed in the sprue or pouring basin (as shown in figures 194 and 195) aids in removing dross from the metal and preventing its entrance into the casting cavity. Agitation of the molten metal while it is being transported to the mold also increases dross formation and gas absorption.

POURING TEMPERATURE

Close control of pouring temperatures is essential to the consistent production of good castings. An immersion pyrometer and an optical pyrometer are furnished for temperature determination. Because of the high temperature involved, the immersion pyrometer is not used for iron or steel. The Chromel-Alumel immersion thermocouples are limited to temperatures of 2,500°F. The optical instrument is impractical for nonferrous metals because their pouring temperatures are too low.

The operation of an immersion pyrometer is a simple matter. The instruments are usually of two types; a self-contained unit or the unit in which the immersion unit is connected by wires to the reading instrument. Before use, the pyrometer should be checked to make sure that the immersion part of the instrument is clean. Lead-wire pyrometers should be checked for any breaks or loose connections in the wire. When taking a reading, the immersion tip should be submerged in the molten metal to a depth of approximately 3 inches and moved slowly back and forth or in a circle. After the temperature reaches a fairly steady reading, it should be recorded. The immersion tip should then be withdrawn from the melt. Immersion pyrometers should be handled with care and periodic inspections made for proper upkeep. Whenever possible, the instruments should be checked and calibrated for good operation.

The optical pyrometer operates by matching the intensity of light from the molten metal with that of a standard light source within the instrument. Exact operating procedures are available with the instruments. Generally, the field of vision will be uniform, as shown in figure 196a, when the instrument is set at the proper temperature. If the instrument is set too high, the inner circle of the field will be brighter as in (b). If the inner circle is darker, the instrument is set at too low a reading. This instrument should be handled with care and given periodic checks and calibrations for proper operation.

Excessive pouring temperatures (that is, temperatures above those required for the proper filling of the mold) result in excessive oxide or dross formation, segregation, rough and dirty casting surface, unnecessarily high liquid shrinkage, coarse-grain metal structure, and increased danger of cavities, tears and porosity. Figure 197 shows the increase in grain size that resulted with increased pouring temperatures for a copper-base alloy. Notice that the high pouring temperature resulted in a very coarse grain structure.

If the pouring temperatures are too low, entrapped gas and dross, misrun castings, or castings with surface laps (cold shuts) are likely to result. Proper pouring temperatures for a given metal vary with the casting size, design, and desired rate of pouring. For this reason, the pouring ranges given below should be taken as a general guide only:

Metal	Pouring-Temperature Range
Steel	2850°F. to 2950°F.
Gray iron	2300°F. to 2600°F.
Aluminum	1250°F. to 1400°F.
Manganese bronze	1875°F. to 1975°F.
Compositions G & M	2000°F. to 2200°F.

In general, thin-walled castings are poured on the high side of the range and thick-walled castings on the low side.

SUMMARY

The important factors in pouring a casting are summarized as follows:

1. Ladle equipment must be kept in good repair.

2. All ladle linings must be rammed uniformly hard and be of the proper thickness.

3. All ladles must be thoroughly dried and at a red heat for some time before use with the high melting point alloys.

4. Ladles should not be filled to more than 3/4 of their capacity.

5. Metal should be skimmed free of all slag or dross before pouring.

6. In pouring, the ladle should be as close to the pouring cup or sprue as possible.

7. Once pouring has started, the stream should not be interrupted. A steady rate of pouring should be used and the sprue should be kept full at all times.

8. The metal should be poured at the correct temperature, neither too high nor too low.

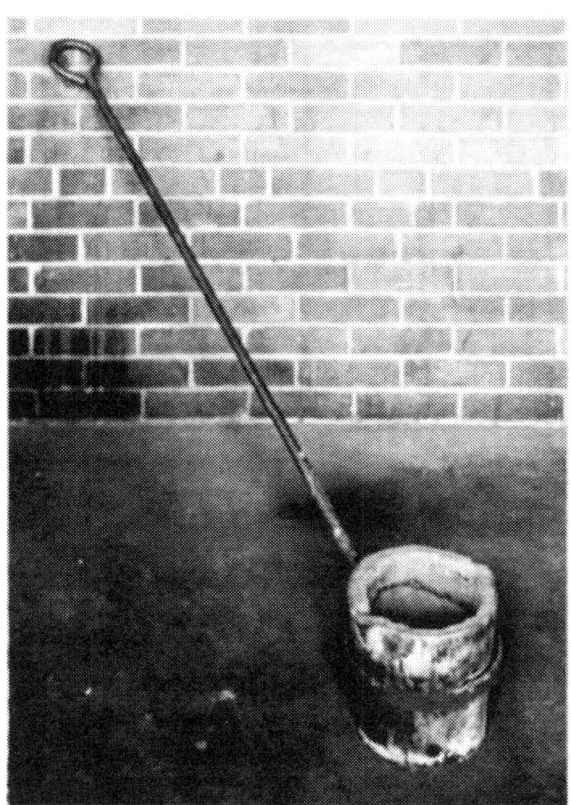

Figure 186. Lip-pouring ladle.

Figure 187. Teapot ladle.

Figure 188. Lip-pouring crucibles.

Figure 189. Teapot crucibles.

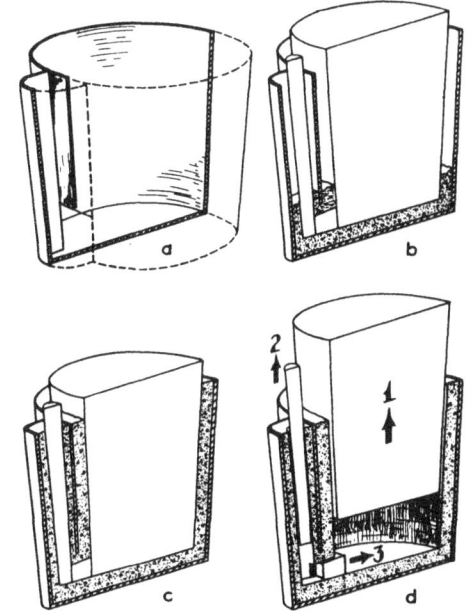

Figure 190. Lining a teapot pouring ladle.

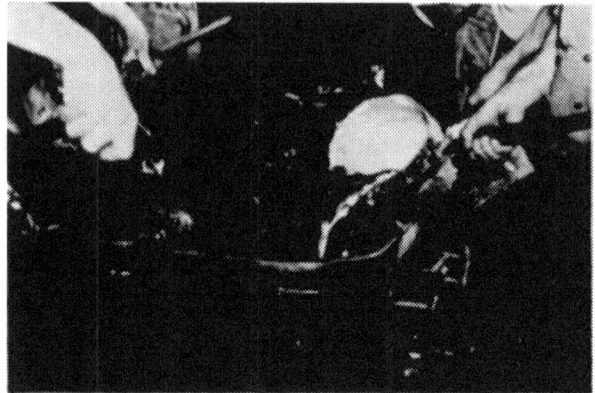

Figure 191. Proper pouring technique.

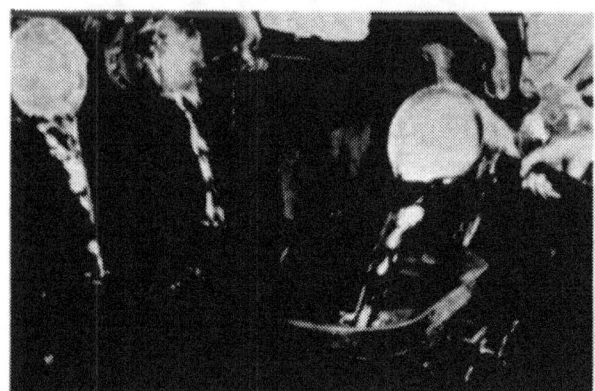

Figure 192. Poor pouring technique.

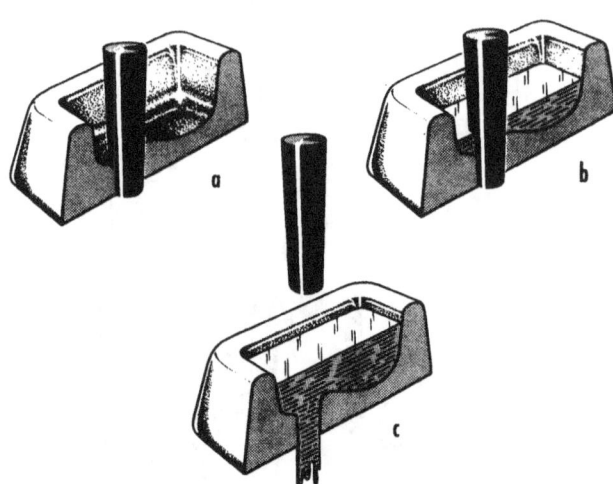

Figure 193. Use of pouring basin and plug.

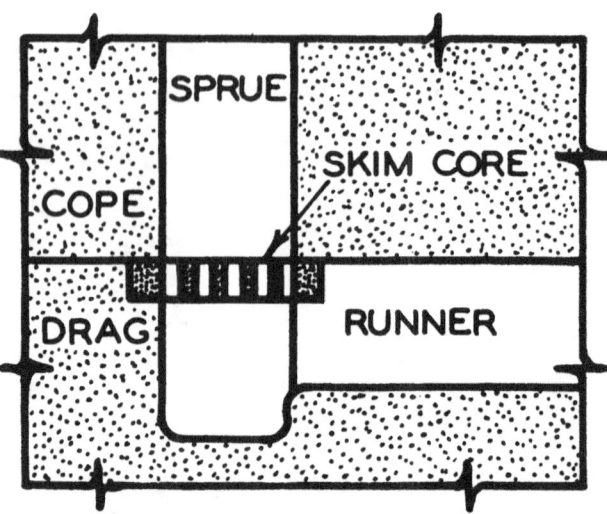

Figure 194. Skim core in down gate.

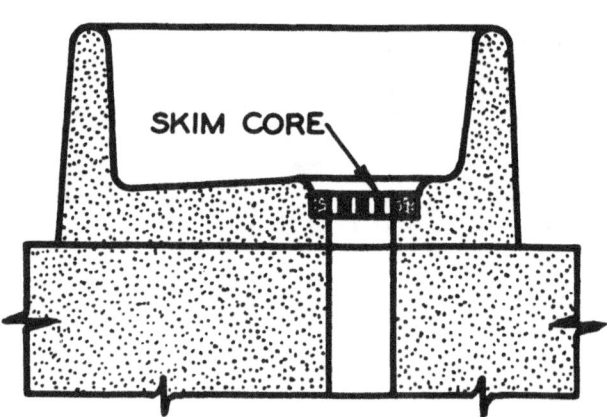

Figure 195. Skim core in pouring basin.

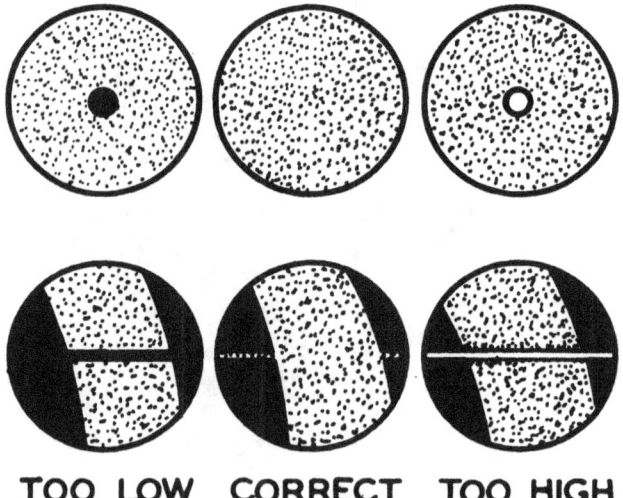

Figure 196. Pyrometer field when at correct temperature, too high a setting, and too low a setting.

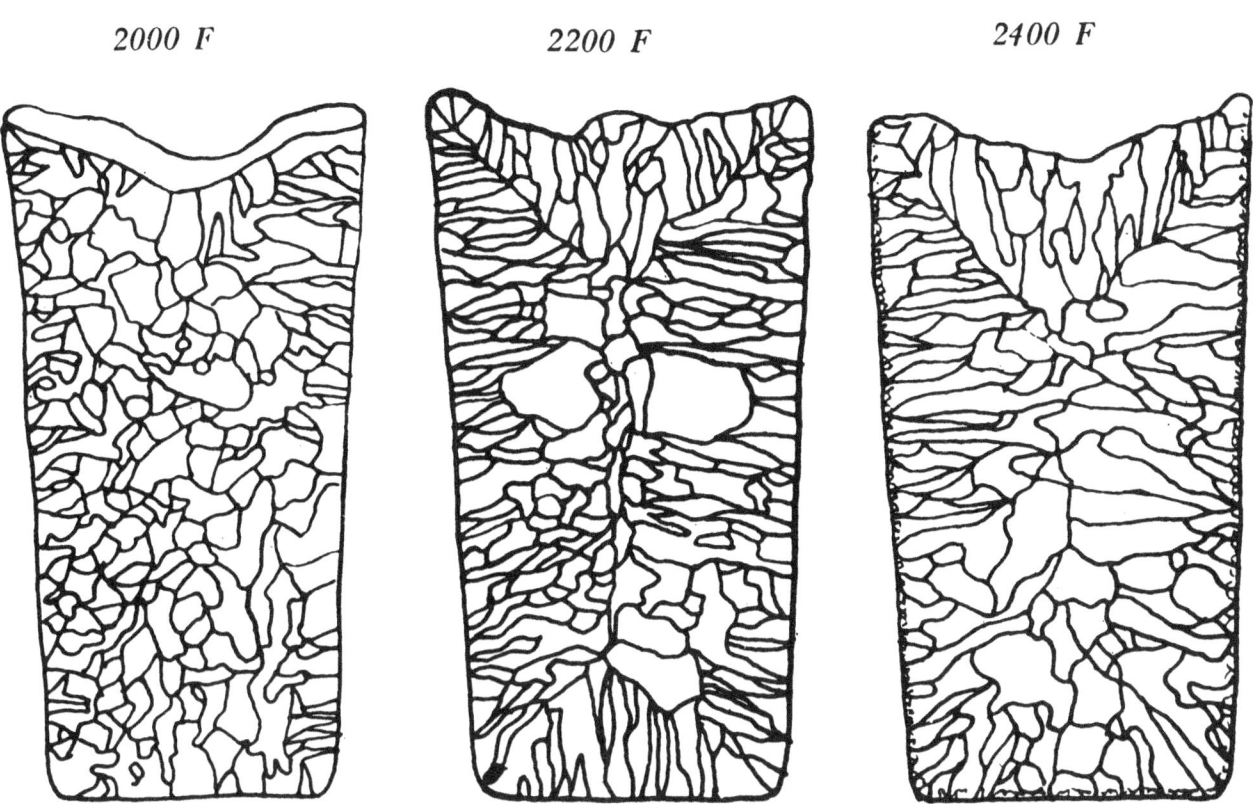

Figure 197. Effect of pouring temperature on grain size.

Chapter X
CLEANING CASTINGS

After the casting has solidified, it should be allowed to cool in the mold until it has reached a temperature which will permit safe handling. The time required for this will vary with the metal, type of mold, and the size and design of the casting. As a general guide, castings should not be shaken out until they have cooled to at least the following temperatures:

Steel	1200°F.
Cast iron	1000°F.
Manganese bronze	1000°F.
Compositions G and M	1000°F.
Aluminum	500°F.

After the casting is shaken from the mold, all adhering sand should be removed with wire brushes or chipping hammers before the casting is cleaned by water, sand, or shot-blasting methods. The casting should not have a lot of excess sand left on it before it is shot blasted. The sand contaminates the metal shot. If the excess sand is removed from the casting, there will also be much less dust to be extracted by the dust-arresting equipment.

REMOVING GATES AND RISERS

The following methods are used to remove gates and risers from castings.

Steel. For Grade B and for low-alloy steel, flame cutting with the oxyacetylene, oxyhydrogen, or oxypropane torch is the best method to use. Thorough cleaning of the casting is important to make it easier to start the cut and to assure a uniform cut. The gates and risers should be cut about three sixteenths or three-eighths of an inch from the casting. The remaining stub is removed by grinding or by chipping hammers.

For stainless steel castings, the gates and risers cannot be removed by flame-cutting. They must be removed by mechanical means such as sawing, chipping, or shearing, with an abrasive cutoff wheel, or by melting off with an electric arc from a welding machine. In melting off, care must be taken to leave a stub of 1/4 to 1/2 inch on the casting to avoid cracking or metallurgical changes in the casting as a result of the high temperature where the cut is made.

If castings show a tendency to crack during cutting, risers should be removed while the castings are at a temperature of at least 400°F. For risers larger than six inches in diameter, it is advisable to preheat to 700°F. or higher. The desired cutting temperature may be that retained during cooling or it may be obtained by reheating the casting in a furnace.

Cast Iron. The gates and risers on cast iron may be removed by flogging, sawing, or chipping. The use of cutting torches is not practical. Flogging is the simplest method and is entirely satisfactory. To flog, the gate or riser is first notched on all sides to keep the break from leading into the casting or a notch is cast into the junction of riser or gate and the casting. The gate or riser should be struck sharply so that the blow is going away from the casting, rather than toward it. This will help to keep breaks from leading into the casting, and will prevent damage to the casting if the hammer misses the gate or riser. Abrasive cutoff wheels can also be used. Sawing with a hacksaw or hand saw is practical if the casting is easy to grip. Stubs remaining on the casting are removed by grinding or chipping.

Brass, Bronze, and Aluminum. The most common methods for removing gates and risers from nonferrous castings are by band saw, high-speed hack saw, abrasive cutoff wheel, or by shearing, depending on the type of equipment available for this purpose.

GRINDING AND FINISHING

Chipping followed by grinding or finishing is used to remove the rough metal remaining on a casting after the gates and risers have been removed. Many times, grinding can be used to salvage a casting which has small fins or localized rough spots on the surface.

When using any type of grinder, the wheel should be protected and the operator should wear goggles. Gloves are a hazard because they may become caught in the wheel. The operator should also avoid loose clothing. Before a grinding wheel is used, it should be struck lightly but sharply with a hammer handle to determine whether the wheel has a high-pitched ring. A dull thud indicates that the wheel may be cracked and may fly apart during use. DO NOT USE A GRINDER UNLESS YOU ARE WEARING GOGGLES.

Grinders available aboard ship are of two types, stand and portable. The portable grinders are air or electric driven. The stand grinder is electric and is used for castings which can be easily manipulated by hand at the face of the grinder. Grinding on large castings must be done with the portable grinders.

Many of the grinding operations done with the stand grinder use bodily contact with the casting to provide the required pressure and stability. Such bodily contact causes the operator to receive a lot of vibration during grinding. Reinforced leather aprons are useful for reducing the physical strain on the operator during grinding. They reduce the vibration transmitted to the operator. For grinding operations requiring a long period of time, a reinforced apron or a similar piece of safety equipment is a necessity. The apron not only serves the purpose of reducing fatigue in the operator, but also may prevent serious injury in case the casting becomes snagged between the tool rest and the grinding wheel.

The portable grinders are normally used for lighter grinding operations, but aboard ship they must be used for heavier grinding on large castings. Small portable grinders are useful for cleaning up minor surface defects in a casting.

When using either type of grinder, an attempt should always be made to use the entire face of the grinding wheel. Moving the work back and forth across the face of a stand grinder, or moving a portable grinder back and forth across the casting, will help in obtaining uniform wear of the wheel face. A grinding wheel is difficult to use when the face becomes grooved because of improper use. It is poor practice to snag a casting between the grinder rest and the wheel in an attempt to apply more pressure and get faster grinding. This causes localized overheating with possible cracking of the casting, unnecessary wear on the grinding wheel, and danger to the operator. This practice or the use of levers to deliver high pressure may cause the wheel to break and injure the operator.

WELDING

Many defective castings may be salvaged by welding. When repairs by welding are required, reference should be made to the "General Specifications for Ships of the United States Navy," Section S9 - 1, "Welding," for general guidance. The actual welding should be done by trained personnel and not attempted by unqualified personnel.

Another use for welding is in the assembly of two or more simple castings into a complicated part. Quite often, an emergency casting can be most simply made by making two or more simple castings and then welding them together. Another scheme is to make some parts of an assembly by casting and to complete the structure by welding wrought metal to the casting.

SUMMARY

Cleaning and grinding of castings is a relatively simple operation compared with the other operations involved in making a casting. It is as important as any of the other operations because carelessness in finishing may ruin an otherwise acceptable casting. The production of good castings depends on the use of correct techniques in all of the operations and not in just a few of them.

Chapter XI
CAUSES AND CURES FOR COMMON CASTING DEFECTS

Defects in castings do not just happen. They are caused by faulty procedure (1) in one or more of the operations involved in the casting process, (2) in the equipment used, or (3) by the design of the part. A casting defect is often caused by a combination of factors which makes rapid interpretation and correction of the defect difficult.

Casting defects arise from many causes and have many names. One of the most prominent causes of defects does not appear on any formal list of defects, it is CARELESSNESS. Its remedy is obvious.

NAMES OF DEFECTS

The table in the summary of this chapter lists the most common types of casting defects, their causes, and their cures. Causes and cures are discussed in more detail later in this chapter. The names of common defects are explained as follows.

A blow or blowhole is a smooth cavity caused by gas in the molten metal. A pinhole is a type of blow that is unusual because of its small size. It is common to find a single large gas hole (or blow) in a casting, but pinholes usually occur in groups.

A shrink or shrinkage cavity is a rough cavity caused by contraction of the molten metal. It is quite often impossible to tell whether a particular hole in a casting is a shrink or a blow. Gas will aggravate a shrink defect, and shrinkage will aggravate a gas defect. The distinction can usually be made that gas pressure gives a cavity with smooth sides (blow) and contraction or lack of feeding gives a cavity with rough sides (shrink). When either a blow or shrink occurs, it is a good idea to correct for both if the cause cannot be determined for sure.

A rat tail, buckle, and scab all originate in the same way and differ mainly in degree. They are caused by uncontrolled expansion of the sand. If the condition is not too bad, a rat tail is formed. The surface of the sand buckles up in an irregular line that makes the casting look as though a rat has dragged his tail over it. If sand expansion is even greater, the defect is called a buckle. If it is still worse so that molten metal can get behind the buckled sand, it is a scab.

A misrun or cold shut occurs when the mold does not completely fill with metal, or when pouring is interrupted so that the metal does not fuse together properly.

Metal penetration causes rough castings. The metal seeps in between the sand grains and gives a rough surface on the casting. Such castings are difficult to clean because sand grains are held by little fingers of metal.

A sticker occurs when sand sticks to the pattern, as the pattern is drawn from the mold.

A runout, bleeder, or breakthrough is a casting in which the mold has failed so that the metal runs out before the casting is solid.

A cut or wash is erosion of the sand by the stream of molten metal. It often shows up as a pattern around the gates and usually causes dirt in some part of the casting.

A swell is an enlarged part of a casting resulting frequently from soft ramming. It is often found in connection with metal penetration.

A crush or drop occurs when part of the sand mold is crushed or drops into the mold cavity. It usually causes dirt in some other part of the casting.

A shift is a mismatching of cope and drag or of mold and cores.

Hot cracks or hot tears are usually irregular and oxidized so that the fracture appears dark. A dark fracture usually shows that the crack or tear occurred while the casting was still hot and contracting. A bright fracture indicates that the break occurred when the casting was cold.

A fin is a thin projection of metal usually found at core prints or parting lines. Fins are common on castings and not too harmful if small. If large, they can cause a runout, or small shrinkage cavities at the junction of the fin with the casting.

Inclusions or dirt are just what the name implies. They are often accompanied by other defects which provide loose sand in the mold.

DESIGN

The most common defects caused by casting design are hot tears and hot cracks. A hot tear is usually recognized by its jagged discolored fracture. It occurs when the stresses in the casting are greater than the strength of metal shortly after the casting has solidified. A hot crack occurs by the same method except that it takes place after the casting has cooled considerably. A hot crack is also recognized by a discolored fracture, but is smooth, as compared with the jagged fracture of the hot tear.

Hot tears and hot cracks both are caused by improper design that does not provide adequate fillets at the junction of sections or that joins sections of different thicknesses without providing a gradual change in section size by tapering. Inadequate fillets (sharp corners) produce planes of weakness at the junctions of the sections and cause failure at these points. Failure from improper joining of heavy and thin sections is caused by the early solidification of the thin section before the heavy section has solidified. The contraction of the thin section produces a stress which is greater than the strength of the partially solidified heavy section. Something has to give; it is usually the heavy section.

The cure for hot cracks and hot tears caused by poor casting design is to provide adequate fillets at all junctions and to use tapered sections where sections of different thicknesses must be joined. Refer to Chapter 2, "Designing a Casting."

Shrinkage cavities, misruns, cold shuts, pinholes, blows, drops, scabs, and metal penetration can also be caused by poor casting design as well as by other factors. Shrinkage cavities may be caused by using fillets large enough to produce a section that cannot be properly fed, or by heavy sections that are so located in the casting that they cannot be properly fed. The latter condition should be corrected by redesigning the casting, the use of chills on heavy sections, or by making the part as two castings which can be welded together.

Misruns and cold shuts are caused by a low pouring temperature for the sections involved, inadequate gating, or inadequate venting of the mold. Redesigning for the use of tapered sections can be used to eliminate these defects. Pinholes can be caused by nonuniform section size. A high pouring temperature necessary to overcome cold shuts and misruns in thin sections may result in pinholes in the heavier sections. This situation requires redesign for uniform section thickness, re-gating to permit lower pouring temperatures, or the use of chills on heavy sections.

Glows due to design can often be traced to insufficient means for the escape of core gas. This may be due to a core print which is too small or inadequately vented. Corrective measures call for an increase in the size of the core print and adequate venting, and the use of core coatings.

Sharp corners in the cope or on protruding sections may become weakened during the drawing of the pattern and cause drops. This is corrected by the use of fillets, increased draft on the pattern, and rounded corners. Deep pockets or overhanging sections in the cope cause drops because of the weight of the sand. If these cannot be overcome by changing the position of the pattern in the flask, reinforcements must be used to give the sand adequate support. Sharp corners also cause scabs because they aggravate the conditions in large flat surfaces, which cause scabs. The use of fillets and round corners will minimize the effect of sharp corners on scab formation. To minimize casting defects caused by improper design, maintain (1) the casting as simple as possible, (2) tapered sections to promote directional solidification, (3) corners rounded or filleted, and (4) plenty of draft on the pattern.

PATTERN EQUIPMENT

The most common defect which can be traced to pattern equipment is the shift. A shift is easily recognized by the mismatching of the cope and drag sections of the casting at the parting. This type of shift is caused by worn pattern equipment. Loose or worn dowel pins in a pattern will permit movement of the pattern parts during molding and cause a shift in the casting. A shift caused by a defective pattern can only be corrected by repairing the pattern. Good pattern maintenance will go a long way toward minimizing the occurrence of shifts due to worn patterns. The recognition of this defect is especially important in repair ship work because the great majority of castings are made with loose patterns.

Another defect frequently caused by a poor pattern is the sticker. A sticker is due to a poor pattern surface, which causes the sand to stick to the pattern. Poor pattern surface can be remedied by smoothing the rough spots and refinishing the pattern. A sticker which is not noticed in the molding operation will have the appearance of a drop in the completed casting.

Other defects that may be caused by pattern equipment include misruns, cold shuts, drops, and metal penetration. Worn pattern equipment, which causes sections to be thinner than designed, may produce misruns and cold shuts in the casting. A drop will be caused by a pattern having insufficient draft. Improper draft will cause the sand to crack when the pattern is drawn, and will cause a drop because of the weakened condition of the sand. Increase in the draft of the pattern is the cure for this defect.

Metal penetration may occur because an irregular parting line has prevented proper ramming of the sand. Metal penetration of this type can be corrected by remaking the pattern with a flatter parting line.

FLASK EQUIPMENT AND RIGGING

Crushes and shifts are the defects most commonly caused by defective flask equipment and rigging. A displacement of sand in the mold after it has been made causes a crush. Improperly aligned flask equipment, warped or uneven flask joints, bad-fitting jackets, and bad bottom boards all result in an unequal pressure on the mold, with the resulting displacement of sand which produces the crush. Properly maintained equipment is the only solution to crushes of this type.

Shifts are also caused by defective flask equipment. Worn pins or defective bushings in a flask allow movement of the cope to occur when closing the mold. Proper maintenance of equipment again is the solution for this defect.

Stickers are often caused by faulty flask equipment. The defective flask prevents a clean pattern draw and, as a result, some sand sticks to the pattern. The sticker shown in figure 198 was caused by loose pins and bushings.

Swells, fins, runouts, bleeders, metal penetration, hot tears, and hot cracks can often be traced to faulty equipment and rigging. Swells and fins are likely to occur when the mold weights are not heavy enough for the casting being poured. Because of the light weight of sand, the molten metal is able to displace the sand and produce a swell in the casting. If this displacement occurs at the parting line or a core print, the molten metal is able to penetrate the joint and a fin is the result. Swells and fins can be remedied by using enough mold weights or clamps to resist the ferrostatic force of the molten metal. Remember that iron, steel, brass, and bronze are heavier than sand, so the cope will tend to float off when these metals are used. For double security, use both mold weights and clamps.

Runouts and bleeders occur when the molten metal penetrates the parting line and reaches the outside of the mold. A breakout may occur anywhere on the mold and may be caused by insufficient sand between the pattern and the flask, or by soft ramming. Runouts and bleeders at the parting line are often caused by uneven matching of the cope and drag. This mismatch may be caused by bad pins and bushings, dirt in the flask joint, bad bottom boards, or uneven clamping. The cures are self-evident.

Metal penetration (rough surface of the casting) is caused by ramming the mold too soft, as when there is too little space between the pattern and the flask. A larger flask will permit harder ramming between the pattern and the flask and reduce penetration of the metal between the grains of sand.

Hot tears and hot cracks can often be traced to a lack of collapsibility in sand which has been excessively reinforced. Excessive reinforcement prevents the sand from collapsing and obstructs free contraction of the casting. Remember that metals contract when they solidify and that the mold must be weak enough to allow the casting to contract. If the mold is too strong, the casting may crack. Reinforcement which is placed too near a sprue or riser has an even greater effect than that mentioned above. The reinforcement in this case restrains the sprue and riser from any free movement with the casting and is almost sure to cause hot tears or hot cracks. If hot tears or hot cracks occur near the point where risers or sprues are attached to the casting, the reinforcement of the mold should be checked as a possible cause.

GATING AND RISERING

Shrinkage cavities, inclusions, cuts, and washes are the defects most frequently caused by gating and risering. If a riser is too small for the section to be fed, there will not be enough metal to feed the section and a shrink will occur in the casting. The gross shrink shown in figure 199 was caused by inadequate feeding. Surface shrinks caused by improper feeding are shown in figures 200 and 201. Improper location of gates and risers for directional solidification can also lead to shrinks. Figure 202 shows an internal shrink which was exposed when the riser was removed. This defect resulted from an improper gating system which resulted in cold metal in the riser. When the casting was gated so as to put hot metal in the riser (as shown in figure 203), the shrink defect was eliminated.

Connections which freeze off too early between a riser and casting produce a shrink by the same method as a small riser because there is no molten metal available to feed the casting. In such a case, the connections should be made larger. The location of the riser with respect to the section it is feeding can also cause a shrink as shown in figure 204. In this casting, the location of the ingate prevented proper feeding of the casting even though the riser contained molten metal.

Inclusions are caused when the gating system permits dirt, slag, or dross to be carried into the casting. The method of eliminating inclusions is to provide a choking action at the base of the sprue by using a tapered sprue of correct cross-sectional area. If it is impossible to provide proper choking action in the gating system, a skim core should be used at the base of the sprue to trap dirt and slag. Dross inclusions in a fractured aluminum casting are shown in figure 205.

Cuts and washes are defects which are also caused by the gating system. If the ingates of a mold are located so that the metal entering the mold impinges or strikes directly on cores or a mold surface, the sand will be washed away by the eroding action of the stream of molten metal. The defect will then appear on the casting as a rough section, usually larger than the designed section thickness. Sand inclusions are usually associated with cuts and washes as a result of the sand eroded and carried to other parts of the casting by the stream of metal.

Improper risering and gating can also cause blows, scabs, metal penetration, hot tears, hot cracks, swells, fins, shifts, runouts, bleeders, misruns, and cold shuts. Blows or gas holes are caused by accumulated or generated gas or air which is trapped by the metal. They are usually smooth-walled rounded cavities of spherical, elongated, or flattened shape. If the sprue is not high enough to provide the necessary ferrostatic heat to force the gas or air out of the mold, the gas or air will be trapped and a blow will result. An increase in the height of the sprue or better venting of the mold are cures for a blow of this type. A similar blow may also occur if the riser connection to the casting freezes off too soon and the metal head in the riser is prevented from functioning properly. To cure this situation, the connections should be made larger, placed closer to the casting, or the connection area should be checked for possible chilling from gaggers or improperly placed chills.

Scabs can be caused by the gating system if the gating arrangement causes an uneven heating of the mold by the molten metal. The cure for a scab from this cause is to regate the casting to obtain a uniform distribution of metal entering the mold.

Metal penetration (rough surface on the casting) occurs if the sand is exposed to the radiant heat of the molten metal for too long a time so that the binder is burned out. An increase in the number of ingates to fill the mold more rapidly will correct this situation. Any gating arrangements which cause the sand to be dried out by excessive radiation will result in metal penetration. A sprue which is too high will cause a high ferrostatic pressure to act on the mold surfaces and cause metal penetration. Metal penetration of this type can be cured by decreasing the height of the cope.

The location of gates and risers can cause hot tears and hot cracks. If the gates and risers restrict the contraction of the casting, hot tears and cracks will occur. If the defects are near the ingates and risers, this cause should be investigated as a possible trouble spot.

Swells, fins, runouts, and bleeders may also be caused by improper gating. Risers which are too high cause an excessive ferrostatic force to act on the mold, with the result that these defects occur. A reduction in cope height will correct defects of this type. Runouts and bleeders may also occur if any part of the gating or risering system is too close to the outside of the mold. In such a situation, there is insufficient sand between the gates, runners, risers, and flask to permit proper ramming. This results in weak sand which cannot withstand the force of the molten metal. The selection of flasks of proper size for the casting being made is the method of overcoming these defects.

Misruns and cold shuts are caused by any part of the gating and risering system which prevents the mold from filling rapidly. Gates or runners which are too small restrict the flow of molten metal and permit it to cool before filling the mold. Improperly located ingates will have the same result. If the pressure head of a casting is too low, the mold will not fill completely and a cold shut will result. Increasing the size of the gating system and relocation of ingates are methods used to correct defects due to the gating system. Increasing the height of the sprue will produce a greater pressure head and help to fill the mold rapidly.

SAND

By itself, the molding sand can cause all of the casting defects that a molder will encounter. This is one of the reasons that it is difficult to determine the cause of some defects. Blows can be caused by sand that is too fine, too wet, or by sand that has low permeability so that gas cannot escape. If the sand contains clay balls because of improper mixing, blows will be apt to occur because the clay balls are high in moisture. A blow caused in an aluminum casting by high moisture content of the sand is shown in figure 206. To remedy this situation, the sand should be mulled to break up the clay balls. If the sand contains too many fines, it will have a low permeability and the moisture or gas will have a difficult time flowing through the sand away from the casting. Fines should be reduced by adding new sands.

Too high a moisture content in the sand makes it difficult to carry the excessive volumes of water vapor away from the casting. Use of the correct moisture contents and the control of moisture content by routine tests with sand testing equipment is the best way to correct this cause. When the permeability of sand is low, it is difficult for even small amounts of moisture to escape through the sand. The addition of new sands and a reduction in clay content serve to "open up" a sand and increase its permeability.

Drops are often caused by low green strength. Such sand does not have the necessary strength to maintain its shape, so pieces fall off. Corrective measures call for an increase in binder, increase in mixing time, or an increase in both binder and mixing time. Don't overlook the possibility of reinforcing a weak section of a mold with wires, nails, or gaggers.

A scab will be caused on a casting when the sand mold cannot expand uniformly when it is heated by molten metal. The individual sand grains have to expand. If the mold does not "give," the surface of the mold has to buckle and cause a scab. An expansion scab is shown in figure 207. The main cause of a sand being unable to expand properly is the presence of too many fines in the sand. These fines cause the sand to pack much harder so that its expansion is restricted. Addition of new sands to properly balance the sand grain distribution and reduce the percentage of fines is used to obtain better sand properties. Another remedy is to add something to the sand to act as a cushion. Cereal flour, wood flour, and sea coal are all used for this purpose.

A molding sand may have acceptable thermal-expansion properties, but a low green strength may also cause an expansion scab. The cure in this situation is to increase the clay content. A high dry strength and a high hot strength can also cause expansion scabs. The sand will be too rigid because of the high strengths, and proper expansion of the sand will be restricted. A reduction in the clay or binders which cause the high strengths will correct scabs due to these causes. If a scab is present on a casting surface where sand shakeout and cleaning was difficult, high hot strength of the sand was probably the cause. The binder should be reduced, cushioning materials added, or fines reduced by adding coarser sand.

The principal cause for cuts and washes is low hot strength. When the sand is heated by the molten metal, it does not have the necessary strength to resist the eroding action of the flowing metal and is washed away. If an increase in the amount of binder does not cure cuts and washes, a different type of binder may be required. An addition of silica flour may also be used to correct low hot strength. A low green strength and a low dry strength may also lead to cuts and washes. These properties are corrected by increasing the binder. A defect closely related to cuts and washes is the erosion scab. It is also caused by a molding sand having a low hot strength. A combination of other factors such as high moisture and hard ramming can also cause an erosion scab such as shown in figure 208. Hard ramming makes the escape of moisture difficult when hot metal is poured into the mold. As a result, the expanding vapor loosens the sand grains and they are washed away by the molten metal. Sand inclusions in some part of a casting are always found when expansion scabs occur.

A metal-penetration defect occurs when the molten metal penetrates into the sand and produces an enlarged, rough surface on the casting. If the metal penetration is not too deep, it may have the appearance of a swell. Coarse sand, high permeability, and low mold hardness are the principal sand properties which cause metal penetration. A sand that is too coarse will have larger openings between sand grains (this accounts for the high permeability). Because of the openings, the molten metal does not have any difficulty in penetrating into the sand. A low mold hardness is caused by soft ramming of the mold. This condition offers a soft surface to the molten metal which, again, can easily penetrate into the sand. To correct penetration due to coarse sand and high permeability, fine sand must be added to the base sand to get a finer sand distribution and reduce the permeability. Harder and improved ramming technique is the cure for metal penetration caused by low mold hardness. If permeability of the sand cannot be reduced, a mold wash may be used to eliminate penetration. An example of metal penetration is shown in figure 209. The left side of the casting has a good surface - the result of using a mold wash to prevent penetration.

Veining is shown in figure 209. It is caused when the sand cracks and the crack is filled by the molten metal. A sand that collapses rapidly under the heat of molten metal will produce veining. This can be corrected by the addition of more binder or silica flour to the sand.

Hot tears and hot cracks are usually caused by poor sand properties. A high percentage of fines and a high hot strength are the principal causes. A high percentage of fine sand grains produces a more closely packed sand, with the result that it cannot contract properly when the casting itself contracts during and after solidification. The reduction of fines can be accomplished by additions of coarse sand. A high hot strength will also prevent the sand from contracting or collapsing properly. A reduction in the content of fines is also a corrective measure for hot cracks and tears due to a high hot strength. A reduction in the binder content may be required to correct a high hot strength. A severe hot tear is shown in figure 210.

Pinholes are caused by a high moisture content in the sand. Pinholes are recognized by their small size and location on the casting surface as shown in figure 211. The cure for pinholes is to use the correct amount of moisture. This can be determined by proper sand testing and control. There are other minor causes of pinholes, but high moisture content in the sand is by far the greatest source of trouble.

An expansion of the sand so that a part of the mold surface is displaced in an irregular line produces a rattail defect. These defects are shown in figures 212 and 213. Rattails do not always occur as severely as shown in these two examples. They may be as fine as hairlines on the surface of a casting. A sand of improper grain size distribution, high hot strength, and that has been rammed hard are the major contributing causes to rattails. To cure this situation, greater care must be taken to ram the mold to make a uniform mold surface of correct hardness. The high hot strength can be corrected by reducing the binder. Better expansion properties can be obtained by proper grain size distribution in the sand, or by adding cushioning materials.

Buckles are similar to expansion scabs and rattails. When an expansion scab is removed from the surface of a casting, an indentation in the casting surface will be revealed. This indentation is a buckle and is shown in figure 214. A rattail is sometimes called a minor buckle. The cure for a buckle is the same as for an expansion scab.

Stickers due to sand are caused by too high a moisture content or by a low green strength. A high moisture content will cause the sand to stick to the pattern. A reduction in the moisture content is necessary to overcome stickers from this cause. If the green strength is low, the sand will not have the necessary strength to permit drawing from a pocket or along a vertical surface. Additions of binder or improvement of the mixing procedure by using a muller to produce a more uniform distribution of binder are corrective measures which can be taken to eliminate stickers of this type. Proper use of parting compounds will minimize sticking.

CORES

The molding sand conditions which contribute to casting defects also apply to cores. Among these conditions are low permeability (which causes blows), low binder content (which leads to cuts and washes), and hot tears (which are caused by cores having low collapsibility). Figure 215 shows a casting that cracked because the core was too hard.

The baking of cores can also cause casting defects. An underbaked core will still contain a large amount of core oil, which may cause a blow when the casting is poured. Such a blow can be cured by baking the core properly and by using the correct amount of binder.

Overbaking of cores causes defects because it results in burned-out binders. An overbaked core will have a weak and soft surface. Cuts, washes, and metal penetration result from overbaking. Correct baking time for the type of binder used and for the size of the core is the method for correcting these defects.

Another contributing factor to the occurrence of hot tears may be over-reinforcement. This is especially true of larger cores where reinforcement is necessary. The use of reinforcing wires and rods only when they are required and only in amounts necessary is the way to overcome hot tears from over-reinforcement.

Core shifts cause runouts, bleeders, misruns, coldshuts, and castings that are dimensionally inaccurate. If cores are improperly fitted in the core print, molten metal can run in between the core and the mold and cause a bleeder or runout. Molten metal may also fill the vent and cause a blow. Incorrectly pasted cores, cores with vents too close to the surface, and cores with inert backing material too close to the surface provide an easy path for the molten metal to run out of the mold. Correct fitting and pasting of cores, relocation of vents toward the center of the core, and central location of inert backing material are the steps required to correct runouts and bleeders from these causes. A core shift may reduce the section thickness of a mold with the result that the section will not be completely filled. A misrun caused by a core shift is shown in figure 216. To correct such a defect, it is necessary to provide better support for the core, either by a better-designed core print or by the use of chaplets.

MOLDING PRACTICE

Molding practice, along with the other operations involved in foundry work, contributes to casting defects if not properly done. Blows are caused by a combination of hot sand and cold cores and flasks. This combination causes moisture to condense and to give a localized concentration of moisture which causes a blow. To prevent this type of blow, sand should be cool before making a mold. Do not use hot sand.

Hard ramming of the sand can cause blows and expansion scabs as shown in figure 217. The blows occur because moisture is prevented from escaping by the closely packed sand. Expansion scabs occur because the hard rammed sand expands and buckles. The casting shown in figure 217 was of such a design that hard ramming could not be avoided. In this case, wood flour additions were made to provide a cushion for the hard rammed sand so that it would expand without buckling.

Figure 218 shows a sticker that was caused by hard ramming in the pockets. Improved technique corrected this defect.

Metal penetration can take place because of soft or uneven ramming, which produces a soft mold surface. Harder and more uniform ramming is the cure for this type of defect. Soft ramming may also result in swells or fins. The sand is soft at the mold surface and cannot retain its shape against the pressure of the molten metal, with the result that the mold cavity is forced out of shape and the defect occurs.

Poor molding practice is probably the major cause for crushes. Careless closing of a mold will result in displaced sand or cores, which in turn result in the crush. If the bottom board is not properly bedded, a high spot of sand on the board will cause pressure against the mold and a displacement of sand in the mold cavity. This again will cause a crush. Incorrectly placed chaplets, or chaplets of incorrect size, will also result in pressure being exerted either on the mold or on the core and cause a crush. Any defect due to poor molding practice can be corrected by only one method; improve the molding technique. Care and attention to the various operations involved can go a long way toward minimizing defects caused by molding practice.

POURING PRACTICE

The defects caused most often by pouring practice are blows, misruns, coldshuts, and slag or dross inclusions. Blows are caused by using green ladles or ladles with wet patches. Severe blows caused by use of a green ladle are shown in figure 219. A defect from this cause is remedied by using ladles which are thoroughly dried after lining and after any patching is done. Misruns and cold shuts are caused by pouring when the metal is too cold or by interrupting the pouring of the mold. With immersion and optical pyrometers in proper operating condition, misruns and coldshuts due to cold metal are minimized. If either of these defects occur when temperature readings indicate hot metal, a defective instrument is indicated. Sometimes more than one ladle of metal will be required to pour a mold. In such a case, pouring with the second ladle should start before the first ladle has been emptied. Otherwise the short interval allowed for the start of pouring from a second ladle is sufficient to chill the metal in the mold and to cause a cold shut, or slag inclusions.

Slow pouring may produce uneven heating of a mold surface by the radiant heat from the molten metal and cause a scab. Faster pouring will fill the mold more rapidly and minimize the radiant heating effects in the mold cavity. Pouring should always be as fast as the sprue will permit. If a slower or faster pouring rate is indicated, a different sprue size should be used.

Pouring from high above the mold results in an increased metal velocity in the mold until the sprue is full and can lead to washing defects. Also, pouring from a ladle which is held high above the mold permits easy pickup of gases by the molten metal, as well as agitation in the stream of metal.

MISCELLANEOUS

The use of rusty or damp chills and chaplets almost always causes blows. The rust on chills and chaplets reacts readily with the molten metal and a large amount of gas is produced in the casting at this point. The localized high gas content cannot escape and a blow is produced. A similar situation is brought about by the use of damp chills or chaplets. The moisture on the chills or chaplets forms steam which results in a blow. Figure 220 shows a blow which was caused in an aluminum casting by using a bad chill.

Careless handling of a mold can result in wasted effort on the part of the molder. Rough treatment may result in drops. The careless placing of mold weights can also result in drops from excess pressure on the cope. A drop due to rough handling of the mold is shown in figure 221. The left side of the figure shows the cope side of the casting. The rough lump of metal which filled the cavity by the displaced sand can easily be seen. The right side of the figure shows the drag side of the casting with the hole at the center core caused by the sand which dropped from the cope. The sand that dropped may float in heavy metal castings and cause a second defect in the cope.

Cracks and tears can be caused by shaking out the casting too early. This causes chilling of the casting and high stresses are produced. The casting usually has a low strength when hot. Dumping of hot castings into wet sand can also cause cracks and tears. Careless grinding of the casting may cause localized overheating, high stresses, and cracks. See Chapter 10, "Cleaning Castings," for correct grinding techniques.

The use of moist, dirty, or rusty melting tools may cause the introduction of moisture into the melt. This source of moisture can result in pinholes in the completed casting. Every effort should be made to maintain good melting practice to prevent the rejection of casting because of carelessness in the melting operation.

SUMMARY

When determining the cause of casting defects, it must always be kept in mind that defects are more often due to a combination of causes rather than to one isolated cause. The use of properly kept records of previous castings, good sand control, and development of a good molding procedure can go far in making the job of eliminating casting defects an easier one. A chart has been included which indicates the causes of the various defects and possible cures and should be used as a convenient reference.

SUMMARY OF CASTING DEFECTS

Defect	Design		Pattern Equipment		Flask Equipment and Rigging		Gating and Risering		Cores		Molding Practice		Pouring Practice		Miscellaneous	
	Cause	Cure	Cause	Cure	Cause	Cure	Cause	Cure	Cause	Cure	Cause	Cure	Cause	Cure	Cause	Cure
Blow	Insufficient core print	Increase print size	—	—	—	—	Low head pressure due to low riser height	Increase height of riser	LOW PERMEABILITY	Open up core sand	HOT FLASKS AND COLD SAND	Make sure sand, cores and flasks are cool enough to prevent any moisture condensation	GREEN, DAMP OR COLD LADLE LIPS AND LINING	Use thoroughly dry ladles	Rusty chills and chaplets	Use only dry, clean chills
	Insufficient venting from internal cores	Provide vent outlet					Low head pressure due to gates freezing too soon	Increase gate size and check for possible chilling in gate area	UNDERBAKED	Use proper baking time for size of core and mix used	HOT SAND AND COLD FLASKS				Damp chills or chaplets	
											HOT SAND AND COLD CORE				Unevenly or incompletely coated chaplets, or chaplets coated with wrong material	Use rustfree, evenly coated chaplets
									Improper venting	Increase size of vents; check for vents closed with paste, etc. Make sure vents are connected	HOT CORE AND COLD SAND					
											Hard ramming	Ease up on ramming in area of blow				
Drop	Sharp corners in cope or projecting surface	Redesign with round corners and fillets	Too little draft	Increase draft	Weak or cracked flasks	Repair or replace flasks	Sand unsupported at riser neck	Provide internal support for sand at riser neck or ram sand harder at this point	—	—	Mold too weak	Use harder ramming	—	—	ROUGH HANDLING OF THE MOLD	Use care in moving mold
	Deep pockets; overhanging or protruding sections	Use internal reinforcement			Jacket too small, cracked or warped	Use proper jacket. Repair or replace defective jacket					Internal supporting members too close to mold cavity	Set reinforcing members farther from mold cavity			Rough handling of weights, dropping them on mold	Use care in the proper placing of weights
					Weights too heavy or uneven	Use proper weights in good condition					Internal reinforcement insufficient	Provide more reinforcement				

- 157 -

SUMMARY OF CASTING DEFECTS (Continued)

Defect	Design		Pattern Equipment		Flask Equipment and Rigging		Gating and Risering		Cores		Molding Practice		Pouring Practice		Miscellaneous	
	Cause	Cure	Cause	Cure	Cause	Cure	Cause	Cure	Cause	Cure	Cause	Cure	Cause	Cure	Cause	Cure
Pinholes	Nonuniform section size requiring a high pouring temperature	Redesign for uniform section size or change gating to produce better metal distribution	–	–	–	–	–	–	–	–	Poor patching	Improve patching technique and keep moisture content of patch at minimum	DAMP LADLE, POOR PATCHING	Thoroughly preheat ladle before use	MOIST MELTING TOOLS	Thoroughly dry tools
															DAMP PATCHING	Dry patches
															MOISTURE IN MELTING ATMOSPHERE	Take steps to protect surface of melt
Inclusions: dirt, slag, and dross	–	–	–	–	–	–	INSUFFICIENT CHOKE TO TRAP SLAG AND DIRT	Increase choke or use skim core	Improperly cleaned cores	Proper cleaning	Poor molding practice, making a dirty mold	Improve molding practice (see chapter 5 "Making a Mold")	Bad ladle lining	Replace ladle lining or improve lining practice	Dirty metal	–
							SPRUE TOO LARGE TO PERMIT CHOKING ACTION	Reduce sprue diameter	Weak splash, skim, or gate cores	Proper core sand mix and core-making procedure			Poor ladle-lining refractory	Use proper refractory		
													Improper skimming	Proper skimming		
													Intermittent pouring	Continuous pouring (especially with high-dross metals)		

SUMMARY OF CASTING DEFECTS (Continued)

Defect	Design Cause	Design Cure	Pattern Equipment Cause	Pattern Equipment Cure	Flask Equipment and Rigging Cause	Flask Equipment and Rigging Cure	Gating and Risering Cause	Gating and Risering Cure	Cores Cause	Cores Cure	Molding Practice Cause	Molding Practice Cure	Pouring Practice Cause	Pouring Practice Cure	Miscellaneous Cause	Miscellaneous Cure
Shrinkage cavities	Abrupt change in section size	Blend sections	—	—	—	—	GATES AND RISERS TOO LARGE	Reduce size of gates and risers	—	—	—	—	Pouring too cold	Increase pouring temperature	—	—
	Fillets insufficient	Use fillets of proper size					RISERS NOT LARGE ENOUGH	Increase size of risers					Pouring too hot for the gating used	Use lower pouring temperature		
	Fillets too large	Use fillets of proper size					GATES AND RISERS DO NOT PROMOTE DIRECTIONAL SOLIDIFICATION	Change gating and risering to obtain directional solidification								
	Heavy sections which cannot be fed	Redesign to reduce section size or permit proper feeding					GATES FREEZING OFF TOO SOON	Change gating system or increase size of gates								
Crush	—	—	Worn patterns and core boxes	Proper maintenance and repair	IMPROPERLY ALIGNED FLASK EQUIPMENT	Repair equipment	—	—	—	—	CARELESS CLOSING	Good molding practice	Improper setting of jackets	Set jackets square and snug	—	—
			Core print too small	Provide proper size core print	WARPED OR UNEVEN FLASK JOINTS				Core too large for core print	Check size of core and provide proper size core print	POOR CLAMPING		Mold weight too heavy	Use lighter weight		
					BAD FITTING JACKETS	Provide jackets of correct size					IMPROPER BEDDING OF BOTTOM BOARDS		Dropping weight on mold	Careful placing of weight		
					BAD BOTTOM BOARDS	Replace					WRONG OR IMPROPERLY SET CHAPLETS	Use proper chaplets and procedure				

SUMMARY OF CASTING DEFECTS (Continued)

Defect	Design		Pattern Equipment		Flask Equipment and Rigging		Gating and Risering		Cores		Molding Practice		Pouring Practice		Miscellaneous	
	Cause	Cure	Cause	Cure	Cause	Cure	Cause	Cure	Cause	Cure	Cause	Cure	Cause	Cure	Cause	Cure
Metal penetration	Thin core surrounded by a heavy section	If possible, eliminate core and machine hole, otherwise use refractory wash on core	Soft ramming due to improper parting line	Locate parting line to permit more uniform ramming	Small flask, causing soft ramming	Use flask with more space around pattern	Mold surface exposed to radiant heat too long	Increase number of gates to fill mold more rapidly	SOFT	Harder ramming	SOFT OR UNEVEN RAMMING	Harder and more uniform ramming	Too high a pouring temperature	Lower pouring temperature	Gray iron – phosphorus too high	Check charge and eliminate source of excess phosphorus
							Riser or sprue too close to vertical wall	Increase space between mold and riser or sprue	Overbaked	Correct baking time; check oven for uniform heat distribution	Poor patch	Make mold over	Too rapid pouring	Reduce pouring rate		
							Too much metal over a given mold surface	Increase number of ingates	Insufficient core wash, improper wash; rough or cracked surface	Make core over and use correct procedures	Not enough facing sand	Increase amount of facing sand				
							Excessive pressure	Reduce riser height								
Hot tears and hot cracks	INADEQUATE FILLETS	Provide proper fillets	Inadequate fillets	Provide proper fillets	Internal reinforcement too near sprue or riser	Increase amount of sand between reinforcement and sprue or riser	Internal reinforcement too near sprue or riser	Increase amount of sand between reinforcement and sprue or riser	Low collapsibility due to too much carbonaceous material, high moisture, high percentage of fines, or too much binder	Check for proper core mix	Too hard ramming	Lighter ramming	Pouring too cold	Increase pouring temperature	Improper removal of gates and risers	Do not remove too hot; cut off any difficult gates or risers
	SECTIONS OF NON-UNIFORM SIZE JOINED	Gradual blending of sections or redesign for uniform section size			Reinforcement too far into deep pockets	Shorten reinforcement	Gating and risering system hindering contraction	Redesign gating system to allow free contraction	Reinforcement too close to surface or over-reinforced	Relocate or reduce reinforcement	Reinforcement too close to mold surface	Place gaggers, wires, etc., farther away from mold surface			TOO EARLY SHAKEOUT CAUSING AIR QUENCH	Leave casting in mold for longer time
															DUMPING CASTINGS IN A WET SAND	Don't
															Cracked in grinding	Grind carefully and do not overheat

SUMMARY OF CASTING DEFECTS (Continued)

Defect	Design		Pattern Equipment		Flask Equipment and Rigging		Gating and Risering		Cores		Molding Practice		Pouring Practice		Miscellaneous	
	Cause	Cure	Cause	Cure	Cause	Cure	Cause	Cure	Cause	Cure	Cause	Cure	Cause	Cure	Cause	Cure
Runouts and bleeders	—	—	—	—	Not enough sand around pattern	Larger flask	Gates, runners or riser too near side or bottom	Increase the amount of sand between gating system and flask	Improperly fitted or pasted core	Fit cores correctly and paste properly	Soft ramming	Harder ramming	Too rapid pouring for gating system	Reduce pouring rate	—	—
					Bad pins and bushings, preventing cope and drag seating properly	Repair or replace pins and bushings	Too much pressure from riser	Relocate riser or reduce in height	Vents too close to surface	Relocate vents	IMPROPER WEIGHTS		Removing weights, clamps, and jackets too soon	Leave weights, clamps etc., on for longer time		
									Inert core material too close to surface	More central location of inert material	BAD CLAMPING	Improve molding practice (see chapter 5, "Making a Mold")				
					Dirt in flask joint line	Clean flask properly					POOR BOTTOM BOARD BEDDING					
					Bad bottom boards	Replace					CARELESS CLOSING OF MOLD					
					Uneven clamping	Correct clamping					VENT TOO CLOSE TO PATTERN					
											JOINTS NOT MUDDED	Make sure joints are closed				
Misruns and cold shuts	Nonuniform section size	Redesign for uniform section size or use proper blending of sections	Worn core boxes and patterns resulting in thin sections	Repair or replace	—	—	Gate, runner or sprue too small	Increase size of gating system	Core shift	Decrease size, use core print that fits the core; provide better support	Any practice resulting in thin sections	Proper molding practice	TOO LOW POURING TEMPERATURE	Increase pouring temperature	—	—
							Gates not properly located	Relocate gates for better distribution of metal in mold			Hard ramming, resulting in back pressure	Ease up on ramming	INTERRUPTED POUR	Pour mold without stopping		
							Pressure head too low	Increase sprue and riser height	Cores too hard	Decrease ramming, use correct baking time; check core mix						

SUMMARY OF CASTING DEFECTS (Continued)

Defect	Design		Pattern Equipment		Flask Equipment and Rigging		Gating and Risering		Cores		Molding Practice		Pouring Practice		Miscellaneous	
	Cause	Cure	Cause	Cure	Cause	Cure	Cause	Cure	Cause	Cure	Cause	Cure	Cause	Cure	Cause	Cure
Expansion scabs, rat tails, and buckles	Large flat surfaces	Use antiscabbing additions in sand	–	–	–	–	Gating arrangement causes uneven heating of the mold	Regate to provide uniform metal distribution into mold	–	–	UNEVEN RAMMING	Uniform ramming	Pouring too slow	Increase pouring rate	–	–
	Sharp corners can aggravate	Use fillets and round corners									Poor patching	Patch only if absolutely necessary, remake mold	Interrupted pour	Avoid stopping pour before mold is full		
											Ramming too hard	Ease up on ramming				
Cuts and washes	Excessive metal flow over a section of mold	Change gates (see gating and risering)	–	–	–	–	Too much metal through a given gate	Change design of gate or use multiple gates	Not enough binder; improperly mixed or of wrong type	Use proper core mix and correct procedure	Soft ramming Hard ramming	Proper ramming	Pouring temperature too high for mold material	Pour at a lower temperature or use a refractory mold wash	–	–
	Jet effects in metal stream						Too little choke in gates	Increase amount of choke in runner or sprue	Damaged surfaces	Careful handling and inspection			Too rapid pouring	Reduce pouring rate		
							Too few gates	Increase number of gates	Vents too close to surface	Relocate vents						
							METAL IMPINGING ON MOLD OR CORE	Relocate gate	Overbaked	Correct baking time; check oven for heat distribution						
							Jet effect in metal stream	Decrease choke or increase size of gate								
							No cushioning of metal by itself	Locate gate so that drop of metal in mold is reduced								

SUMMARY OF CASTING DEFECTS (Continued)

Defect	Design Cause	Design Cure	Pattern Equipment Cause	Pattern Equipment Cure	Flask Equipment and Rigging Cause	Flask Equipment and Rigging Cure	Gating and Risering Cause	Gating and Risering Cure	Cores Cause	Cores Cure	Molding Practice Cause	Molding Practice Cure	Pouring Practice Cause	Pouring Practice Cure	Miscellaneous Cause	Miscellaneous Cure
Swells, fins, and sags	Long cores sagging without proper support	Redesign for proper core support	Worn or improper patterns, resulting in fins	Repair or replace pattern	Worn strike-off bar and bad bottom boards, resulting in swells or fins	Straight strike-off bar and new bottom boards	Lack of relief vents	Provide relief vents	Not reinforced properly, resulting in sag	Provide sufficient reinforcement	Large, flat copes insufficiently supported lead to sagging	Proper use of gaggers	Pouring too rapidly	Reduce pouring rate	–	–
					Weights too light or improper fitting jackets permitting swells	Heavier weights and correct fitting jackets	Risers too high	Reduce riser height or relocate	Core too small for core print, producing fin	Provide proper core; check size of both core and print	SOFT RAMMING, IMPROPER TUCKING AND PEENING RESULT IN SWELLS AND FINS	Good molding technique (see chapter 5, "Making a Mold")	Ladle too high above sprue	Ladle lip as close as possible to sprue		
Shifts, core shift, and raise	Not enough core support	Support	COPE AND DRAG PATTERNS NOT IN LINE	Repair pattern	Loose or worn pins and bushings	Repair or replace	METAL FROM GATE STRIKING ON CHAPLET AND MELTING IT TOO EARLY, CAUSING A RAISE	Relocate ingate or chaplet	Not enough reinforcement in core	Increase reinforcement	Excessive rapping of loose patterns, producing oversize core print	Rap minimum amount	Bumping molds with ladle	Proper pouring procedure	–	–
	Core print too small	Redesign for proper use	Loose or worn flask pins and guides		Defective jackets	Replace with good jackets			Core sagged during baking	Correct core mix and reinforcement	Reversing cope and drag	Care in setting cope	Pouring temperature too high for core materials	Reduce pouring temperature or produce a more refractory core mix		
											Improper chaplets	Chaplets of correct size				
											Bad clamping	Careful clamping				
											Omission of chaplets	Check mold before closing				

SUMMARY OF CASTING DEFECTS (Continued)

Defect	Design		Pattern Equipment		Flask Equipment and Rigging		Gating and Risering		Cores		Molding Practice		Pouring Practice		Miscellaneous	
	Cause	Cure	Cause	Cure	Cause	Cure	Cause	Cure	Cause	Cure	Cause	Cure	Cause	Cure	Cause	Cure
Sticker	Back draft or not enough draft	Correct draft	POOR SURFACE FINISH ON PATTERNS	Smooth over rough spots on patterns	Bad pins and bushings	Repair or replace	Gate or runner too close to pattern	Provide more space between pattern and gating system	Ram-up cores that fit too tight against pattern	Ram-up cores with more clearance	Hard ramming in pockets	Lighter ramming and reinforcement	–	–	–	–
											Rough handling of pattern while drawing	Rap lightly and draw pattern with smooth, steady movement				

- 171

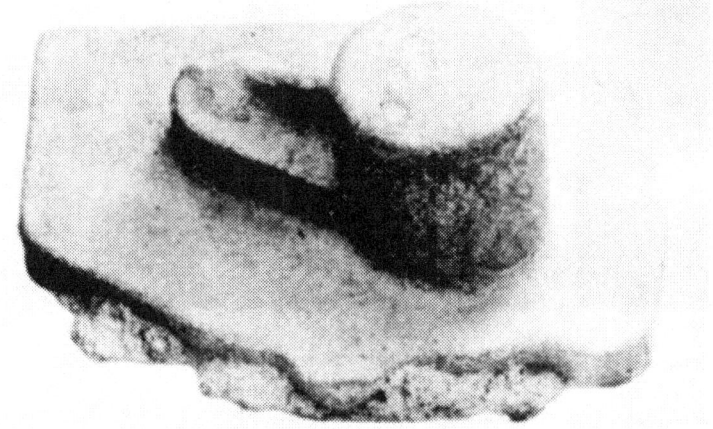

Figure 198. Sticker. (Caused by loose pins and bushings)

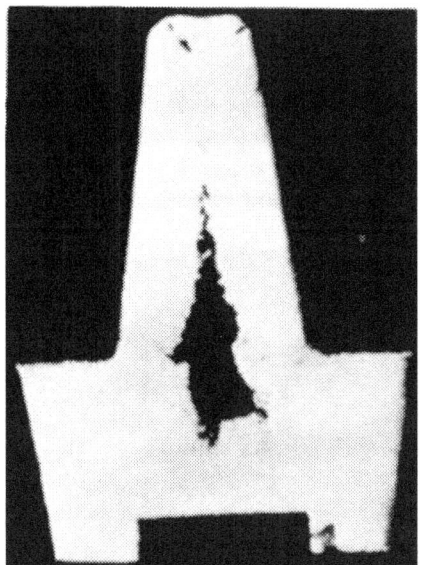

Figure 199. Gross shrink. (Caused by inadequate feeding)

Figure 200. Surface shrink. (Caused by improper feeding)

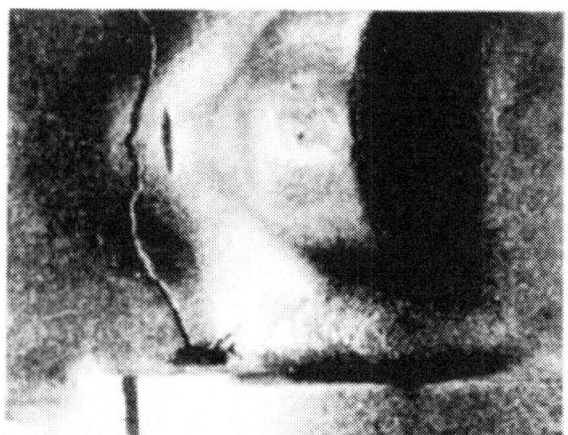 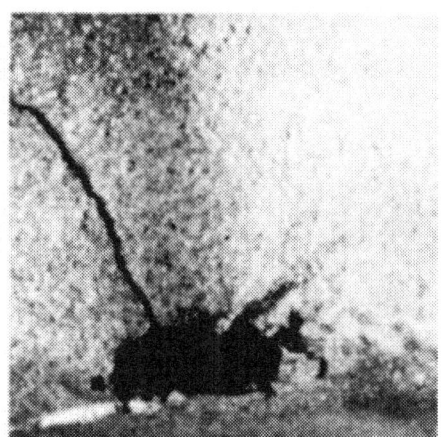

Figure 201. Surface shrink. (Caused by improper feeding.)
(Crack resulted from breaking the casting for examination.)

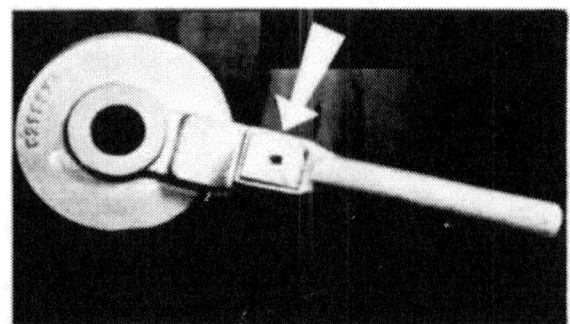

Figure 202. Internal shrink.
(Caused by cold-metal riser arrangement)

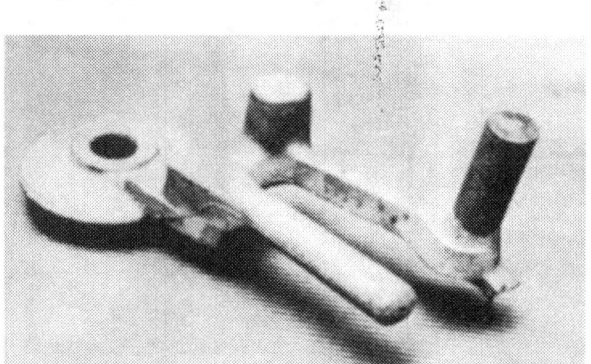

Figure 203. Gating and risering that corrected internal shrink in figure 202.

Figure 204. Gross shrink. (Caused by improper location of ingate.)

Figure 205. Gross inclusions.
(Revealed by fractured aluminum casting)

Figure 206. Blow. (Caused by high moisture content)

Figure 207. Expansion scab. (Caused by too many fines in the sand)

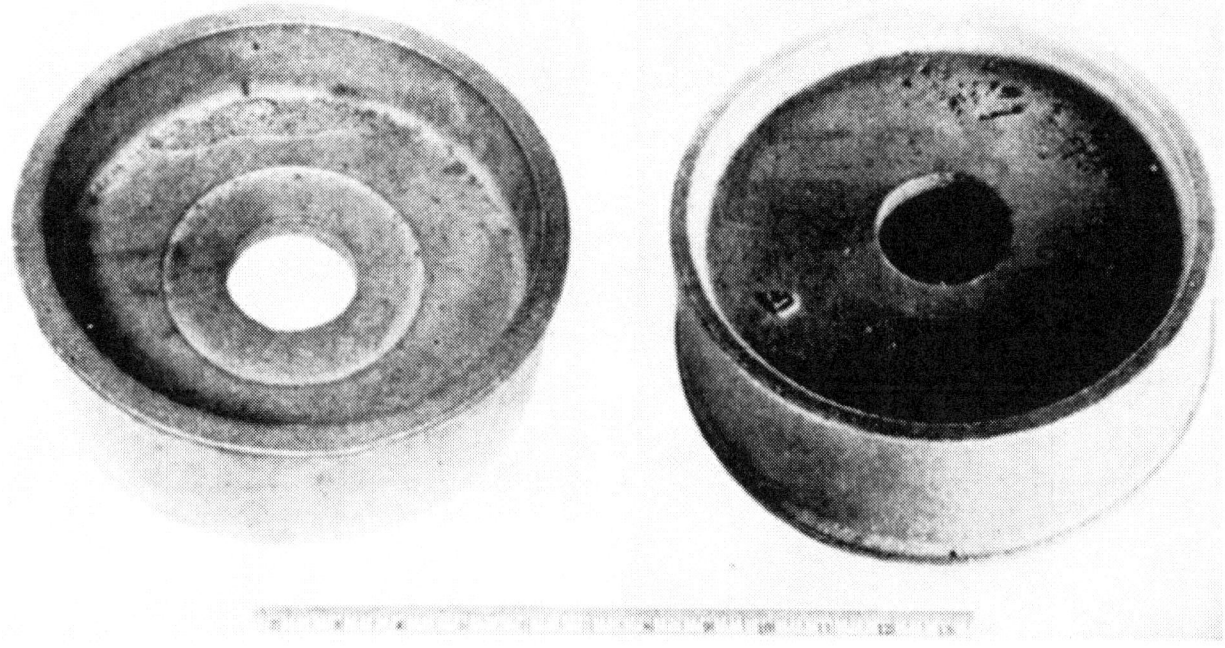

Figure 208. Erosion scab and inclusions.

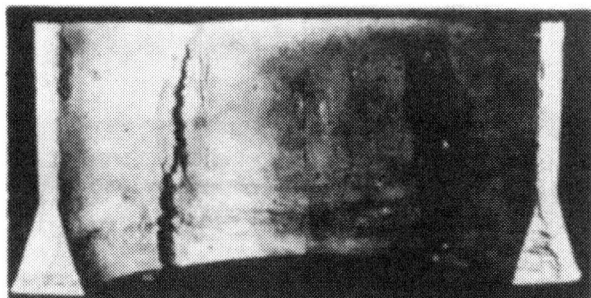

Figure 210. Hot tear. (Caused by too high hot strength of the molding sand)

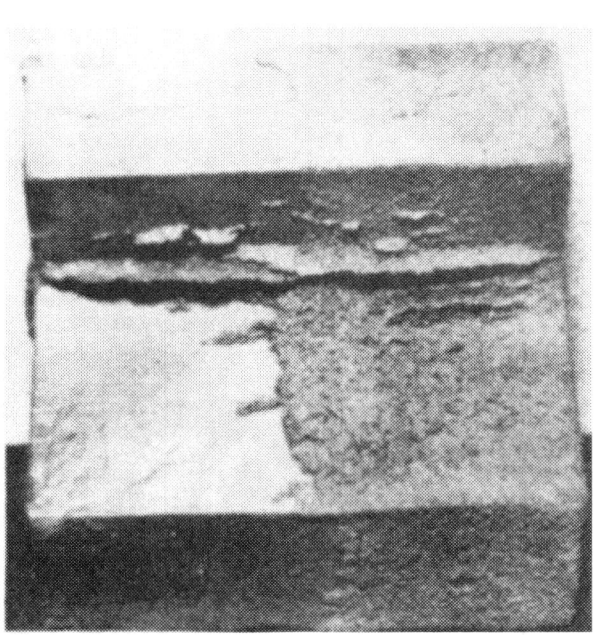

Figure 209. Metal penetration and veining. (Penetration caused by an open sand veining caused by metal penetration into cracked sand)

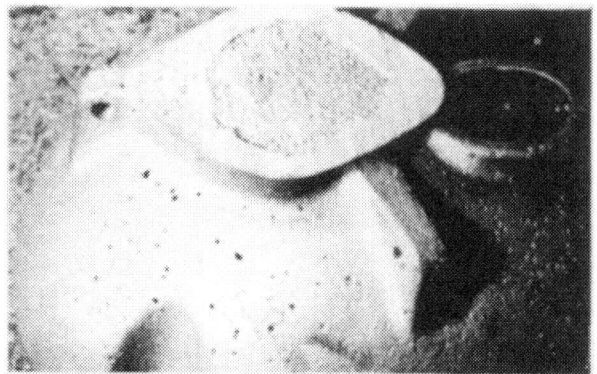

Figure 211. Pin holes. (Caused by high moisture content of the sand)

Figure 212. Rattails. (Sand lacked good expansion properties)

Figure 213. Rattails. (Cause same as for Fig. 207)

Figure 214. Buckle.

Figure 215. Cracked casting. (Caused by a hard core)

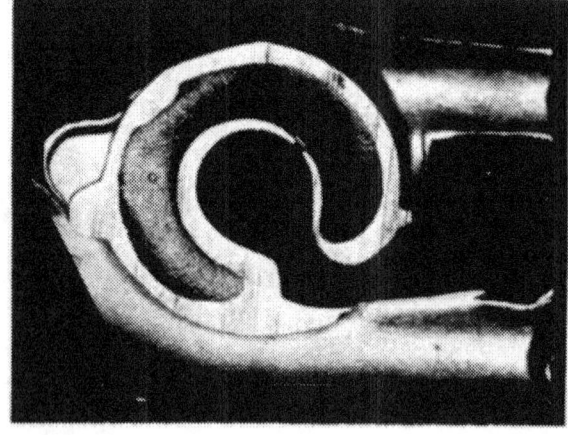

Figure 216. Misrun. (Caused by a core shift)

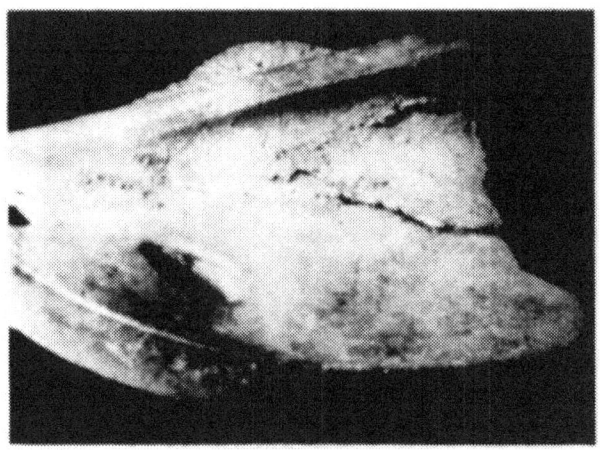

Figure 217. Blow and expansion scab. (Caused by hard remming of the sand)

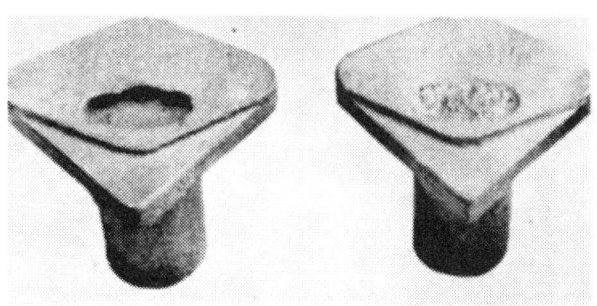

Figure 218. Sticker. (Caused by hard remming in pockets)

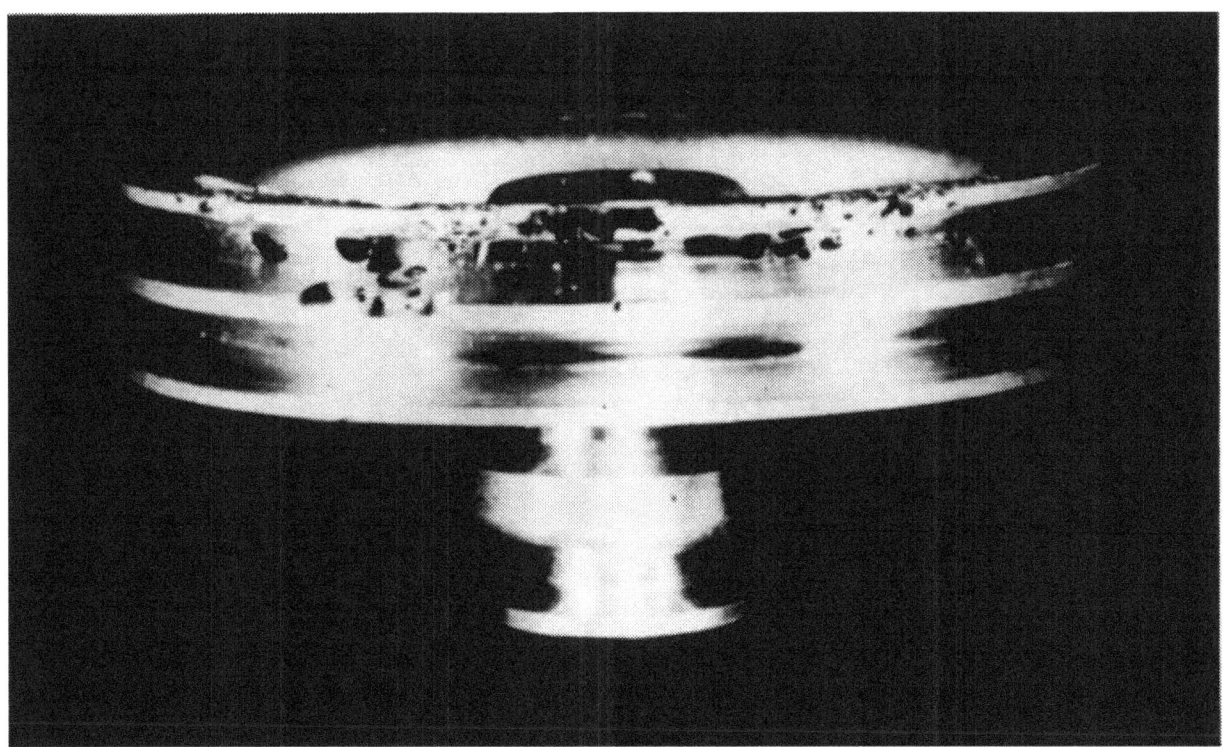

Figure 219. Blows. (Caused by moisture pickup from a damp ladle)

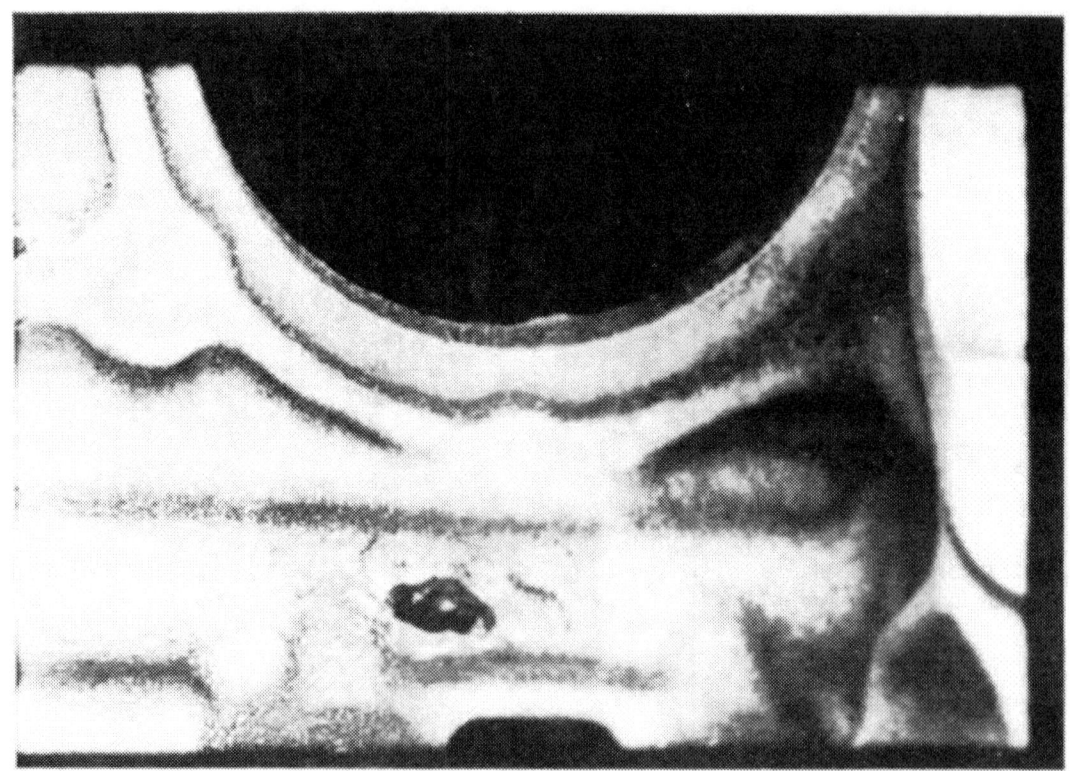

Figure 220. Blow. (Caused by a bad chill)

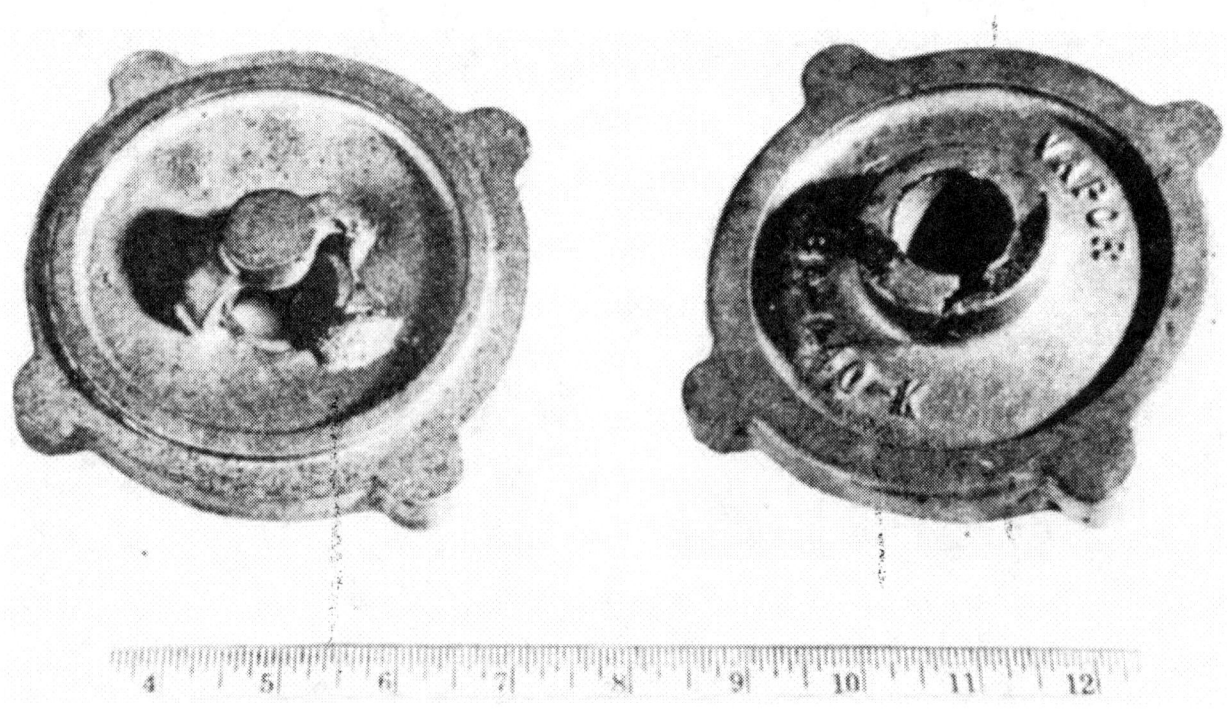

Figure 221. Drop. (Caused by rough handling of the mold)

Chapter XII
HEAT TREATMENT OF CASTINGS

The most common types of heat treatment applied to castings are as follows:

IRON AND STEEL CASTINGS

STRESS-RELIEF ANNEAL

Low-temperature treatment to improve dimensional stability and increase available strength by decreasing residual stresses. Can be applied to all castings. Requires slow cooling in the furnace. Usually has little effect on hardness.

FULL ANNEAL

High-temperature treatment to homogenize the cast structure. Improves mechanical properties or machinability. Required for steel castings that are not normalized. Desirable for cast iron where "chill" is present and castings must be machined. Requires slow cooling in the furnace.

NORMALIZE

High-temperature treatment for steel castings that are not annealed. Improves structure and ductility. Requires cooling in air.

TEMPER

Low-temperature treatment following normalizing or quenching. Similar to stress-relief anneal but involves cooling in air. Results in softening of normalized steel or iron castings.

QUENCHING

Fast cooling of castings from high temperatures by immersing them in quenching oil, water, or brine. Results in hardening of iron and steel castings. Involves considerable risk of cracking the casting. Details of quenching are outside the scope of this manual. Quenching must almost always be followed by tempering.

NONFERROUS CASTINGS

STRESS-RELIEF ANNEAL

Low temperature treatment to improve dimensional stability and increase the available strength by decreasing residual stresses. Can be applied to all castings. May increase the hardness. Temperature and time of treatment depends on the particular alloy.

SOLUTION HEAT TREATMENT

High temperature treatment. Usually for aluminum for prolonged periods, generally just below the solidus temperature to homogenize the structure, followed by quenching in warm water. Temperature and time of treatment depends on the particular alloy. This treatment produces the softest most ductile condition. It is generally followed by artificial aging.

AGING

A low temperature treatment following solution heat treatment to produce maximum hardness and yield strength.

REASONS FOR HEAT TREATMENT

There are several reasons for heat treating castings. The properties of some alloys such as heat-treatable bronzes, heat-treatable aluminum alloys, various steels, and cast irons can be improved by heat treatment. Improvement of mechanical properties is the main reason for the heat treatment of castings.

A second important reason for heat treatment is for the removal of residual stresses that are the result of casting design, solidification, or lack of free contraction because of sand properties. See Chapter 1, "How Metals Solidify," Chapter 2, "Designing a Casting," and Chapter 4, "Sand for Molds and Cores." Stress-relief heat treating involves heating followed by slow cooling.

Another reason for heat treating is to make ferrous castings softer than they were in the as-cast condition, so that they will be easier to machine. Also, it may be necessary to reduce the hardness that may have been caused by chilling, such as may occur with gray irons. Such heat treatment is called annealing.

Because of the slow cooling in sand, many castings will have a coarse grain structure that does not provide the best properties. This can be corrected by a heat treatment that will cause the solidified metal to recrystallize and form smaller grains. This recrystallization will produce improved properties in the casting.

In many alloys, it is difficult to produce a uniform structure in the casting because of the alloy itself, or because of the conditions controlling solidification of the casting. In such a case, heat treatment can be used to obtain a uniform homogeneous structure.

All of the changes in properties obtained by heat treating of metals or alloys depend on the movement of the atoms of metal. When a metal or alloy solidifies, it forms in a definite pattern of atoms. This pattern determines its crystal structure. Heat treatment often produces a rearrangement of the atoms to produce the desired properties. Movement of the atoms to accomplish this rearrangement is called diffusion.

Diffusion takes place in a metal or alloy when it is heated to a certain critical temperature which permits the easy movement of the various atoms. This critical temperature at which rapid diffusion starts is known as the activation temperature. Below this temperature, diffusion does take place but is so slow that heat treatment at these lower temperatures would be impractical. The old-fashioned practice of "aging" castings for months by storing them at room temperature is an example of a low temperature heat treatment which requires a long time, but which can be speeded up to a matter of several hours by proper heating of the casting.

A simple example of diffusion and the effect of temperature can be shown with water and a dye. If the water is frozen into ice and a drop of dye placed on its surface, the dye will maintain its drop shape and possibly penetrate a very short distance into the ice. With the application of heat, the ice will reach its melting point. As it melts, the dye can be seen moving through the water and tinting it. This movement of the dye is diffusion. The melting temperature of the ice (which resulted in water) is the activation temperature.

The process by which the atoms diffuse to produce smaller grains is known as recrystallization. The formation of the new crystal takes place at nuclei which may be centers of high concentration of a particular element, impurities in the metal, ideal atom arrangements, or even centers of high stress in the casting. Whatever their cause, they serve the same purpose as the nuclei described in Chapter 1, "How Metals Solidify." They act as centers of crystal growth. The only difference between recrystallization and solidification, so far as crystal growth is concerned, is that during solidification the atoms form the crystals from a liquid state and in recrystallization, the new crystals are formed by the diffusion of the atoms through the solid metal.

In the heat treatment of most alloys, a preliminary step known as solution treatment is necessary. Solution treatment means to change the state of the alloy into that of a solid solution. A solid solution alloy is one in which the alloys are soluble in each other in the solid state. Under a microscope, a solid-solution alloy would have the appearance of a pure metal. It would be uniform, without any indication of the presence of more than one metal. This would be in contrast to other alloys in which the presence of more than one alloy is shown by two or more characteristic phases.

BRASS AND BRONZE

There are only a few copper-base alloys that can be heat treated to improve their mechanical properties. For most copper-base alloys only a stress-relief annealing treatment is used to remove residual stresses.

STRESS-RELIEF ANNEAL

Tin-bronzes can be stress relieved by heating at a temperature between 700°F. and 800°F. for at least 1 hour. The castings are then cooled in air from this temperature. Where extreme dimensional stability is required, the casting should be rough machined before stress relieving.

Manganese-bronze can be stress relieved by heating in a temperature range of 600-800°F. (1 hour for each inch of casting thickness) and air or furance cooling.

Copper-base castings in general can be stress relieved by heating at 700°F. to 800°F. (one hour for each inch of casting thickness) and then cooled in air.

SOFTENING AND HARDENING

Certain aluminum-bronze alloys can be heat treated to obtain higher strength and hardness. This heat treatment consists of softening the alloy by a solution heat treatment, followed by an aging treatment that hardens it to the desired strength. Classes 2, 3, and 4 aluminum-bronze alloys fall into this group. The exact heat-treating cycle is determined by the composition of the alloy and the properties desired. Castings should be heated to a temperature between 1,600°F. and 1,650°F., water quenched, and aged between 1,000°F. and 1,150°F. A good general rule for holding time at the heat treating temperature is to hold for 1 hour for each inch of section thickness.

ALUMINUM

STRESS-RELIEF ANNEAL

Aluminum alloys are not usually heat treated for removal of casting stresses. When it is desirable to reduce the residual stresses in an intricate casting so as to make it stronger, less susceptible to cracking, or more stable dimensionally, heat the casting to about 600°F., hold it from two to seven hours, and then air cool. (T2). Additional data is given under "General Heat Treatments."

GENERAL HEAT TREATMENTS

Aluminum alloys in classes 1, 3, 4, 7, and 8 can be heat treated. The heat treatment involves a softening of the alloys by a solution heat treatment followed by quenching in warm water followed by an appropriate ageing treatment to harden it.

Exact information on heat-treating procedures should be obtained from the Bureau of Ships.

Some typical heat treatments of aluminum casting alloys are as follows:

GENERAL HEAT TREATMENT OF ALUMINUM ALLOYS

Alloy Type	Condition Desired[a]	Solution Heat Treatment			Aging		Annealing	
		Time at Temp., hours	Temp.[b] F.	Quenching Media	Time, hours	Temp.[b] F.	Time, hours	Temp.[b] F.
Aluminum-Silicon	T2	(Not heat treatable and seldom used)			--	--	2 to 4	600
Aluminum-Copper (or Al-Cu-Si)	T4	12	960	Boiling water	--	--	--	--
	T6	12	960	Ditto	3 to 5	310	--	--
Aluminum-Magnesium	T4	12	810	Ditto	--	--	--	--
Aluminum-Magnesium-Silicon	T5	--	--	--	--	--	7 to 9	440
	T6	12	980	Boiling water	3 to 5	310	--	--
	T7	12	980	Ditto	--	--	7 to 9	440

[a]T2 Annealed.
T4 Solution heat treated for maximum softness and ductility.
T5 Stress relieved.
T6 Solution heat treated and aged for maximum strength and hardness.
T7 Solution heat treated and overaged for maximum dimensional stability combined with good strength and hardness.

[b]Furnaces must generally be air-circulating type and controlled to ±3 F.

IRON AND STEEL

STEEL

Cast steel has a coarse microstructure. Castings placed in service in this condition may fail because of the brittle nature of the material. Therefore, steel castings must be heat treated to refine the grains by recrystallization, to homogenize the structure, and to improve the physical properties. Two types of heat treatment are (1) annealing, and (2) normalizing followed by tempering. Steel castings that have been welded or worked should be stress relieved before use. The details of these treatments are as follows:

Class A castings and Class B castings under Specification MIL-S-15083 shall be annealed or heat treated in accordance with the specification requirements. Heat treatment should be accomplished with adequate and calibrated pyrometric equipment. Castings should be charged so that the lighter castings will be shielded from the heat of the furnace by the heavier castings, placing the casting loosely and, if possible, a few inches off the floor of the furnace, so that the hot gases will have free circulation throughout the charge. Large castings must be properly supported under heavy sections to minimize distortion.

Annealing. Navy specifications for Class B and low-alloy steel castings require that the castings shall be placed in a furnace, the temperature of which is not more than 500°F. above that of the castings and shall be uniformly heated at a controlled rate to a temperature of 1600°F. or above. The castings shall be held at the annealing temperature for a period of at least 1 hour per inch of thickest cross section, but in no case less than 1 hour. The temperature difference between the hottest and coldest part of the charge during the holding period shall not be greater than 75°F. The castings shall be cooled slowly in the furnace from the heat-treating temperature. When the temperature of the hottest part of the charge has fallen to 500°F. above the ambient temperature, the castings may be removed from the furnace and cooled in still air. Where time is a factor, the cooling rate of small castings, such as pipe fittings or those where dimensional stability is not a controlling factor, may be accelerated when the temperature of the casting has fallen to 1000°F.

Normalizing. "The castings shall be heated and held at temperature***" as described under annealing. They shall then be removed from the furnace and permitted to cool through the critical range in still air. The difference between annealing and normalizing should be made clear. In annealing, the casting is cooled slowly in the furnace and is softened. In normalizing, the casting is cooled in air and may be hardened. Normalizing must be followed by tempering. Annealed castings are not tempered.

Tempering Heat Treatment of Normalized Castings. Castings which have been normalized shall be given a tempering heat treatment by heating in accordance with the requirements given under annealing until the temperature of 1000° to 1250°F. is reached. The temperature of the charge shall remain within this range for a period of not less than 1 hour for each inch or fraction thereof of maximum thickness of section. The castings shall be cooled in air from the heat-treating temperature.

Certain steels develop a brittleness characterized by loss of ductility and impact strength when held for excessive times or slow cooled through temperature ranges of about 400° to 700°F. and 850° to 1100°F. The causes of this "temper brittleness" are not well understood. Fine-grained aluminum-killed steels are less susceptible to temper brittleness than coarse grained silicon-killed steels.

Stress-Relief Heat Treatment. Castings which have been subjected to cold straightening, welding, or forming shall be given a stress-relief heat treatment identical with the tempering heat treatment of normalized castings described.

Bend Test. A bend test may be used as a shop test for checking the quality of a steel and the effectiveness of heat treatment. Bend-test specimens are machined to about 6 inches long with a rectangular cross section of 1 by 1/2 inches and with the long edges rounded to a radius of not over 1/16 inch.

A rectangular bar which has been adequately gated and risered to ensure soundness and which is suitable for machining to the above dimensions should be poured from each heat of steel. The bend-test bar should be given the same heat treatment as that used for the castings and then machined.

If the steel has been properly made and heat treated, the specimen should withstand cold bending around a 1-inch-diameter pin through an angle of approximately 120°F. without cracking. This is an indication of satisfactory ductility.

CAST IRON

Stress-Relief Annealing. When gray iron castings are removed from the mold, they contain temperature gradients caused by nonuniform cooling of uneven section thicknesses within the casting. This uneven cooling causes internal stresses in the castings. These residual stresses should be relieved by heat treatment, particularly for gray iron castings which require good dimensional stability. A proven practice is to heat slowly and uniformly to 800° to 950°F., hold at temperature for 1 hour per inch of thickness (heaviest section), and cool slowly in the furnace. This treatment, if carried out properly, will not appreciably affect the strength or hardness of the castings. For machinery castings where dimensional stability is important, it may be advisable to rough machine before stress relieving. Then, the heat treatment will relieve internal stresses which may have been introduced as a result of severe rough machining operations.

Annealing. Occasionally, an "off-analyses" heat or a chilled casting may be produced which will be difficult to machine. As an emergency measure, annealing may be used to soften the casting and improve the machinability. The annealing temperatures will vary. If chilled or white iron interferes with machinability, it may be necessary to anneal the casting at 1700°F. and cool it slowly in the furnace to restore machinability. If chill is not present, machinability can be improved by annealing at 1400°F. and cooling slowly in the furnace.

CAUTION: Annealing gray iron castings will cause softness and lower physical properties.

MONEL

It may be necessary to soften "S" Monel to permit easier machining. Annealing consists of heating at a temperature of 1600°F. for approximately 1 hour and air cooling to 1300°F., in a reducing atmosphere followed by either quenching in water or oil. The castings should be heated rapidly to the desired temperature and held for a minimum time to prevent excessive grain growth. CAUTION - The furnace atmosphere must be reducing and free of sulfur, and quenching oil must not contain over 0.5 percent sulfur. Sulfur is very easily picked up by the monel at the high temperatures, and it is detrimental to the physical properties. Severe intercrystalline attack will occur if the furnace atmosphere fluctuates between reducing and oxidizing. Very intricate castings should be quenched in oil.

Water quenching produces a slightly softer condition than oil quenching. If a casting is simple in shape and maximum softness is desired, it is sometimes possible to quench directly from the 1600°F. annealing temperature.

After machining, the casting can be age-hardened by heating at a temperature of 1100°F. for 4 to 6 hours and cooling in air. This treatment produces a hardness as high, or higher, than the as-cast alloy.

SUMMARY

In the heat treatment of castings, it must be remembered that the section size of the casting has an important effect on the properties obtained. (Refer to Chapter 2, "Designing a Casting.") The best mechanical properties are usually obtained in thin sections. A step-bar casting is useful for determining the effect of section size on the properties obtained by heat treatment.

Slow heating, holding for the correct length of time at temperature, and control of the heat-treating temperature by the use of properly operating pyrometers are the most important steps to watch in the heat treating of castings. A good casting can be ruined by slipshod heat treatment. For general information, the various temperatures for heat treatment are summarized in table form.

Alloy	Heat Treatment*			
	Stress Relief (To relieve residual stresses)	Anneal (To homogenize and to soften reduces strength)	Solution Treatment (softens)	Aging Treatment (hardens)
Copper base (general) Tin-bronze	Hold[1] at 700°F. to 800 F., air cool.	--	--	--
Manganese bronze	Hold[1] at 600°- 800°F. air or furnace cool.	--	--	--
Aluminum bronze class 2 and 4			Hold[1] at 1600°F. to 1650°F., water quench.	Hold[1] at 1100°F. to 1150°F., water quench.
class 3			Hold[1] at 1600°F. to 1650 F., water quench.	Hold[1] at 1000°F., air cool.
Nickel bronze			1400°F., oil quench[2].	500°F., to 600°F. (5 hours).
Aluminum base (general)	600°F. to 800°F. (1 hour).	--	--	--
Class 3, 4, and 7 (Exact treatment can be obtained from BuShips)	--	--	960°F. to 1000°F., quench[2].	310°F. to 475°F.
"S" Monel		Hold[1] at 1600°F., air cool to 1200 F., water or oil quench[2].		Hold[1] at 1100°F. 4 to 6 hours air cooled.

Alloy	Heat Treatment*			
	Stress Relief (To relieve residual stresses)	Anneal (To homogenize and to soften reduces strength)	Solution Treatment (softens)	Aging Treatment (hardens)
Cast iron	Hold[1] at 800°F. to 950°F., furnace cool.	Hold[1] at 1700°F., furnace cool (for chill removal). Hold[1] at 1400°F., furnace cool (improve machinability).	--	
Steel, Class B and low alloy.	Hold[1] at 1100°F. to 1250°F, furnace cool.	Hold[1] at 1600°F. or higher, furnace cool.	Hold[1] at 1650°F. to 1750°F., cool in still air.	Hold[1] at 1100°F. to 1250°F., furnace cool.

*CAUTION: The temperatures listed are only approximate and intended as a guide.
[1] Hold 1 hour for each inch thickness of heaviest section. Minimum time is 1 hour.
[2] Exact procedures should be obtained from the Bureau of Ships. Poor practice will result in cracked castings. Aluminum is generally quenched in boiling water to minimize danger of cracking.

Chapter XIII
COMPOSITION OF CASTINGS

The selection of the metal or alloy to be used for a casting is usually specified by the ship requesting the work. Whenever there is any doubt about the metal or alloy to be used, the blueprints of the part found aboard the ship originating the work order should be referred to. The blueprints specify the metal or alloy which provides the best combination of properties for that particular casting. If blueprints or other specifications are not available, and if the selection of the metal for a casting must be made locally, the metal should be selected on the basis of information given in this chapter and in the detailed table given at the end of this chapter.

An alloy is a metallic mixture of two or more elements (of which at least one is a metal). The definition of an alloy must be broad and general so that it will include many nonmetallic elements used in alloys. Examples of nonmetallic elements in iron and steel that may be added to obtain desired properties (or which may be present as impurities) are carbon, sulfur, phosphorus, oxygen, and hydrogen.

Many times an alloy will be referred to as a copper-base alloy, nickel-base alloy, etc. Such a reference means that the particular metal mentioned is the principal metal in the alloy. For example, brass is a copper-base alloy; steel is an iron-base alloy. Through usage in the foundry trade, some names have deviated from their intended meanings. This is particularly true with the brass and bronze alloys. Some alloys called brasses, such as red brass, will have a composition which should fall into a bronze classification, and some other alloys called bronzes, such as manganese bronze, are actually brasses.

Definitions of the alloys commonly used in brass and copper foundry work are as follows:

Copper is a commercially pure metal or one which is alloyed with not more than approximately 1 percent of other elements.

Brass is a copper alloy in which zinc is the principal alloying element. Brass usually contains small quantities of other alloying elements. Because of their appearance or some other property, some brasses have become known as bronzes. Examples are leaded bronze and manganese bronze both of which are actually brasses. Nickel silver is also a brass alloy which has been renamed because of its silvery appearance. A nickel silver is a brass in which nickel has been substituted for part of the zinc.

Bronze was originally the term applied to copper alloys having tin as the principal alloying element. Pure bronzes have been modified with other elements to obtain specific properties, and present-day names often include the secondary alloying element in the name (such as phosphor bronze). In some cases, the tin has been replaced as the principal alloying element, but the alloy is still called a bronze. Aluminum bronze, for example, has aluminum as the principal alloying element, and silicon bronze has silicon as the principal alloying element.

SPECIFICATIONS

If the composition of the alloy desired in a casting is not specified by the ship requesting the part, the molder may have a difficult choice to make in selecting the proper alloy for a new casting. A broken casting or another casting of the same type may give valuable information suggesting the alloy to be used. Aluminum is readily distinguished from brass, bronze, iron, and steel by its weight and color. Most brasses and bronzes have a distinctive color. Most irons and steels, except some stainless steels, are magnetic, while nonferrous alloys (and some stainless steels) are nonmagnetic. (Most irons and steels) rust, nonferrous castings and stainless steels do not. An alert observer will distinguish other differences that will be helpful in identifying an alloy in an emergency. After the general type of alloy (cast iron, steel, bronze, or aluminum), has been established, refer to the appropriate chapters (14 to 18) on the selection of the proper alloy. The only proper selection can be made from blueprint information.

Help in selecting the proper alloy for a particular casting may be obtained from tables A and B, in which are listed the common alloys that the repair ship might be called on to cast.

Table A. Compiled Chart of Selected Military and Navy Specifications for Cast Nonferrous Alloys.

Table B. Compiled Chart of Selected Military and Navy Specifications for Cast Ferrous Alloys.

SELECTION OF METAL MIXTURES

When the foundryman has a choice of the alloy to be used for a casting, he should consider using the simplest and most easily cast alloy that will do the job adequately. As a general rule, the alloy that has the lowest tensile strength of a group will be the easiest to cast. For example, hydraulic bronze (ounce metal) with a tensile strength of 30,000 p.s.i. is easier to cast than manganese bronze with a tensile strength of 65,000 p.s.i. Ordinary cast iron is easier to cast than high-strength cast iron. The more simple alloy should be rejected only when it is inadequate because of requirements for strength, corrosion resistance, or special service.

COPPER-BASE ALLOYS

The copper-base alloys that will be of most universal use on shipboard are hydraulic bronze (also known as red brass and ounce metal), gunmetal (a tin bronze known as Composition G), and valve bronze (a leaded tin bronze known as Composition M). The following description will illustrate the common applications for these materials and will show where some of the other copper-base alloys may be required.

Red Brass, also called hydraulic bronze or ounce metal, is suitable for general service and easy to cast. It can be used for valves, small gears, ornamental work, machine parts, and any other parts that do not involve use as a bearing. It can be used for pressure castings up to 350 p.s.i. (MIL-B-16444).

Gunmetal, also called Composition G, is a tin bronze used for parts requiring medium strength and resistance to salt-water corrosion. It has fair machinability. Typical uses are for valve boxes, expansion joints, flanged pipe fittings, gear wheels, condenser heads, water chests, struts, safety valves, and stop valves. It also may be used for bearings and bushings for high-duty work (MIL-M-16576).

Valve Bronze, also called Composition M, is a leaded tin bronze. The lead, as for all leaded brasses and bronzes, is added to improve machinability. This alloy is similar to Composition G except that the tin content has been reduced and lead increased. This results in strength slightly lower than G metal. Composition M may be used in place of G if the physical properties permit. It is a general-purpose bronze used where medium strength and resistance to salt-water corrosion is necessary. It is used for draft gauges, hose couplings, hose fittings, propeller-shaft sleeves, stuffing boxes, and low and medium-pressure valves (MIL-B-16541).

Yellow Brass is a simple high-strength alloy of about 60 percent copper and 40 percent zinc. It has good strength, is tough, and has moderately good resistance to corrosion. Yellow brass is particularly useful in torpedo-tube castings (MIL-B-17512).

Commercial Brass can be used where strength and corrosion resistance are not important (as in handrail fittings, instrument cases, name plates, oil cups, and trim). This alloy is essentially a leaded yellow brass to which tin has been added to improve the physical properties and lead added to improve the machinability (MIL-B-17668).

Naval Brass has a composition similar to that of the commercial brass but the tin and lead contents may be lower. It is used as a general purpose brass for applications such as door fittings and frames, pipe flanges, rail and ladder stanchions, and tarpaulin hooks. It has a corrosion resistance slightly higher than that of commercial brass (MIL-B-17511).

Ornamental Bronze is actually a leaded semi-red brass, which because of its color has become known as a bronze. It has excellent machinability and is used for threaded pipe, ornamental and hardware fittings requiring a high finish, and electrical fittings, and may be used for low-pressure valve bodies (MIL-B-18343).

Phosphor Bronze is used for castings that require a medium strength and are subject to salt-water corrosion. Typical applications are for gears, bushings, bearings, expansion joints, pump pistons, and special pipe fittings (MIL-B-16540).

Tin-Nickel Bronze is used where there is a need for an alloy having a lower tin content but a corrosion resistance comparable to hydraulic bronze (MIL-B-17528).

Tinless Bronze, formerly known as X-1 Metal, has good casting properties and excellent machinability. It can be used where good corrosion resistance or good bearing properties are needed but the use of tin must be minimized (MIL-B-16358).

Manganese Bronze is a popular and useful alloy that can be produced in two levels of strength. Low-tensile manganese bronze has a good combination of high strength and good corrosion resistance in salt water. The designation "low tensile" refers to its minimum tensile strength of 65,000 p.s.i., which is low when compared with other manganese bronzes. Typical applications are for crosshead slipper shoes, engine framing, gypsies and capstans for submarines, nonmagnetic structures, periscope supports, propeller blades, hubs, and worm wheels (MIL-B-16443).

High-tensile manganese bronze contains more aluminum to give it greater strength. Alloys of this type have strengths comparable to those of mild steels. They are used where strength, toughness, and resistance to corrosion by salt water is required. Typical applications are in framing, gears, and worm wheels (MIL-B-16522).

Aluminum Bronze has the highest strength, wear resistance, and toughness of any copper-base alloy. By proper attention to casting procedures and heat treatment, tensile strengths of more than 100,000 p. s. i. can be produced. Aluminum and iron are the principal alloying elements. Aluminum bronze can be used in applications similar to manganese bronze. Its uses include gears, pinions, propeller blades, and worm wheels (MIL-B-16033).

Silicon Bronze (copper-silver alloy) has a combination of good casting properties and homogeneous structure. It is used where high strength, toughness, and resistance to corrosion are required. It may be used for pump bodies, marine hardware, and machine parts, and makes excellent bells (QQ-C-593).

Nickel Silver (also known as copper-nickel-zinc alloy) is a copper-base alloy with nickel as the principal alloying element. It also contains additions of tin, zinc, and lead. This alloy has a white color, good corrosion resistance, good mechanical properties, and good tarnish resistance. A typical application is in graduated sight drums for fire-control instruments. It is also used for hospital equipment (MIL-C-17112).

Copper Nickel is a 70 copper - 30 nickel alloy that provides a high grade material for pipe and tube fittings (MIL-C-20159).

Bearing Bronzes are generally special high-leaded tin bronzes. Grade I bearing bronze contains 18 to 21 percent lead. It is soft and is used principally for bearing liners under conditions where the bearing metal is required to deform locally to conform to irregularities of motion or imperfection in fitting. It also supplies reasonable service where it is difficult to lubricate. Typical applications are in bearings for winches and conveyors (MIL-B-16261).

Grade II bearing bronze, which is stronger than Grade I, has a lead content of 7 to 9 percent and a tin content of 7 to 9 percent. It is good for general bearing surfaces and fairly good for structural purposes. It is suitable where bearings are cast as a part of supporting or enclosing structures (MIL-B-16261).

Grade III bearing bronze is a hard bearing bronze (tin bronze) also known as hard-gear bronze. Examples of its use are for bushings in ammunition hoists, turret turning gear, winches, and rudder bearing rings (MIL-B-16261).

Grade IV bearing bronze has a lead content of 13 to 16 percent, a tin content of 6.25 to 8 percent, and a zinc content of 0.0 to 0.75 percent. Examples of its use are in low pressure valves and fittings, general hardware, and plumbing supplies (MIL-B-16261).

Grade V bearing bronze has a lead content of 23 to 26 percent and a tin content of 4.5 to 6 percent. Examples of its use are bearings under light loads and high speeds (MIL-B-16261).

Grade VI bearing bronze is a high-leaded tin bronze that contains up to 0.5 percent nickel. It is used wherever high strength, hardness, or shock resistance is required. Bearings made from this alloy are finely finished and installation must be true running for the heavy loads carried. Lubrication is necessary (MIL-B-16261).

Grade VII bearing bronze has a lead content of 14 to 16 percent and a tin content of 12 to 14 percent. Examples of its use are for general purpose, low-speed, moderate pressure, bearings (MIL-B-16261).

COPPER CASTINGS

Copper castings are used only for special parts where high electrical or high heat conductivity are required. Castings requiring an electrical conductivity greater than 85 percent must be made from a high-purity copper and only small amounts of deoxidizers may be used. Electrical conductivity is greatly reduced by even small amounts of tin, silicon, magnesium, or aluminum. Copper castings must be heavily risered because they are subject to high shrinkage during solidification.

NICKEL-BASE ALLOYS

Monel is a nickel-base alloy with copper as the principal alloying element. It has very high corrosion resistance and high strength. It is particularly useful where high corrosion resistance and high strength are needed at elevated temperatures. Typical applications are shaft nuts, shaft caps, high-pressure valves, valve trim, and fittings (QQ-N-00288).

Monel modified by additions of silicon is useful where nongalling and antiseizing properties are required. It is also used for propellers (QQ-N-00288).

ALUMINUM-BASE ALLOYS

Aluminum-base casting alloys are widely used because of their light weight. The alloys used for casting aboard ship are of two general groups: (1) those used in the as-cast condition, and (2) those which must be heat treated to obtain the best properties.

Class 2 and Class 5 alloys are used in the as-cast condition. The Class 2 alloy is for general use where maximum corrosion resistance is required, or where an intricate leakproof casting is necessary. Class 5 alloy is used where good tensile strength and corrosion resistance is required. The corrosion resistance is obtained at a sacrifice of tensile strength (MIL-A-17129).

Aluminum alloys in Classes 1, 3, 4, 6, 7, and 9, require heat treatment to obtain the best properties. See chapter 12 for information on the heat treatment of these alloys. The Class 1 alloy is for general use where high strength, ductility, and resistance to shock are necessary. For castings requiring high quality and excellent fluidity, the Class 3 alloy is used. Its properties include pressure tightness in complex castings, strength, and resistance to corrosion. The Class 4 alloy has higher tensile strength at a sacrifice in corrosion resistance. Typical applications are for ammunition stowages, ladder treads, nonmagnetic structures, and sprocket guards. Castings requiring high strength at elevated temperatures can be made from the Class 7 alloy (MIL-A-17129).

LEAD AND TIN-BASE ALLOYS

Babbitt or antifriction metal is a lead or tin-base alloy with antimony or copper as the main alloying element. There are four grades of antifriction alloys generally used in shipboard foundries. Grade 1 is a medium-hard babbitt metal intended for use in aircraft-engine bearings. Grade 2 is a true babbitt metal intended for general use for all bearing surfaces requiring a hard ductile white-metal alloy. Grade 3 is intended for diesel-engine bearings when specifically required. For diesel-engine bearings where loads are excessive and impact is not severe, Grade 4 is used (QQ-T-390).

STEEL

Steel is an iron-base alloy containing a small amount of carbon. Control of the carbon content is vitally important. Low-carbon steels (for example, 0.20 percent carbon) develop little strength but are highly ductile when properly annealed (see chapter 12). High-carbon steels (for example, 0.70 percent carbon) develop higher strength but can be very brittle and difficult to weld if not properly handled. It is best to keep the carbon content low unless there are special requirements for strength or hardness. Low-carbon alloy steel can be one of the most difficult alloys for the foundryman to cast and heat treat.

Class B Steel is a general-purpose cast steel having a medium tensile strength and high ductility. It has good machinability and good resistance to vibration and shock. Typical applications are for motor bedplates, turbine castings, hoist drums, pipe fittings, struts for shafting, and safety valves. This steel should not be used for temperatures exceeding 650°F (MIL-S-15083).

Class A Steel is made in four grades of higher strength and quality than Class B. The improved strength is obtained by a slight increase in the carbon content, by heat treatment, and in certain cases by alloying. Proper heat treatment and tensile testing are requirements for the production of Class A steel. As with Class B steel, the Class A steels are usually not alloyed unless necessary. If the castings are to be welded, hardening alloys such as molybdenum and chromium should be avoided. Class A-70 steel is used for general structural parts. Class A-80 steel is used for castings which are subject to compressive stresses or surface wear (such as may be found in chain pipes, fair leads for anchor cable, followers for piston valves, guides, hawse pipes, and strong-backs). Where greater strength and fair ductility are required, Class A-90 steel should be used. Some typical applications are bearings for turret-turning pinions, carriages for ammunition hoists, rudder crossheads for surface ships, and thrust blocks for turret worm gears. Class A-100 steels are used where maximum strength, hardness, abrasion, and wear resistance are required. The ductility is low (MIL-S-15083).

ALLOY STEELS

Alloy steel is used for special purposes where some special property is desired. It should not be used except where specified or where it is known that a special property must be developed. These steels are tricky to cast and to heat treat. Improper practice can easily produce an alloy steel which is less satisfactory than a common unalloyed steel.

Molybdenum alloy steel is a special low-carbon steel for certain high-pressure hydraulic services as specifically approved. It is useful in steam applications at temperatures up to 850°F., but is not intended for general service (MIL-S-870).

Chromium-molybdenum alloy steel is made in two classes for steam service up to 1,050°F.

It is a specialty steel and is not intended for general use (MIL-S-15464).

Hadfield manganese steel is a special purpose high-carbon steel in which manganese is used as an alloy to obtain high abrasion resisting properties without the use of nickel or chromium. In the as-cast condition, it is extremely brittle and castings require a solution heat-treatment followed by water quenching to develop good ductility. The high manganese content also makes the steel virtually nonmagnetic. This steel has a unique property in that it becomes very hear and wear resistant as it is worked (that is, it work hardens). If the steel is severely cold worked, it becomes magnetic. It has good resistance to corrosion. Typical uses include anchors, aircraft arresting hooks, and gypsy heads. It is a difficult steel to cast, exceedingly difficult to machine, and should be used only where called for (MIL-S-17249).

STAINLESS STEELS

Stainless steels are made up of the class 300 and class 400 steels. The 300 stainless steels are called austenitic steels, and the 400 class are called ferritic steels.

Austenitic alloy steels are the so-called 18-8 stainless steels. They are highly alloyed with about 18 percent chromium and 8 percent nickel. Their carbon content is low. Higher carbon contents than specified will ruin their properties. The main features of these steels are their high resistance to corrosion and oxidation (rusting) and the fact that they are generally nonmagnetic in the annealed condition. They are also known as corrosion resistant steels. They are specified in three grades of MIL-S-867 and three grades of MIL-S-17509. All six grades are rather difficult to machine.

Ferritic alloy steels are chromium alloy steels that are generally magnetic. The type specified for shipboard use contains 12 percent chromium. It is a special corrosion resistant steel. Properties of this steel can be improved by heat treatment (MIL-S-16993).

CAST IRON

There are three types of gray cast iron used in making castings for shipboard use. They are: (1) ordinary cast iron, (2) high-test cast iron, and (3) scale-resisting alloyed cast iron. As with all metals, the simplest (lowest strength) alloy should be used unless there is a real need for another material. As the strength or alloy content increases, the iron becomes more difficult to cast and more difficult to control.

Ordinary gray iron has a low tensile strength, about the same as bearing bronze or some as-cast aluminum-base alloys. The tensile strength is greatly dependent on the section thickness of the casting. Thinner sections have a higher tensile strength than thicker sections. Ordinary cast iron has little impact strength and should not be used where shock is encountered. It is stiff and has high rigidity, but is brittle. Gray iron is one of the easiest metals to cast and can be used for intricate shapes, which cannot be cast in other metals. It is one of the easiest metals to machine. It has good heat resistance, but should not be used above 425°F. Gray iron has good bearing qualities and the ability to absorb vibrations. Gray iron differs from steel mainly in its carbon content (about 3.0 percent carbon in cast iron and 0.30 percent carbon in steel). Ordinary cast iron is used in crankcases, cylinder blocks, piston rings, pistons for reciprocating pumps, and rotors for rotary pumps (QQ-I-652).

High-test gray iron is similar to ordinary gray iron except that higher strengths are developed by controlling the structure. It is more rigid and harder than ordinary gray iron (QQ-I-652).

Scale-resisting gray iron is a carefully controlled alloy used for resistance to scaling, warpage, and growth at high temperatures. A typical use is for galley range tops and furnace parts. It is also used for resistance to acid, caustic, and salt solutions (MIL-G-858).

RAW MATERIALS AND CALCULATION OF CHARGES

One of the major problems in producing a heat of metal for casting is the necessity of making the desired composition. This situation has been made easier by commercial smelters who supply various copper-base and aluminum alloys in ingot form. A good meltdown practice is all that is necessary to produce a heat very close to the desired analysis. For this reason, it is advisable for repair ship foundries to stock alloys such as G metal, M metal, and valve bronze in the ingot form. Any other alloys which are used extensively should also be stocked in ingot form. Some melting losses will occur and should be compensated for in the charge or during melting.

CALCULATION OF CHARGES FOR NONFERROUS MELTS

The calculation of a charge for an alloy is a matter of simple arithmetic. It is usually desirable to use some scrap metal. Hence, the weight of each element in the scrap must be considered individually. A standard form for calculation of charges is recommended so that records may be kept of all heats. This form is based on the following calculations which are illustrated later in two examples:

1. (Weight of metal desired) times (desired percent of element in metal) equals (the weight of element required to produce metal of proper analysis).

2. (Weight of element required in metal) times (percent loss of element) equals (weight loss of element).

3. (Weight loss of element) plus (the desired weight of element in metal) equals (the weight of element which must be added as scrap, new element, or master alloy).

4. (Weight of element which must be added) minus (weight of element present in scrap) equals (weight of element which must be added as new metal or master alloy).

The form shown in figure 222 contains all of the necessary information in compact form and makes it easy to check any errors in calculating a charge.

A charge for ounce metal is computed as an example. The desired composition is 85.0 percent copper, 5.0 percent tin, 5.0 percent zinc, and 5.0 percent lead. The weights of each element needed in the melt are determined by using Calculation 1 as listed above and are entered as line 2 of the form. Remember that 85 percent means 85/100 or 0.85, and 5 percent means 5/100 or 0.05.

The analysis of the scrap is used to determine the weights of the various elements available from the scrap. If analyses are not available, an estimate of the composition will have to be made. These weights are determined by using Calculation 1 and entering the results on line 3. Weights of the various elements available in the ingot are determined in the same way and entered on line 4. The weights of the various elements available in the charge are then added in their respective columns and entered in the Sub-Total as line 6. Reference to these figures shows that 0.7 pound of tin and 2.3 pounds of zinc are required to raise these elements to the desired composition. However, because there will probably be a 1 percent melting loss for lead and a 2 percent melting loss for zinc, these losses must also be added in. Because 25 pounds of lead are required in the final melt, the melting loss will be 0.01 x 25 = 0.25 pound. The zinc loss will be 0.02 x 25 = 0.5 pound. One-half pound of zinc is required to make up the estimated melting loss. The total lead addition will be the 0.7 pound required to get the analysis, plus the 0.25 pound for the melting loss. The zinc addition is the 2.3 pounds required to get the analysis, plus the 0.5 pounds for the melting loss, and is 2.8 pounds. The additions are made in the proper columns, added, and entered in the Total as line 10. The total of the weight column (vertical) should equal the total of the various elements in the total (horizontal) column. If these weights are not the same, an error has been made in computing the charge.

THE USE OF SCRAP METALS

One of the most difficult problems of the emergency foundry will be the classification of scrap alloys. There is no easy way to distinguish between G, M, or leaded bronzes by any means other than chemical analyses. This is not serious for these particular bronzes if lower mechanical properties from lead contamination can be tolerated. If, however, silicon bronze or manganese bronze should be alloyed with a leaded bronze for pressure castings, or if aluminum bronze is alloyed with Composition G, M, or hydraulic bronze, the pressure tightness and mechanical properties would probably suffer to a considerable extent. Aluminum in any form must be carefully avoided when melting the tin bronzes, as it is harmful to the quality of the casting.

In order to classify scrap without chemical analyses, it is a good plan to segregate it according to its color, weight, and use. Oxide films, paint, or dirt must be removed by filing. Bronzes are usually reddish in color and brasses yellowish. Segregation can be effectively accomplished by putting all the tin bronze valve bodies in one bin, manganese bronze propellers and rudders in a second, lead bronze bushings in a third, sheet copper, bus bars, and other copper of electrical-conductivity grade in a fourth. Copper for electrical purposes is usually as pure as virgin ingot. This same general plan can be followed for all scrap material. In making charges, it is safe to assume that old bushings are reasonably good scrap for new bushings, old valve bodies for new ones, etc. Gates, risers, and excess metal which has been poured into pigs are also usually of known composition. If virgin metals are scarce, they should be used only for making minor adjustments in composition.

It must be remembered that many of the nominal compositions recommended in this chapter are the product of several decades of experimentation by a number of investigators. The bronzes have been used for hundreds of years. Changes in composition of any of the alloys should never be attempted on the spur of the moment.

The correct composition obtained by proper segregation of scrap is a major factor in the production of sound castings having good mechanical properties.

CALCULATION OF A CAST IRON CHARGE

The calculation of a charge for cast iron or steel follows the same procedure as that previously described for nonferrous heats. The properties of cast iron and steel are affected by very small changes in the various elements. It is necessary, therefore, to know the analysis of materials making up a charge in order to produce metal having the desired properties.

Should it be necessary to produce 100 pounds of gray iron for a casting of the following chemical specifications, a calculation should be made as shown in figure 223.

	Percent
T.C.	3.15-3.25
Mn	0.80-0.90
Si	1.70-1.90
P	Less than 0.20
S	Less than 0.12
Ni	1.00-1.10

The proportions of raw materials to be used should be estimated and entered in the appropriate columns. This is determined by trial and error and sometimes two or three estimations are necessary before the desired analysis is achieved. (Remember that 3.20 percent means 3.20/100=0.032.)

The silicon content of the raw materials in lines 3, 4, 5, and 6 should be calculated at less than the desired analysis, to permit a ferrosilicon (line 8) addition to the molten bath. Additions of ferronickel, ferrochromium, and ferromolybdenum may be made with the cold charge. It is preferred to add the ferromanganese (line 7) with the ferrosilicon to the molten bath. For example purposes, graphite additions are made (line 10) to adjust the carbon content to show the mechanism of handling this material.

The weight in pounds contributed by each element of each raw material in the proportions used should be calculated. For example, the percentage of carbon in steel scrap is 0.20 percent. Thus, 0.20/100 times 15 equals 0.03 pound of carbon contributed by the steel scrap charge. The manganese contributed would be 0.40/100 times 15 equals 0.06 pound. These calculations should be made for all the constituents of the charge and the figures entered in the appropriate columns. It will only be necessary to carry calculations out to the third decimal place.

The weight in pounds contributed by each element should be added and entered as the Sub-Total (line 6). It will be noted in figure 223, that carbon, silicon, and manganese are now below the desired analysis, and that nickel is missing entirely. Additions must be made to meet the desired analysis and to compensate for melting losses. Table 21 shows the average melting losses which can be expected in an indirect-arc furnace. The manganese melting loss is 10 percent. The manganese addition must be 0.10 x 0.84 = 0.084 pound, plus the 0.106 pound required to meet the analysis. The manganese is added as an 80 percent ferromanganese, which means that the alloy contains 80 percent manganese, with the balance iron. The amount of alloy required is obtained by dividing the required weight of the element by the percent of the alloy, which in this case is 0.190/0.80 equals 0.24 pound of ferromanganese. The silicon, carbon, and nickel additions are calculated in a similar manner.

TABLE 21. AVERAGE MELTING LOSSES IN THE INDIRECT-ARC FURNACE

	Percent
T.C.	Nil
Mn	10
Si	3
P	Nil
S	Nil
Ni	Nil
Cr	Nil
Mo	Nil
Graphite	20

CALCULATION OF A STEEL CHARGE

The mechanism for calculating the final composition for a steel charge is exactly the same as outlined for cast iron. However, very accurate predetermined carbon analyses cannot be calculated from the average analyses specified in the cast iron section, as it is necessary to consider the carbon content of the alloy additions. Therefore, in some instances, it might appear as though there were a slight gain in carbon content when actually this is not the case.

Losses. The melting losses will vary and are functions of the physical characteristics of the charge, tapping temperature, and the length of time held at the superheating temperature prior to tapping. The losses for each element will be considered separately as follows:

(a) <u>Carbon</u>. When the dead-melting method or producing steel is used, there is an average loss of 0.02 percent carbon from the steel when consideration is given to the carbon present in the ferroalloys added. When the boiling method is used, there should be a loss of 0.10 percent carbon, if freedom from porosity is to be obtained.

(b) **Manganese.** The losses for manganese range from 0.30 to 0.40 percent with the dead-melting practice and from 0.40 to 0.50 percent with the boiling method.

(c) **Silicon.** There may be a considerable pick up of silicon when the steel is melted in a fire clay or sillimanite lining (as in the direct-arc electric rocking furnace). Therefore, the maximum calculated silicon (including additions) should be 0.43 percent for the Class B castings and 0.38 percent for the carbon-molybdenum castings to keep the silicon within the ranges indicated by each specification.

(d) **Phosphorus, Sulfur, Molybdenum, and Nickel.** There is practically no loss of these elements.

To assist in estimating compositions of raw materials, some typical compositions of various raw materials are listed in table 22. Whenever analyses are available, either from supply sources or from actual analyses, they should definitely be used in preference to the analyses in table 22.

(For the melting of Remelt No. 1 and Remelt No. 2, see Chapter 17, "Cast Iron.") Also, to simplify the calculation of ferrous charges, weight charts for various charge materials are listed in tables 23 and 24.

TABLE 22. AVERAGE COMPOSITIONS OF RAW MATERIALS

	Composition, percent							
	Total Carbon	Manganese	Silicon	Phosphorus	Sulfur	Nickel	Chromium	Molybdenum
Pig Iron, Grade A	3.63	0.84	2.75	0.55	0.027			
Low Phosphorus Pig Iron	4.26	0.78	1.40	.026	.016			
Structural Steel Scrap	0.20	0.40	.04	.02	.03			
Remelt No. 1 (soft)	3.30-3.40	0.80	2.00	0.10	.02			
Remelt No. 2 (hard)	3.05-3.15	0.80	1.60	0.10	.02			
FeSi (50%) (lump)			50.00					
FeSi (95%) (granular)			95.00					
FeMn		80.00						
FeNi						94.00		
FeCr							70.00	
FeMo								60.00

TABLE 23. WEIGHT CHARTS FOR USE IN CALCULATION OF CAST IRON HEATS

Percent of Total Charge	Structural Steel Scrap					Percent of Total Charge	"A" Foundry Pig				
	Weight in Pounds per 100 Pounds						Weight in Pounds per 100 Pounds				
	TC	Mn	Si	P	S		TC	Mn	Si	P	S
5	0.010	0.02	0.002	0.001	0.0015	5	0.182	0.042	0.138	0.0275	0.00135
10	.020	.04	.004	.002	.0030	10	.364	.084	.276	.0550	.00270
15	.030	.06	.006	.003	.0045	15	.546	.126	.414	.0825	.00405
20	.040	.08	.008	.004	.0060	20	.728	.168	.552	.1100	.00540
25	.050	.10	.010	.005	.0075	25	.910	.210	.690	.1375	.00675
30	.060	.12	.012	.006	.0090	30	1.092	.252	.828	.1650	.00810
Low Phosphorus Pig Iron						Remelt No. 1					
5	0.213	0.039	0.070	0.0013	0.0008	20	0.6700	0.160	0.40	0.020	0.004
10	.426	.078	.140	.0016	.0016	25	.8375	.200	.50	.025	.005
15	.639	.117	.210	.0039	.0024	30	1.0050	.240	.60	.030	.006
20	.852	.156	.280	.0052	.0032	35	1.1725	.280	.70	.035	.007
25	1.065	.195	.350	.0065	.0040	40	1.3400	.320	.80	.040	.008
30	1.278	.234	.420	.0078	.0048	45	1.5075	.360	.90	.045	.009
35	1.491	.273	.490	.0091	.0056	50	1.6750	.400	1.00	.050	.010
40	1.704	.312	.560	.0104	.0064	55	1.8425	.440	1.10	.055	.011
45	1.917	.351	.630	.0117	.0072	60	2.0100	.480	1.20	.060	.012
50	2.130	.390	.700	.0130	.0080	65	2.1775	.520	1.30	.065	.013
55	2.343	.429	.770	.0143	.0088	70	2.3450	.560	1.40	.070	.014
60	2.556	.468	.840	.0156	.0096	75	2.5125	.600	1.50	.075	.015
65	2.769	.507	.910	.0169	.0104	80	2.6800	.640	1.60	.080	.016
70	2.982	.546	.980	.0182	.0112	85	2.8475	.680	1.70	.085	.017
75	3.195	.585	1.050	.0195	.0120	90	3.0150	.720	1.80	.090	.018

TABLE 23. WEIGHT CHARTS FOR USE IN CALCULATION OF CAST IRON HEATS—Continued

Percent of Total Charge	Structural Steel Scrap Weight in Pounds per 100 Pounds					Percent of Total Charge	"A" Foundry Pig Weight in Pounds per 100 Pounds				
	TC	Mn	Si	P	S		TC	Mn	Si	P	S
Remelt No. 2											
20	0.620	0.160	0.320	0.020	0.004	60	1.860	0.480	0.960	0.060	0.012
25	.775	.200	.400	.025	.005	65	2.015	.520	1.040	.065	.013
30	.930	.240	.480	.030	.006	70	2.170	.560	1.120	.070	.014
35	1.085	.280	.560	.035	.007	75	2.325	.600	1.200	.075	.015
40	1.240	.320	.640	.040	.008	80	2.480	.640	1.280	.080	.016
45	1.395	.360	.720	.045	.009	85	2.635	.680	1.360	.085	.017
50	1.550	.400	.800	.050	.010	90	2.790	.720	1.440	.090	.018
55	1.705	.440	.880	.055	.011	95	2.945	.760	1.520	.095	.019

TABLE 24. WEIGHT CHARTS FOR USE IN CALCULATION OF FERROUS HEATS

Percent of Alloy Additions	Weight of Element Contributed per 100 Pounds of Charge						
	FeMn (80%)	FeSi (50%)	FeSi (95%)	FeNi (94%)	FeCr (70%)	FeMo (60%)	FeVa (35%)
0.20	0.160	0.100	0.1900	0.1880	0.1400	0.120	0.0700
.25	.200	.125	.2375	.2350	.1750	.150	.0875
.30	.240	.150	.2850	.2820	.2100	.180	.1050
.35	.280	.175	.3325	.3290	.2450	.210	.1225
.40	.320	.200	.3800	.3760	.2800	.240	.1400
.45	.360	.225	.4275	.4230	.3150	.270	.1575
.50	.400	.250	.4750	.4700	.3500	.300	.1750
.55	.440	.275	.5225	.5170	.3850	.330	.1925
.60	.480	.300	.5700	.5640	.4200	.360	.2100
.65	.520	.325	.6175	.6110	.4550	.390	.2275
.70	.560	.350	.6650	.6580	.4900	.420	.2450
.75	.600	.375	.7125	.7050	.5250	.450	.2625
.80	.640	.400	.7600	.7520	.5600	.480	.2800
.85	.680	.425	.8075	.7990	.5950	.510	.2975
.90	.720	.450	.8550	.8460	.6300	.540	.3150
.95	.760	.475	.9025	.8930	.6650	.570	.3325
1.00	.800	.500	.9500	.9400	.7000	.600	.3500
1.05	.840	.525	.9975	.9870	.7350	.630	.3675
1.10	.880	.550	1.0450	1.0340	.7700	.660	.3850
1.15	.920	.575	1.0925	1.0810	.8050	.690	.4025
1.20	.960	.600	1.1400	1.1280	.8400	.720	.4200
1.25	1.000	.625	1.1875	1.1750	.8750	.750	.4375
1.50	....	.750	1.4250	1.4100	1.0500	.900	.5250
1.55	....	.775	1.4725	1.4570	1.0850	.930	.5425
1.60	....	.800	1.5200	1.5040	1.1200	.960	.5600
1.65	....	.825	1.5675	1.5510	1.1550	.990	.5775
1.70	....	.850	1.6150	1.5980	1.1900	1.020	.5950
1.75	....	.875	1.6625	1.6450	1.2250	1.050	.6125
2.00	....	1.000	1.9000	1.8800	1.4000	1.200	.7000
2.25	....	1.125	....	2.1150	....	1.350	.7875
2.50	....	1.250	....	2.3500	....	1.500	.8750
2.75	....	1.375	....	2.5850	....	1.650	.9675
3.00	....	1.500	....	2.8200	....	1.800	1.0500

SUMMARY

The selection of the proper alloy to be used for a particular casting requires considerable know-how. The only recommended method for selection is to refer to blueprints of the part to be cast. If this is impossible, make the part from an alloy used for similar parts. Remember that the simplest alloy (or alloy of lowest strength) will usually give the best casting with the least danger of rejects. Use special materials or unfamiliar alloys only when they are specified on blueprints or required for some special application where it is known that the difficulties involved in using a special alloy are justified.

The success of making a heat that will produce a casting of the desired strength and other properties depends on how accurately the charges are calculated. Aboard repair ships, it is impossible to have chemical analysis on every heat, as are made in commercial foundries. The use of melting records for obtaining estimated melting losses based on various melting practices is invaluable in determining proper charges. Whenever possible, chemical analyses should be obtained at Navy Yards or from other sources. Records should be maintained on all materials purchased and analyses should be obtained from the supply source.

Scrap will sometimes accumulate until the estimated analysis is very much in doubt. In such cases, it is best to melt the scrap and pig it. Chemical analyses should then be obtained for the pigged metal before using it for casting production.

Comparison of melting practice with actual chemical analyses is the best source of information when making an estimated analysis on scrap material.

Another point that cannot be overlooked in the preparation of charges is that of proper segregation of scrap and proper marking of all melting stock. Mixed scrap should be discarded or used for making practice castings.

COMPILED CHART OF SELECTED MILITARY AND NAVY SPECIFICATIONS FOR CAST NONFERROUS ALLOYS

Chemical and Mechanical Requirements and Uses

Specification Number	Material	Class, Grade, or Alloy	Cu	Sn	Zn	Pb	Fe	Ni	P	Al	Mn	Si	Mg	Sb	Ti	Cr	Other, Total	Condition	Tensile, p.s.i.	Yield Strength, p.s.i.	Elongation, % in 2 in.	BHN	Uses
MIL-B-16033	Bronze, aluminum; castings	1	86.0[1]	–	–	–	2.5-4.0	–	–	8.5-9.5	–	–	–	–	–	–	0.50	As cast	65,000	25,000[2]	20	–	High tensile, hard, resists corrosion, has good wearing qualities, suitable for purposes similar to manganese bronze. Typical applications – gears, pinions, propeller blades, worm wheels.
		2	86.0[1]	–	–	–	0.75-1.5	–	–	9.0-11.0	–	–	–	–	–	–	0.50	As cast / Heat treated	65,000 / 80,000	25,000[2] / 40,000[2]	20 / 12	–	Same as Class 1.
		3	83.0[1]	–	–	–	3.0-5.0	2.5	–	10.0-11.5	0.50	–	–	–	–	–	0.50	As cast / Heat treated	75,000 / 90,000	30,000[2] / 45,000[2]	12 / 6	–	Same as Class 1.
		4	78.0[1]	–	–	–	3.0-5.0	3.0-5.5	–	10.0-11.5	3.5	–	–	–	–	–	0.50	As cast / Heat treated	90,000 / 110,000	40,000[2] / 60,000[2]	6 / 5	–	Same as Class 1.
MIL-B-16522	Bronze, aluminum-manganese; castings	1	60-68	0.50	Remainder	0.20	2.0-4.0	–	–	3.0-7.5	2.5-5.0	–	–	–	–	–	–	–	110,000	60,000[2]	12	–	Intended for use where strength, toughness, or resistance to corrosion by sea water is required. Typical applications – framing, gears, worm wheels.
		2	60-68	0.50	Remainder	0.20	2.0-4.0	–	–	3.0-7.5	2.5-5.0	–	–	–	–	–	–	–	90,000	45,000[2]	18	–	Same as Class 1.
		3	66-68	0.20	Remainder	0.20	2.0-4.0	–	–	4.5-5.5	3.0-5.0	–	–	–	–	–	–	–	90,000	45,000[2]	18	–	Same as Class 1. Also for use where stress corrosion is a factor.
MIL-B-16261	Bronze, bearing; castings	1	74-77	5.0-6.5	0.75	18-21	0.15	1.0	0.05	–	–	–	–	0.75	–	–	–	–	22,000	–	7	–	Suitable for bearing liners; for operation under conditions such that the bearing metal is required to deform locally to conform to irregularities of motion or imperfection of fitting. Also provides reasonable service where lubrication is difficult. Typical applications – bearings for winches and conveyors.
		II	82-85	7-9	0.75	7-9	0.15	1.0	0.50	–	–	–	–	0.50	–	–	–	–	25,000	–	8	–	Good for general bearing surfaces. Also reasonably good structural bronze, and is suitable material for members in which the bearings are integral with the supporting or enclosing structure.
		III	84-86	13-15	1.50	0.20	0.10	0.75	0.05	–	–	–	–	–	–	–	–	–	30,000	–	1	–	A hard bearing bronze, also known as hard gear bronze. Typical applications – bushings for ammunition conveyors, ammunition hoists, booms, cranes, davits, steering gear bearings, turret turning gear, winches, rudder bearing rings; sleeves for steering gear link pins.
		IV	75-79	6.25-8.0	0.75	13-16	0.15	0.75	0.25	–	–	–	–	0.75	–	–	–	–	20,000	–	10	–	Same as Grades I and II.
		V	68.5-72.5	4.5-6.0	0.50	23-26	0.15	0.50	0.05	–	–	–	–	0.75	–	–	–	–	21,000	–	7	38-41[3]	Same as Grade I.
		VI	81-85	6.25-7.5	2-4	6-8	0.20	0.50	0.15	–	–	–	–	0.35	–	–	–	–	30,000	–	18	–	Suitable for use wherever strength, hardness, or shock resistance is required. Suitable for heavily loaded, true running, finely-finished bearings which are lubricated.
		VII[4]	69-72	12-14	0.50	14-16	0.25	0.50-1.0	0.05	–	–	–	–	0.35	–	–	–	0.35	–	–	–	80[3] / 50[3]	Chill cast / Sand cast — Used for torpedo engine and tail bushings.
MIL-B-16358	Bronze, copper-lead-phosphorus; castings	–	Remainder	1.0	1.0	8-12	0.10	0.75-1.5	3-4	–	–	–	–	–	–	–	–	–	30,000	–	6	–	A tinless bronze formerly called X-1 Metal. Has good casting properties, excellent machinability, good corrosion resistance, and good bearing properties.

Footnotes appear on last page of table.

COMPILED CHART OF SELECTED MILITARY AND NAVY SPECIFICATIONS FOR CAST NONFERROUS ALLOYS

Chemical and Mechanical Requirements and Uses

(Continued)

Specification Number	Material	Class, Grade, or Alloy	Cu	Sn	Zn	Pb	Fe	Ni	P	Al	Mn	Si	Mg	Sb	Ti	Cr	Other, Total	Condition	Tensile, p.s.i.	Yield Strength, p.s.i.	Elongation, % in 2 in.	BHN	Uses
MIL-B-16444	Bronze, hydraulic; castings (ounce metal)	A	84-86[5]	4-6	4-6	4-6	0.30	1.0[5]	0.05	–	–	–	–	–	–	–	–	–	30,000	–	20	–	Standard composition for general service, also known as red brass. It is satisfactory for use with pressures up to 350 pounds. Miscellaneous castings not of a bearing nature.
MIL-B-16443	Bronze, manganese; castings	1	55-60	1.0	Remainder	0.40	0.40-2.0	–	–	0.50-1.5	1.5	–	–	–	–	–	–	–	65,000	–	20	–	Relatively high tensile, medium elongation, highly resistant to salt water corrosion; used where strength and corrosion resistance are primary requirements. Typical applications – Crosshead slipper shoes, deck sockets, engine framing, gypsies and capstans for submarines, nonmagnetic structures, periscope supports, propeller blades and hubs, worm wheels.
MIL-B-16343	Bronze, castings, ornamental		78-82[74]	2.25-3.5	7-10	6-8	0.40	1.0	0.02	–	–	–	–	–	–	–	–	–	25,000	–	12	–	A brass, used for threaded pipe, ornamental, or electrical fittings.
MIL-B-16540	Bronze, phosphor; castings	A	85-89	7.5-9.0	3-5	1.0	0.25	1.0	0.50	–	–	–	–	–	–	–	–	–	35,000	–	18	–	Medium tensile and elongation; used for castings which require medium strength, particularly where exposed to the action of salt water, or where good bearing qualities are essential, such as gears, in general. Grade A may be used for applications requiring good strength and resistance to sea water corrosion. Typical applications – bearings, bushings, expansion joints, gears, pump pistons and castings, special pipe fittings.
		B	85-89	7.5-9.0	3-5	1.0	0.25	1.0	0.50	–	–	–	–	–	–	–	–	–	30,000	–	12	–	See first sentence of Grade A. Grade B bronze in intended for use where stresses are low or structural strength is not required.
		C	78-86	2.5-6.0	4-10	4-8	0.40	1.0	0.05	–	–	–	–	–	–	–	–	–	29,000	–	16	–	See first sentence of Grade A. Grade C is a grade requiring refining for use as a constituent of any of the stronger alloys.
MIL-B-16542	Bronze castings (for screw pipe fittings)	A	79-82	4-6	10-16	2-3	0.35	0.75	0.05	–	–	–	–	–	–	–	–	–	28,000	–	15	–	Cast composition fittings for screwed pipe in general.
MIL-B-17528	Bronze, tin-nickel; castings	1	86-89	5-6	1-3	0.50	0.15	4.5-5.5	0.03	–	–	–	–	–	–	–	–	–	40,000	–	20	–	For use where a lower tin alloy having corrosion resistance in sea water equivalent to that of hydraulic bronze is required. Has higher strength than Alloy No. 2.
		2	86-89	5-6	3-5	1-2	0.25	2.5-3.5	0.03	–	–	–	–	–	–	–	–	–	35,000	–	20	–	See Alloy No. 1. Has better machinability than Alloy No. 1.
MIL-B-16541	Bronze, valve, castings	A	86-89	5.5-6.5	3-5	1-2	0.25	1.0	0.05	–	–	–	–	–	–	–	–	–	34,000	–	22	–	A general purpose bronze, used for parts not highly stressed where the physical properties permit. Typical applications – cocks draft gauges, hose couplings and fittings, macomb strainers, manifolds for stem tubes and propeller shaft sleeves, stuffing boxes, low and medium pressure valves.
MIL-W-16376	Gun metal; castings	A	86-89	7.5-9.0	3-5	0.30	0.15	1.0	0.05	–	–	–	–	–	–	–	–	–	40,000	–	20	–	Used for parts requiring medium strength and resistance to salt water corrosion. Typical applications – air pump castings, buckets, and valve boxes, expansion joints, flanged pipe fittings, gear wheels, grease extractors, condenser heads, shapes, and water chests, stem tube, strut, and spring bearings, safety and stop valves.

COMPILED CHART OF SELECTED MILITARY AND NAVY SPECIFICATIONS FOR CAST NONFERROUS ALLOYS

Chemical and Mechanical Requirements and Uses
(Continued)

Specification Number	Material	Class, Grade, or Alloy	Cu	Sn	Zn	Pb	Fe	Ni	P	Al	Mn	Si	Mg	Sb	Ti	Cr	Be	Co	Other, Total	Condition	Tensile, p.s.i.	Yield Strength, p.s.i.	Elongation, % in 2 in.	BHN	Uses
MIL-B-17512	Brass castings	–	59-63	(6)	Remainder	(6)	(6)	(6)	–	0.50-1.0	(6)	–	–	–	–	–	–	–	0.20(7)	–	55,000	–	25	–	Intended for use for torpedo tubing castings, and other applications where strength, toughness, and/or moderate resistance to corrosion is required.
MIL-B-17668	Brass, commercial; castings	1	65-70	1.5	Remainder	1.5-3.75	0.75	–	–	0.30	–	–	–	–	–	–	–	–	–	–	–	–	–	–	Used where cheap brass will serve the purpose and strength and resistance to corrosion are not important. Typical applications – fittings in general, handrail fittings, instrument cases, nameplates, oil caps, trim.
		2	70-74	0.75-2.0	Remainder	1.5-3.75	0.60	–	–	–	–	–	–	–	–	–	–	–	–	–	–	–	–	–	Same as Grade 1.
MIL-B-17511	Brass, naval; castings	–	60-65	0.50-1.5	Remainder	0.75-1.5	0.75	–	–	0.50	–	–	–	–	–	–	–	–	–	–	40,000	14,000(8)	15	–	Low tensile strength and elongation; used as a general purpose brass. Relatively less affected by corrosion than commercial brass. Typical applications – belaying pins, door fittings and frames, pipe flanges, rail and ladder stanchions, scuttle frames, tarpaulin hooks.
MIL-C-20159	Copper-nickel alloy castings (70-30)	(9)	Remainder	–	1.0	0.03	0.25-0.60	28-32	–	0.50	1.25	0.70	–	–	0.25	–	–	–	–	–	60,000	32,000(10)	20	–	A high grade material suitable for pipe and tube fittings.
MIL-C-17112	Copper-nickel-zinc alloy castings (nickel-silver)	–	63-67	2-4	8-12	0.5-1.0	–	18-22	–	–	–	–	–	–	–	–	–	–	–	–	–	–	–	–	For uses where good mechanical properties, white color, and/or excellent resistance to corrosion and tarnishing are desired. Typical applications – graduated sight drums on fire control instruments.
QQ-C-593a	Copper-silicon alloy castings	a	Remainder	1.0	12.0-16.0	0.5	2.5	–	–	1.5	1.5	2.5-5.0	–	–	–	–	–	–	0.50	–	50,000	21,000	18	–	Intended for purposes requiring sound homogeneous castings of high strength, toughness, workability, and resistance to corrosion.
		b	Remainder	1.0	5.0	0.5	2.5	–	–	1.5	1.5	1.0-5.0	–	–	–	–	–	–	0.50	–	45,000	18,000	20	–	Same as Class a.
MIL-C-19464	Copper	1	Remainder	0.01	0.01	0.01	0.10	0.25(25)	–	0.10	–	0.15	–	–	–	0.01	0.50-0.75	2.35-2.75(26)	–	A B	40,000(29) 95,000	8,000(27) 55,000	25(28) 8	45(27,28,29) 92(29)	Moderate strength and hardness with good electrical conductivity; used for switches and switch gear, circuit breakers, and other current carrying devices.
	Beryllium	2	Remainder	0.10	0.10	0.02	0.25	0.20(25)	–	0.15	–	0.20-0.35	–	–	–	0.10	1.90-2.15	0.35-0.65(26)	–	A B	55,000(27) 155,000	20,000(27) 115,000	30(28) 0	75(27,28,29) 38(30)	High strength and good wear and corrosion resistance; used for bushings, cams, bearings, gears, safety bolts, etc.
	Alloy	3	Remainder	0.10	0.10	0.02	0.25	1.00-1.50	–	0.15	–	0.15	–	–	–	0.10	2.35-2.55	–	–	B	155,000	130,000	1	39(30)	High strength and hardness; general purpose.
	Castings	4	Remainder	0.10	0.10	0.02	0.25	0.20(25)	–	0.15	–	0.20-0.35	–	–	–	0.10	2.50-2.75	0.35-0.65(26)	–	A B	75,000(27) 150,000	30,000(27) 110,000	5(28) 0	90(27,28,29) 42(30)	Used where high tensile strength and yield strength with maximum hardness are required.

COMPILED CHART OF SELECTED MILITARY AND NAVY SPECIFICATIONS FOR CAST NONFERROUS ALLOYS
Chemical and Mechanical Requirements and Uses
(Continued)

Specification Number	Material	Class, Grade, or Alloy	Cu	Sn	Zn	Pb	Fe	Ni	Al	P	Mn	Si	Sb	Mg	Ti	Cr	Be	Co	Other, Total	Condition	Tensile Strength, p.s.i.	Yield Strength, p.s.i.	Elongation, % in 2 in.	BHN	Uses
QQ-N-00288	Nickel-copper and Nickel-copper-silicon alloy castings	A	26-33	–	–	–	2.5	62-68	0.5	–	1.5	2.0	–	–	–	–	–	–	–	–	65,000	32,500	25	125-150[14,28]	"Monel", medium tensile, high elongation, high corrosion-resisting alloy, used where high corrosion resistance to strength at elevated temperatures is required. Typical applications – shaft nuts and caps, high-pressure valves, valve trim and fittings. See manual.
		B[12]	27-33	–	–	–	2.5	61-68	0.5	–	1.5	2.7-3.7	–	–	–	–	–	–	–	–	100,000	60,000	10	240-290[14,28]	Used for applications involving nongalling and antiseizing characteristics coupled with moderately high hardness and relative ease of machinability.
		C	27-31	–	–	–	2.5	60[1]	0.5	–	1.5	3.3-4.3	–	–	–	–	–	–	–	–	120,000	80,000	10	250-300[14,28]	High strength, high corrosion and high abrasion resistance; used for ship propellers.
		D	27-31	–	–	–	2.5	60[1]	0.5	–	1.5	3.5-4.5	–	–	–	–	–	–	–	–	–	–	–	300[7,14]	Intended for use where nongalling and antiseizing characteristics are desired. Similar to composition B, but harder.
		E[12]	26-33	–	–	–	3.5	60[1]	0.5	–	1.5	1.2	–	–	–	–	–	–	–	–	65,000	32,500	25	125-150[14,28]	Similar to composition B, but weldable.
MIL-A-17129	Aluminum alloy castings (sand)	1	0.40	–	5-7	–	1.0	–	Remainder	–	0.30	0.30	–	0.45-0.7	0.15-0.25	0.40.6	–	–	0.15[15]	Aged	32,000[16]	–	3[16]	–	For general use where high strength, ductility, and resistance to shock are required.
		2	0.25	–	0.30	–	0.80	–	Remainder	–	0.30	4.5-6.0	–	0.05	0.20	–	–	–	(15)	As cast	17,000	–	3	–	For general use with maximum corrosion resistance, and for leakproof castings of intricate design.
		3	0.20	–	0.30	–	0.60	–	Remainder	–	0.30	6.5-7.5	–	0.2-0.4	0.20	–	–	–	0.15[15]	Solution heat-treated	25,000[17]	–	3[17]	–	For use with complex castings where castability, pressure tightness, strength, and resistance to corrosion are required. Will respond to heat treatment to improve strength. For applications requiring high casting quality and excellent fluidity.
																				Artificially aged	23,000	–	3	–	
																				Solution heat-treated and artificially aged	30,000	–	3	–	
		4[18]	4-5	–	0.30	–	1.0	–	Remainder	–	0.30	1.5	–	0.03	0.20	–	–	–	0.15[15]	Solution heat-treated	29,000[17]	–	6[17]	–	High tensile, with less corrosion resistance than Classes 1,3,5,7, and 8. Heat treatment required. Typical applications – ammunition stowages and hoist finger trays, frames and sills for joiner doors, ladder treads, nonmagnetic structures, sprocket guards.
																				Solution heat-treated and partially aged	32,000	–	3	–	
																				Solution heat-treated and aged	36,000	–	–	–	
																				Solution heat-treated and stabilized	29,000	–	3	–	
		5	0.10	–	0.10	–	0.50	–	Remainder	–	0.60	0.30	–	3.5-4.5	0.20	–	–	–	0.15[15]	As cast	22,000	–	6	–	For use wherever good tensile strength and relatively high resistance to corrosion are required. Heat treatment not required. For applications similar to Class 4 but requiring resistance to corrosion at a sacrifice of tensile properties.

- 201 -

COMPILED CHART OF SELECTED MILITARY AND NAVY SPECIFICATIONS FOR CAST NONFERROUS ALLOYS
Chemical and Mechanical Requirements and Uses
(Continued)

Specification Number	Material	Class, Grade, or Alloy	Chemical Composition, percent maximum or range indicated															Mechanical Requirements, minimum				Uses
			Cu	Sn	Zn	Pb	Fe	Ni	P	Al	Mn	Si	Mg	Sb	Ti	Cr	Be	Co	Other, Total	Condition	Tensile, p.s.i.	Yield Strength, p.s.i.	Elongation, % in 2 in.	BHN	
MIL-A-17129 (Continued)	Aluminum alloy castings (sand)	6	1.5-2.25	1.25-1.75	0.5-1.25	–	0.80	–	–	Remainder	0.30	0.25	0.5-1.0	–	0.1-0.25	0.15-0.3	–	–	0.15(15)		30,000	–	4	–	Moderate strength and corrosion resistance. For general uses where excellent machinability is required.
		7	1.0-1.5	–	0.30	–	0.60	–	–	Remainder	0.30	4.5-5.5	0.4-0.6	–	0.20	0.25	–	–	0.15(15)	Solution heat-treated and partially aged	32,000	–	2	–	For general use, where high strength and corrosion resistance are required. Also retains strength at elevated temperature.
																				Artificially aged	25,000	–	–	–	
																				Solution heat-treated and stabilized	35,000(19)	–	–	–	
		8	0.4-1.0	–	7-8	–	1.0	0.10	–	Remainder	0.60	0.30	0.2-0.5	–	0.20	0.30	–	–	0.20(20)	Aged	32,000(21)	–	3(21)	–	Same as Class 1.
		9	0.20	–	0.10	–	0.3	–	–	Remainder	0.10	0.20	9.5-10.6	–	0.20	–	–	–	0.15(15)	Solution heat-treated	42,000	–	12	–	For castings requiring maximum strength, elongation, and resistance to shock; requires special founding practice.
QQ-T-390	Antifriction metal castings	1(22)	4.0-5.0	90-92	0.005	0.35	0.08	–	–	0.005	–	–	–	4.0-5.0	–	–	–	–	0.10	–	–	–	–	–	A medium hard babbitt metal intended for use in aircraft engine bearings.
		2(22)	3.0-4.0	88-90	0.005	0.35	0.08	–	–	0.005	–	–	–	7-8	–	–	–	–	0.10	–	–	–	–	–	Genuine babbitt metal. Intended for general use for all bearing surfaces requiring a hard, ductile, white metal alloy.
		3(22)	7.5-8.5	83-85	0.005	0.35	0.08	–	–	0.005	–	–	–	7.5-8.5	–	–	–	–	0.10	–	–	–	–	–	A rather hard babbitt metal which may be used for bearings subject to moderately heavy pressures or severe reciprocating motion.
		4(22)	5-6	80.5-82.5	0.005	0.25	0.08	–	–	0.005	–	–	–	12-14	–	–	–	–	0.10	–	–	–	–	–	Harder than Grades 1, 2, and 3. Intended for use involving very heavy pressure and high speed.

NOTE: The above listed alloys are those normally used by the Bureau of Ships. For other compositions, refer to the specification.

(1) Minimum.
(2) Extension under load 0.005 inch per inch.
(3) 500 kg.
(4) Sulfur content – 0.25 percent maximum.
(5) In determining compliance with minimum copper, copper may be computed as copper plus nickel.
(6) Manganese, tin, nickel, iron and lead total – 1.00 percent maximum.
(7) Each.
(8) 0.5 percent extension under load.
(9) Carbon content – 0.15 percent maximum.
(10) Yield point.
(12) Carbon content – 0.30 percent maximum.
(13) Proof-stress.
(14) 3000 kg.
(15) Other elements – 0.05 percent maximum each.
(16) Obtainable after 21 days at room-temperature aging or after artificially aging at 356°F. for 10 hours.
(17) Obtainable when tested after 48 hours and before 96 hours of room-temperature aging.
(18) Total of all constituents other than copper and aluminum, 3.0 percent maximum.
(19) Applies to castings which are water quenched after heating to the solution temperature. In the case of castings which are of such complicated design that water quenching results in cracks, and quenching in an air blast is used, the minimum tensile strength shall be 32,000 p.s.i.
(20) Other elements – 0.10 percent maximum, each.
(21) Obtainable after 21 days at room-temperature aging or after artificially aging at 250°F. for 10 to 16 hours.
(22) Arsenic content – 0.10 percent maximum.
(23) On a 20-gram sample.
(24) Copper plus nickel.
(25) Nickel is a residual element and is not intentionally added to the melt.
(26) Nickel plus cobalt.
(27) Maximum.
(28) For information; not required.
(29) Rockwell B.
(30) Rockwell C.

COMPILED CHART OF SELECTED MILITARY AND NAVY SPECIFICATIONS FOR CAST FERROUS ALLOYS

Chemical and Mechanical Requirements and Uses

Specification Number	Material	Class, Grade, or Alloy	Chemical Composition, percent maximum or range indicated									Mechanical Requirements, minimum					Uses
			C	Mn	P	S	Si	Cu	Ni	Mo	Cr	Tensile, p.s.i.	Yield Point, p.s.i.	Elongation, % in 2 in.	R.A., %	Cold Bend, degrees	BHN, 3000 kg.	
MIL-S-15083	Steel castings	CW	0.30(1)	0.70(1)	0.07	0.06	-	-	-	-	-	(2)	(2)	(2)	(2)	(2)	-	Intended for applications where strength is not of prime importance, but where welding may be required.
		B(3)	0.30(1)	0.60(1)	0.05	0.05	0.60(4)	0.30	0.50(5)	-	0.20	60,000	30,000	24	35	120(6)	-	Medium tensile, high elongation, readily machinable; used as a general purpose steel wherever cast steel is required or permitted, particularly where resistance to vibration or shock is necessary. Not satisfactory for temperatures exceeding 650°F. Typical applications: bedplates for motors, turbine castings, drums for hoists, foundations, pipe fittings, struts for shafting, safety valves.
		A70	0.35	-	0.05	0.05	-	-	-	-	-	70,000	35,000	22	30	-	-	Class A70 castings and those of higher strength are intended for special structural services. If these castings are to be welded, the composition should be selected to minimize the use of hardening elements insofar as practicable.
		A80	-	-	0.05	0.05	-	-	-	-	-	80,000	40,000	18	30	-	-	Used for important parts subjected to compressive stresses of surface wear. Typical applications: chain pipes, fair leads for anchor cable, followers for piston valves, guides, hawsepipes, pistons, strongbacks.
		A90	-	-	0.05	0.05	-	-	-	-	-	90,000	55,000	18	30	-	-	Used in special cases similar to Class A80 cast steel where greater strength combined with ductility is required. Typical applications: bearings for turret turning pinions, carriages for ammunition hoists, rudder crossheads for surface ships, sprockety wheels, thrust blocks for turret worm gears.
		A100	-	-	0.05	0.05	-	-	-	-	-	100,000	60,000	15	30	-	-	Same as A90.
MIL-S-870	Steel, molybdenum alloy castings	-	0.25(1)	0.50-0.75	0.05	0.05	0.20-0.50	0.30(3)	1.00(3)	0.40-0.60	0.20(1,3)	65,000	35,000	20	30	120(6)	-	Used for steam applications, for temperatures up to 850°F., and for certain high pressure hydraulic services as approved.
MIL-S-15464	Steel, chromium-molybdenum alloy castings	1	0.20	0.50-0.80	0.05	0.05	0.20-0.60	0.50(3)	0.50(3)	0.40-0.60	1.00-1.50	70,000	40,000	20	35	-	-	Used with superheated steam at temperatures of 850°F. to 1050°F.
		2	0.18	0.40-0.70	0.05	0.05	0.20-0.60	0.50(3)	0.50(3)	0.80-1.10	2.00-2.75	70,000	40,000	20	35	-	-	Used with superheated steam at temperatures of 950°F. to 1050°F.
		3(8)	0.18	0.40-0.70	0.05	0.06	0.60	0.50(3)	0.50(3)	0.40-0.60	1.00-1.50	70,000	40,000	20	35	-	-	Same as Class 2.
MIL-S-867	Steel, corrosion-resisting austenitic castings	1	0.08(9)	1.50	0.05	0.05	2.00	-	8.00-11.0	-	18.0-21.0	70,000	28,000	35	-	-	-	Used for castings exposed to various combinations of corrosive and high temperature conditions.
		II(10)	0.08	1.50	0.05	0.05	2.00	-	9.00-12.0	-	18.0-21.0	70,000	30,000	30	-	-	-	Same as Class I.
		III	0.08	1.50	0.05	0.05	2.00	-	9.00-12.0	2.0-3.0	18.0-21.0	70,000	30,000	30	-	-	-	Same as Class I.
MIL-S-16993	Steel castings, 12 percent chromium	1	0.15	1.0	0.05	0.05	1.5	-	1.00	0.5	11.5-14.0	90,000	65,000	18	30	-	-	Intended for use in load carrying applications for elevated temperature (up to 1200°F.) service. At higher temperatures the alloy has good resistance to oxidation, but is not suitable for stressed applications. Typical applications: pump castings, compressor housings, jet engine parts.
		2	0.15	1.0	0.05	0.05	0.5	-	0.65-1.00	0.50-0.70	11.5-14.0	90,000	65,000	18	30	-	-	Intended for ship propellers and other applications requiring resistance to impact at low temperatures.

Footnotes appear on last page of table.

COMPILED CHART OF SELECTED MILITARY AND NAVY SPECIFICATIONS FOR CAST FERROUS ALLOYS

Chemical and Mechanical Requirements and Uses

(Continued)

Specification Number	Material	Class, Grade, or Alloy	C	Mn	P	S	Si	Cu	Ni	Mo	Cr	Tensile, p.s.i.	Yield Point, p.s.i.	Elongation, % in 2 in.	R.A., %	Cold Bend, degrees	BHN, 3000 kg	Uses
MIL-S-17249	Steel castings, Hadfield manganese (low magnetic permeability)	A and B	1.00-1.35	12.00-14.00	0.06	--	0.40-1.00	--	1.00[3]	0.50[3]	0.75[3]	100,000	45,000	25	--	--	--	A nonstrategic material used for nonmagnetic applications. Typical applications: anchors, aircraft arresting hooks, gypsy heads.
MIL-S-17509	Steel castings, austenitic, chromium-nickel (low magnetic permeability)	I	0.15-0.30	1.50	0.05	0.05	1.50	--	8.0-11.0	0.50	17.0-20.0	70,000	30,000	30	--	--	--	Intended for use where austenitic corrosion resisting steel having low magnetic permeability is required.
		II[11]	0.10	1.50	0.05	0.05	1.50	--	11.0 min.	0.50	17.0-20.0	65,000	30,000	30	--	--	--	Same as Class I. May be used in steam applications at temperatures not exceeding 1050°F.
		III	0.08	1.50	0.05	0.05	1.50	--	10.0 min.	0.50	17.0-20.0	65,000	28,000	30	--	--	--	Same as Class I.
QQ-I-652	Iron castings, gray	20	--	--	0.25	0.15	--	--	--	--	--	20,000	--	--	--	--	--	Ordinary cast iron, low tensile, no ductility, classification as to physical characteristics is dependent on the size of the casting as per specification; not satisfactory for temperatures exceeding 425°F. Used for machinery parts where neither weight, strength, vibration, shock resistance, nor high rigidity are important considerations, but where cheapness of production is required, or for wearing surfaces which may be easily and cheaply renewed. Typical applications: cylinder liners, pistons, and piston rings for internal combustion engines, furnace fittings. See manual.
		25	--	--	0.25	0.15	--	--	--	--	--	25,000	--	--	--	--	--	Same as Class 20.
		30	--	--	0.25	0.15	--	--	--	--	--	30,000	--	--	--	--	--	Same as Class 20.
		35	--	--	0.25	0.15	--	--	--	--	--	35,000	--	--	--	--	--	Higher tensile cast iron, practically no ductility, classification as to physical characteristics is dependent on size of casting as per specification; not satisfactory for temperatures exceeding 450°F. or where strength is a requirement, unless specifically approved. Used for purposes similar to ordinary cast iron where greater strength, rigidity, or resistance to wear is necessary. Typical applications: crankcasts and cylinder blocks for internal combustion engines, liners for steam cylinders on reciprocating pumps, piston rings and pistons for reciprocating pumps, rotors for rotary pumps.
		40	--	--	0.25	0.15	--	--	--	--	--	40,000	--	--	--	--	--	Same as Class 35.
		50	--	--	0.25	0.15	--	--	--	--	--	50,000	--	--	--	--	--	Same as Class 35.
		60	--	--	0.25	0.15	--	--	--	--	--	60,000	--	--	--	--	--	Same as Class 35.
		Special	--	--	(14)	(14)	--	--	--	--	--	(14)	--	--	--	--	--	Same as Class 35.
MIL-G-858	Gray iron castings, scale-resisting	1	2.60-3.00	1.0-1.5	0.20[12]	0.10[12]	1.25-2.20	5.5-7.5	13.5-17.5	--	1.8-3.5	25,000	--	--	--	--	120-180	Intended for use for resistance to scaling, warpage, and growth in high temperature service, such as galley range tops; also for resistance to corrosion of acid, caustic, and salt solution.
		2	2.60-3.00	0.80-1.30	0.20[12]	0.10[12]	1.25-2.20	0.50	18.0-22.0	--	1.75-3.50	25,000	--	--	--	--	120-180	Same as Class 1.

NOTE: Recommend insertion of transverse test properties. See Table I of specifications.

(1) For each reduction of 0.01 percent carbon under the maximum specified, an increase of either 0.04 percent manganese or 0.04 percent chromium above the maximum specified will be permitted but in no case shall the manganese content exceed 1.00 percent or the chromium content exceed 0.40 percent.

(2) Determination of mechanical properties will not ordinarily be required of Class CW steels. For design purposes, the following properties may be considered to be the minimum for this class:

Tensile Strength	55,000 p.s.i.
Yield Point	27,000 p.s.i.
Elongation	15 percent in 2 inches
Reduction in area	25 percent in 2 inches

(3) Unless otherwise noted, the limits shown for molybdenum, copper, nickel, and chromium, are permissible residual elements and shall not be added

(4) Minimum silicon 0.20 percent.

(5) For Ordnance castings a maximum of 1.00 percent nickel is permitted.

(6) Not required if reduction in area is 40 percent or higher.

(7) Class A70 is a plain carbon steel, however, for Ordnance castings, a maximum of 1.50 percent nickel is permitted.

(8) Vanadium content — 0.15-0.25 percent.

(9) If chromium is over 20 percent and nickel is over 10 percent, a maximum carbon content of 0.12 percent will be permitted.

(10) Columbium, tantalum, or titanium content. Columbium or columbium plus tantalum shall be not less than 10 times the carbon content and not more than 1.10 percent (tantalum shall not exceed 0.4 times the sum of the columbium and tantalum content), or titanium content shall be not less than 6 times the carbon content and not more than 0.75 percent.

(11) Columbium or tantalum content — Columbium or columbium plus tantalum shall be not less than 10 times the carbon content and not more than 1.20 percent (tantalum shall not exceed 0.4 times the sum of the columbium and tantalum content).

(12) Unless otherwise specified, the sulphur and phosphorus limits for galley range top castings shall be 0.20 and 0.70 percent respectively.

(13) Maximum.

(14) As specified.

Heat No. _____ Alloy _____ Date _____

Nominal Composition __85-5-5-5__

Total Charge Weight __500__ Pounds

Line	Composition 85.0Cu, 5.0Sn, 5.0Zn, 5.0Pb	Weight, lb	Element							
			Sn		Zn		Pb		Cu	
1			Percent	Lb	Percent	Lb	Percent	Lb	Percent	Lb
2	Desired Analysis	500	0.05	25.0	0.05	25.0	0.05	25.0	0.85	425.0
	Charge									
3	Scrap	200	0.045	9.0	0.04	8.0	0.05	10.0	0.865	173.0
4	Ingot	300	0.51	15.3	0.049	14.7	0.05	15.0	0.85	255.0
5										
6	Sub-Total			24.3		22.7		25.0		428.0
7	Additions Virgin Lead	0.25						0.25		
8	Virgin Zinc	2.80				2.8				
9	Virgin Tin	0.60		0.60						
10	Total	503.75		25.00		25.5		25.25		428.0

Figure 222. Example of charge calculation for ounce metal.

Heat No. _____ Alloy __Cast Iron__ Date _____

Nominal Composition __3.15-3.25C, 1.30-1.90Si__

Total Charge Weight __100__ Pounds

Line	Composition 3.20C, 0.84Mn, 1.75Si, 1.00Ni, less than 0.20P, less than 0.12S	Weight, lb	Element								
			C		Si		Mn		Ni		P
			Percent	Lb	Percent	Lb	Percent	Lb	Percent	Lb	Percent	Lb
2	Desired Analysis	100	0.032	3.20	0.0175	1.75	0.0084	0.84	0.01	1.00	0.002	0.2
3	Charge Steel Scrap	15	0.002	0.030	0.0004	0.006	0.004	0.060			0.002	0.003
4	Low Phosphorus Pig	30	0.0426	1.278	0.0140	0.420	0.0078	0.234			0.00026	0.008
5	Remelt No. 2	55	0.031	1.705	0.016	0.880	0.008	0.440			0.001	0.055
6	Sub-Total			3.013		1.306		0.734				0.066
7	Additions FeMn (80%)	0.25					0.80	0.20				
8	FeSi (50%)	1.00			0.50	0.50						
9	FeNi (94%)	1.20							0.94	1.03		
10	Graphite (80%)	0.30	0.80	0.240								
11	Total			3.253		1.806		0.934		1.03		0.066

Figure 223. Example of charge calculation for gray iron.

Chapter XIV
COPPER-BASE ALLOYS

SELECTION OF ALLOY

COMPOSITION G

Composition G, or gun metal, is a tin bronze which has good resistance to salt-water corrosion and to dezincification. These properties make it a useful alloy for castings required aboard ship. It is often used in valves and steam fittings.

COMPOSITION M

Composition M, or valve bronze, is a tin bronze to which lead has been added to improve the machinability. It can be used in place of Composition G if its lower strength is adequate.

HYDRAULIC BRONZE

This alloy, the familiar 85-5-5-5 alloy, is a leaded red brass, also known as ounce metal. It is a general-purpose alloy having good corrosion resistance. Castings subjected to hydraulic pressures up to 350 pounds can be made from this alloy.

MANGANESE BRONZE

Manganese bronze castings are strong, ductile, and are corrosion resistant to sea water, sea air, waste water, industrial wastes, and other corroding agents. Typical uses are in propeller hubs, propeller blades, engine framing, gun-mount castings, marine-engine pumps, valves, gears, and worm wheels. Manganese bronze has an excellent combination of corrosion resistance, strength, and ductility that makes it very useful for the various marine castings. It has the disadvantage of having a high solidification shrinkage and a comparatively high drossing tendency. These disadvantages, however, can be overcome by proper design and casting procedures.

YELLOW BRASS

Castings which are not subjected to air or water pressure but which require corrosion resistance are made from yellow brass. It is a general-purpose alloy used for fittings, name plates, and similar applications. Naval brass is a yellow brass which has a higher corrosion resistance than commercial yellow brass.

ALUMINUM BRONZE

Strength, hardness, ductility, and corrosion resistance for the properties which make aluminum bronze a desirable alloy for shipboard use. Higher strength and hardness can be obtained in some of the alloys by proper heat treatment. Typical uses are for worm gears, bearing sleeves, pinions, and propeller blades.

70-30 CUPRO-NICKEL

Cupro-nickel has an excellent resistance to salt water corrosion. Fittings such as couplings, tees, ells, pump bodies, and valve bodies are cast from this alloy.

NICKEL SILVER

Nickel silver provides good mechanical properties and excellent resistance to corrosion and tarnish. It has a pleasing white color that makes it useful where the appearance of the cast part is important.

HOW COPPER-BASE ALLOYS SOLIDIFY

Copper-base alloys solidify by the nucleation and growth of crystals as described in Chapter 1, "How Metals Solidify." There are two types of alloys so far as solidification is concerned: (1) those which have a short solidification range, and (2) those which have a long solidification range. A long solidification range means that an alloy solidifies slowly over a wide range of temperature. Ordinary solder is such an alloy, and solidification over a long range is shown by the fact that the alloy remains mushy for quite a while during solidification. Alloys with short solidification ranges do not show this mushy behavior. Manganese bronze, aluminum bronze, and the yellow brasses have a short solidification range, which cause their high solidification shrinkage. Composition G, Composition M, and hydraulic bronze have a long solidification range, which permits extensive growth of the dendrites. These alloys have a tendency toward interdendritic shrinkage and microporosity, with the result that piping in the riser is not so pronounced.

Solidification of all the alloys begins at the mold and core surfaces. The part of the alloy having the highest solidification temperature (the copper-rich material) solidifies first and a crystalline structure is formed. This is the exterior shell of the casting. As the molten metal continues to cool, the parts of the alloy with the lower freezing temperature will crystallize on the already-growing dendrites. This process continues until the metal is completely solid. The composition of the alloy within the dendrites will vary from the center to the outer edges. The center of the dendrite (first part to solidify) will have a composition corresponding to the high-freezing part of the alloy, while the outer parts of the dendrite will have a composition corresponding to the last part of the alloy to solidify. Any lead or other insoluble material in the alloy will be trapped between the dendrites. Failure to properly feed alloys having a long solidification range results in microporosity and leakage under pressure.

PATTERNS

Patterns for copper-base alloy castings should be constructed with the following points in mind:

(1) Patterns should be parted in such a manner that the important machined surfaces are in the drag.

(2) Heavy sections should be cored or altered in such a way that sections will be of uniform thickness or will be gradually tapered.

(3) Where there is a possibility of casting distortion, tie-bars or brackets should be placed on the pattern as required.

(4) Follow boards and "stop off" bracing should be used to avoid pattern breakage or warpage.

(5) Core boxes should be constructed to permit adequate venting of the cores.

(6) In constructing cylindrical bushings or circular patterns, annular risers should be made as part of the pattern.

(7) Parting lines should be made as even as possible. Flat-back patterns are preferred.

MOLDING AND COREMAKING

SAND MIXES

Various sand mixes and properties are given here as a guide for preparing sand for the different metals poured. As has been mentioned in Chapter 4, "Sands for Molds and Cores," the best properties for any sand mixture can only be obtained through proper mixing by the use of the sand muller. Also, the maintenance of properties of a sand mixture can be accomplished only through continuous and correct sand-testing procedures.

Most copper-base alloys will be cast in the all-purpose sand described in chapter 4. When time and materials are available, better results will be obtained by using the following recommended sand mixtures.

Compositions G and M. Sand used for these alloys should have properties within the following limits, depending on the size of the casting.

Grain Fineness Number	100-140
Clay Content, percent	5 for synthetic sand to 20 for natural sand
Permeability, AFS units	10-50
Green Compressive Strength, p.s.i.	4-9
Moisture, percent	5-6

The table of properties given below for hydraulic bronze can also be used as a guide for Compositions G and M.

For some types of work, a molasses water spray for the mold surface may be used. Facing sands are not generally used, but for heavy castings a plumbago coating may be used.

Hydraulic Bronze. Sand properties for castings of various weights and section thickness are given below:

Casting Weight, pounds	Section Thickness, inches	Permeability, AFS units	Green Compressive Strength, p.s.i.	Moisture, percent
Up to 1	1/2	20	7	6.5
1 to 10	1	30	7	6.0
10 to 50	2	40	7	6.0
50 to 100	3	50	8	5.5
100 to 200	4	60	10	5.5
200 to 250	5	80	12	5.5

The preceding sand properties should be used as a guide in obtaining similar properties with the all-purpose sand.

Manganese Bronze. Sand which has too high a moisture content usually results in damaged castings. To overcome this condition, the sand should be worked with moisture content on the low side. This is a good general rule to follow when casting any alloy, because water is almost always harmful to any alloy.

Typical properties of sand for manganese bronze are as follows:

Casting Weight, pounds	Section Size, inches	Permeability, AFS units	Green Compressive Strength, p.s.i.
1 to 100	1/4 to 3/4	20	8
Up to 250	3/4 to 1-1/4	40	8
500	1-1/4 to 1-3/4	60	10

Moisture contents of 5 to 6 percent are normally used for manganese bronze castings. Because of the high strength of manganese bronze, strong cores may be used with this alloy without causing excessive strains in the casting. A graphite core wash may be used to make core removal from the casting easier.

Yellow Brass. A sand having a permeability of 20, green strength of 7 p.s.i. and a moisture content of 6 percent is suitable for the majority of small castings made from yellow brass. Castings up to 50 pounds with wall thicknesses up to 1/2-inch can be made with sand having these properties.

Aluminum Bronze. Aluminum bronze alloys are difficult to cast in green sand molds because of the high drossing tendency of the alloys and the possibility of surface pinholes and porosity in the finished castings. The defects caused by high moisture in the green sand molds can be minimized by using dry sand molds. It is recommended that dry sand molds be used as described in chapter 4. Property ranges for sand mixtures are as follows:

Grain Fineness	100-160
Clay Content, percent	10-20
Permeability, AFS units	20-50
Green Compressive Strength, p.s.i.	5-12
Moisture, percent	3-6

Cupro-Nickel and Nickel Silver. The 70-30 cupro-nickel alloys and the nickel silver alloys should have a sand with a permeability between 40 and 60 and a moisture content between 4.5 and 5.5 percent. The sand grain size should be about 95 Fineness Number, with an 18 percent clay content. Nickel silvers are sensitive to gas from organic binders. Such binders, therefore, should not be used.

PROCEDURES

The procedures for coremaking, molding, and the use of washes are the same as the practices described in previous chapters. Certain procedures are repeated here to stress their importance to copper-base alloy castings.

Coremaking. Cores used in making copper-base alloy castings should be strong and well vented. Many of the castings are of such a design that considerable pressure is placed on the core during the pouring of the casting. Care must be taken, however, not to make the cores too hard or hot cracks and tears will result. Refer to Chapter 4, "Sands for Molds and Cores," for representative core mixes, and to Chapter 6, "Making Cores," for coremaking techniques.

Molding. Good molding practice as described in Chapter 5, "Making Molds," is the principal requirement when making molds for copper-base alloy castings. Extra precautions should be taken to ram the molds as uniformly as possible. Uneven ramming will cause localized hard spots and agitation of the molten metal at these points because of nonuniform permeability. In high-zinc alloys, this will cause zinc to boil out and produce rough surfaces at the areas of agitation. Alloys containing aluminum will form dross at these areas and the castings will be dirty.

Washes. Washes for copper-base alloy castings are used primarily to prevent metal penetration. The wash most generally used in plumbago. Molasses water is sometimes sprayed on the mold surface to provide a stronger bond in the surface sand. A typical molasses-water mix contains one part of molasses thoroughly mixed with 15 parts of water.

GATING

"Gating Principles," described in Chapter 7, "Gates, Risers, and Chills," should be used in the gating of Compositions G and M. Because these two alloys are tin bronzes and subject to interdentritic shrinkage, the gating system should be designed to make maximum use of directional solidification.

Heavy bushings, or "billets," can be cast by two methods to obtain directional solidification. They can be molded horizontally, as shown in figure 224, by using a thin gate to provide a choking action and causing the metal to enter the mold quietly. The mold should be tilted with the riser end lower during pouring to provide an uphill filling of the mold. The mold is then tilted with the riser up to provide maximum gravity feeding. A second method of gating bushings is to use a circular runner with pencil gates. This method, if used with a pouring temperature on the low side, will provide the best conditions for directional solidification. The cold metal will be at the bottom of the mold and the hottest metal at the top, where it will be available to feed solidification shrinkage. A third method may be used, which utilizes a tangential gate as shown in figure 225. This method permits the metal to enter the mold with the least amount of agitation, but is not so good for directional solidification of the metal.

A gating system that has proved successful for the production of valve bodies uses a sprue diameter of 1/2 inch or 5/8 inch, depending on the casting size. The runner is placed in the cope, and a reduction of at least 20 percent in cross-sectional area from the sprue to runner to gate is used.

Manganese bronze castings can be gated successfully by using a reverse horn gate into the riser and gating from the riser into the casting is illustrated in figure 226. When a number of small castings of manganese bronze or red brass are made in the same mold and gated from the same sprue and runner system, a gating system such as shown in figure 227 can be used. The runner is placed in the cope and the ingates in the drag. The castings are gated through small blind risers. The dimensions indicated for the runner and ingates show the range in sizes that can be used and depend on the size of the castings.

A gate with a large cross section (as shown in figure 228) is used for thin castings made in nickel silver. This is a plate cast in nickel silver. Notice the many ingates to permit rapid filling of the mold and also the large size of the runner. The gating system should also give uniform distribution of the metal in the mold. A gating system which resulted in a defective cupro-nickel check valve is shown in figure 229. The improved method of gating that produced a pressure-tight casting is shown in figure 230.

RISERING

Risering of copper-base alloy castings follows the same principles as described in Chapter 7, "Gates, Risers, and Chills." Castings made from Composition G, Composition M, and hydraulic bronze may give some difficulties in feeding because of their long solidification ranges that permit strong dendritic growth and make feeding of heavy sections difficult. Risers for these alloys will have to be made larger to obtain proper feeding. In connection with the risering of these alloys, it is only through experience and records of successful risering practice that correct risering procedures can be developed.

The correct placement of risers as well as the correct size is important. Good and bad risering practices are shown in figures 231, 232, 233, and 234. Figure 231 shows a globe valve that was poured without risers on the flange sections. Porosity in the casting caused low physical properties. A revised procedure is shown in figure 232. Notice that risers were used on all of the flange sections. This risering arrangement resulted in a casting with greatly improved physical properties because of the improved soundness. Similarly, the lack of risers on a high-pressure elbow, as shown in figure 233, resulted in microshrinkage in the flange section with low physical properties. The revised risering system shown in figure 234 resulted in improved physical properties in the casting.

Risers for nickel silver and cupro-nickel alloys must be large to provide enough molten metal to feed heavy sections and to compensate for the high solidification shrinkage. An example of a suitable riser is shown in figure 235. The casting is a 4-inch check valve body cast in cupro-nickel.

CHILLS

The use of chills to aid in directional solidification is described in Chapter 7, "Gates, Risers, and Chills." The procedures are generally the same for all copper-base alloy castings. Alloys having a long solidification range (such as G metal, M metal, and hydraulic bronze) require a stronger chilling action than the metals with shorter solidification ranges (manganese bronze, aluminum bronze, and yellow brass). Stronger chilling action means that if two identical castings are made, one with a long solidification-range alloy and one with a short solidification-range alloy, the long solidification-range alloy will require larger chills or chills with higher heat capacity in order to obtain the same amount of directional solidification as the short solidification-range alloy.

Recent studies have shown that special chills may be used to produce strong directional solidification in G metal. The chills used are wedge-shaped, as shown in figures 236 and 237. Their use is recommended to produce the desired directional solidification in G-metal castings. The size of the tapered chills must conform to the size of the casting. Chills 24 inches long were cut into two 12-inch pieces to prevent warping of the chills. Fifteen-inch chills were used in one piece. A general idea of the placing of the chills for flat castings can be obtained from figure 236. Figure 237 shows the use of chills on a bushing casting. Notice that the casting was top poured. The chills should not extend to the riser, because this would cause heat extraction from the riser and nullify the desired directional solidification. It is suggested that records be kept on the use of chills of this type so that effective use can be made of experience gained with their use.

VENTING

Venting procedures used for other metals and alloys are applicable to copper-base alloys.

MELTING

Copper-base alloys can be melted in any of the melting units that are to be found aboard repair ships. The melting procedures for all the units are essentially the same. Oil-fired furnaces require closer attention during melting because of the need for maintaining the proper furnace atmosphere.

When melting copper-base alloys, it is important to develop some means of determining the quality of the melt. This is best done by the use of a fracture test. Refer to Chapter 21, "Process Control," for details on developing a fracture test.

OIL-FIRED CRUCIBLE FURNACES

Any copper-base alloys that are melted in an oil-fired crucible furnace should be melted under a slightly oxidizing atmosphere. This means that at all times there must be a slight excess of air in the combustion chamber. An easy method for checking the nature of the furnace atmosphere is to hold a piece of cold zinc in the furnace atmosphere for 2 or 3 seconds. If the zinc shows a black carbon deposit when it is removed, the atmosphere is strongly reducing and more air is required. If the zinc is straw colored, the atmosphere is slightly reducing. If the zinc remains clean, the atmosphere is oxidizing. It is good practice to check the furnace atmosphere before any metal has been charged into the crucible.

General Procedure. When charging a crucible, the remelt material such as gates, risers, sprues, and scrap castings should be charged first. Ingot material may be charged on top if there is sufficient room in the crucible. Under no circumstances should any of the charge material extend above the crucible. Such conditions will permit direct flame impingement with resulting high oxidation losses and gas pick up by the metal. If the crucible is not large enough to accommodate the entire charge, the first part of the charge should be melted and the remainder added after the initial meltdown. Any ingot material that is added to the melt should be thoroughly dried and preheated.

No fluxes, glass-slag covers, or charcoal should be used at any time during melting. Experience has shown that any of these practices may lead to poor quality metal.

Procedure for Tin Bronzes. Melting should be done under oxidizing conditions as described under "General Procedure." The melt should be superheated only 25° to 50°F. above the pouring temperature and thoroughly skimmed before removal from the furnace. The melt is skimmed again if it is transferred to a pouring ladle. The melt is then flushed by plunging a piece of zinc (4 ounces for each 100 pounds of melt) deep below the surface of the melt. A phosphorizer or a pair of refractory-coated tongs are used for this purpose. Extreme care should be taken to insure that any tools used for this purpose are thoroughly dry. Moisture on the tools not only causes undesirable gassing of the metal, but also causes severe splashing of metal with danger to personnel.

The melt is allowed to stand for 2 or 3 minutes and come into equilibrium with the surrounding atmosphere. It is then deoxidized by plunging phosphor-copper into the melt (2 to 3 ounces for each 100 pounds of melt). The same precautions must be observed as when flushing with zinc. The melt is then ready for pouring into the molds.

Procedure for Manganese Bronzes and Yellow Brasses. The melt should be brought up to a temperature of about 1,800°F. to 2,000°F. in a slightly oxidizing atmosphere, the temperature at which the flaring of zinc occurs, and allowed to flare for a few minutes under a good ventilating system. The purpose of the flaring is to flush the melt with the aid of the escaping zinc vapor. Under normal operation, a flaring period of 3 to 5 minutes will result in a zinc loss of approximately one percent. Care should be taken **not** to overheat these alloys, because the zinc loss and resulting zinc fume will be a serious health hazard. After the flaring is finished, the melt should be skimmed. The crucible is then removed from the furnace and the melt skimmed again or, if the melt is poured into a ladle, it is skimmed after the transfer. Enough zinc should then be added to replace that lost by flaring. The melt should be allowed to cool to the desired temperature and poured.

Procedure for Aluminum and Silicon Bronzes. These two alloys are also melted under oxidizing conditions. The control of the furnace atmosphere is more critical than for the previously described alloys. Aluminum and silicon oxidize very easily and form dross and surface films. Therefore, the atmosphere must not have too much excess air or the dross formation and oxidation losses will be high. The melt should be superheated at least 25 to 50°F. above the pouring temperature and skimmed before removal from the furnace.

If any zinc additions are required, they are added at this time. The melt is then allowed to cool to the desired temperature and poured.

Procedure for Cupro-Nickel. Electro-nickel, electrolytic copper, 97 percent metallic silicon, and low-carbon ferromanganese are used in making up charges of cupro-nickel. Nickel-copper shot is used to make nickel additions to the base charge. Up to 50 percent remelt in the form of gates and risers can be used in the charge. This scrap should be clean. Borings and turnings should not be used.

The meltdown procedure is the same for all types of equipment available aboard ship. The nickel, copper, and iron are charged first and melted down. With the oil-fired furnace, an oxidizing atmosphere should be maintained. The melt is then deliberately oxidized with 1-1/4 ounces of nickel oxide or 3-1/2 ounces of copper oxide for each 100 pounds of virgin metal. This addition may be placed in a paper bag and stirred vigorously into the melt. If the scrap is heavily oxidized, this procedure is not necessary. After the deliberate oxidation treatment, the remelt scrap is added, melted down, and the heat brought to the desired temperature. Manganese and silicon additions are made as part of the deoxidation practice.

If the charge consists of new metal, 1-1/4 pounds of manganese and 9 ounces of silicon should be added for each 100 pounds of new metal. Final deoxidation is made with 0.025 to 0.05 percent of magnesium.

Procedure for Nickel Silver. Charges for nickel silver can be made from virgin metals such as electro-nickel, ingot copper, tin, lead, and zinc; 50-50 nickel-copper alloy, ingot copper, tin, lead, and zinc; from commercially prepared ingot.

For a virgin metal heat, the copper is charged first, the zinc next, and the nickel last. Remelt may be added on top or added to the heat as it settles during melting. In crucible melting, a charcoal or glass-slag cover may be used. The heat is brought to the desired temperature and the remainder of the zinc is added and stirred into the melt. The lead is then added, followed by the tin. All additions should be thoroughly stirred into the melt. The heat is then ready for deoxidation.

The recommended deoxidation practice for nickel silver is to add 0.10 percent of manganese (1-1/2 ounces for each 100 pounds of melt) 5 to 7 minutes before pouring. This is followed by 0.05 percent of magnesium (3/4 ounce for each 100 pounds of melt) 3 to 5 minutes before pouring, and 0.02 percent of phosphorus, as 15 percent phosphorcopper (2 ounces for each 100 pounds of melt) immediately before pouring. The phosphorus deoxidation may be done in the pouring ladle. If phosphorus is used, a check should be maintained on the scrap, if at all possible, to make sure there is not a buildup of phosphorus in the circulating scrap.

INDIRECT ELECTRIC-ARC FURNACE AND RESISTOR FURNACE

The melting procedures in these furnaces are the same as far as the handling of the melt is concerned. The indirect electric-arc furnace requires much closer control than the resistor furnace. Poor arc characteristics in the indirect-arc furnace will cause a highly reducing atmosphere that causes silicon to be picked up from the furnace lining and contaminate the melt. The furnace must be maintained in proper condition at all times when melting copper-base alloys. Refer to Chapter 8, "Description and Operation of Melting Furnaces." A smoky operation is sure to be reducing and will produce metal of low quality.

A factor that is of major importance in both types of furnaces is the proper drying of linings and patches. Copper-base alloys are very easily gassed and moisture in the lining is a major source of gassing troubles.

General Procedure. Any scrap material charged into the indirect electric-arc or resistor furnaces should be as free as possible of dirt and sand. Preferably all scrap should be sand blasted to clean it. Sand in particular will cause a slag blanket that will increase the melting time and make handling of the heat more difficult. Heavy pieces of scrap should be charged to the rear of the barrel with ingots on top and close to the arc.

Additions of zinc, tin, and lead should be made as new metals in the order mentioned approximately 3 to 5 minutes prior to tapping. The additions should compensate for any shortages in the desired analysis and any melting losses of zinc and lead. One quarter of one percent (0.25 percent) of the total charge for zinc and lead is usually sufficient to compensate for melting losses.

When melting tin bronzes, aluminum bronzes, or silicon bronzes in these furnaces, it is important to maintain the proper amount of oxygen in the bath in order to prevent gassiness caused by hydrogen. Opening of the charging door or blowing of air into the furnace is poor practice because this causes increased electrode consumption. A better method of obtaining oxygen in the melt is to use copper oxide.

Deoxidation Procedures. These are the same as described under the procedure for the oil-fired crucible furnace.

ELECTRIC INDUCTION FURNACE

The melting procedure is essentially a crucible process. The heat is generated entirely in the charge itself, melting is rapid, and there is only a slight loss of the oxidizable elements. Furthermore, on account of the rapidity of operation, preliminary bath analyses are not usually made. The charge is preferably made up of carefully selected scrap and alloys of an average composition to produce as nearly as possible the composition desired in the finished metal. Final additions are made to deoxidize the metal or to adjust composition, as for the other melting methods just described.

General Procedure. The heavy scrap is charged first and as much of the charge as possible is packed into the furnace. The current is turned on and, as soon as a pool of molten metal has formed in the bottom, the charge sinks and additional scrap is introduced until the entire charge has been added. The charge should always be made in such manner that the scrap is free to slide down into the bath. If the pieces of the charge bridge over during melting and do not fall readily into the molten pool, the scrap must be carefully moved to relieve this condition. Rough poking of the charge must be avoided at all times, however, because of danger or damaging the furnace lining. Bridging is not serious if carefully handled but, if allowed to go uncorrected, overheating of the small pool of metal may damage the lining seriously and will have an undesirable effect on the composition of the metal. The molten metal in the crucible below the bridged charge will become highly superheated with a resulting loss in the lower melting metals such as zinc and lead. There is no way of determining the metal loss when such a condition occurs. When loosening the bridged charge material, extreme caution should be observed, and the charge should never be forced down into the crucible in an effort to loosen the bridge. Forcing the charge may result in a cracked or broken crucible with a resulting run-out of molten metal and damage to the furnace coil.

Safety precautions should always be observed when holding or melting molten metals. Protective eye and face shields and safety clothing should be worn at all times.

The compactness of the charge in the furnace has an important influence on the speed of melting. The best charge is a cylindrical piece of metal slightly smaller in diameter than the furnace lining. This will draw very close to the full current capacity of the equipment. Two or three large pieces with considerable space between them will not draw maximum current because the air cannot be heated by induction. The charge should not be so tightly packed that it cracks the crucible or lining when it expands during heating.

As soon as the charge is completely melted and refining or superheating operations finished, further necessary additions of alloys or deoxidizers are made. The furnace is then tilted to pour the metal over the lip. If the entire heat is poured into a large receiving ladle, the power is turned off before tilting. If, however, the metal is taken out in small quantities in hand ladles, reduced power may be kept on while pouring. This maintains the temperature of the bath and facilitates slag separation by keeping it stirred to the back of the bath. When the heat is poured, the furnace is scraped clean of adhering slag and metal and is then ready for the next charge.

It is important that only similar metals be melted in the same lining or crucible. When melting cast iron or steel, the lining absorbs iron. Brass or bronze melted in the same lining will become contaminated with iron. The reverse will also be true. Cast iron or steel can become contaminated with copper, tin, or zinc. If it ever becomes necessary to melt different metals in the same furnace, a wash heat similar in composition to the next heat planned can be used to cleanse the crucible. It is always better practice to have separate furnaces or crucibles for different types of metals that may be required.

Most metals have a tendency to absorb gas and oxidize upon heating. Gas absorption and oxidation increase with time and temperature, with the largest increase occurring at the melting point of the metal, and continuing to increase as the temperature increases. The possibilities of gas absorption in induction-melted heats are less than for other types of furnaces because combustion products are absent. Nevertheless, to minimize hydrogen pickup it is important that metals be melted as rapidly as possible and be held no longer than necessary in the molten state after the desired temperature is attained.

By selective arrangement of the charge, melting time and metal loss can be kept at a minimum. The usual procedure when using new virgin metals is to charge the base metal or the higher melting metals first. Best results are obtained when the crucible is filled to capacity with the larger pieces placed on the bottom. If scrap makes up a proportion of the charge, it may be added with the base metal. Otherwise, it should be charged after the base metal is melted to cut down metal loss and prevent overheating. Scrap melts faster because it usually has a lower melting point than the pure base metal and also has greater electrical resistance. Alloying additions should be made gradually in amounts small enough to allow rapid solution.

Alloying. Elements may be added as commercially pure metals or as master alloys (hardeners). The lower melting metals used for alloying present little difficulty because they are molten at temperatures near those of the base metals. However, such metals as iron, manganese, silicon, nickel, and copper present a problem because of their relatively high melting points. In most cases, it is undesirable to heat the base metals to temperatures necessary to effect rapid solution of the higher melting metals. For this reason, master alloys with their lower melting ranges are used. Because of the ease with which castings can be made from prealloyed ingots of the proper composition, it is a good idea to obtain special ingots of the compositions that will probably be needed.

PROCEDURES FOR MELTING BRONZES

G Bronze. The nominal composition of this bronze is 88 percent copper, 8 percent tin, and 4 percent zinc. A slightly oxidizing atmosphere is preferred during melting.

The copper is melted and heated to approximately 2,000°F. Tin is then added. In adding the zinc, the power should be turned down, the melt cooled nearly to freezing, and the zinc held beneath the surface of the bath with an iron rod to prevent excessive loss. It is customary to add from 2 to 5 percent more zinc than is desired in the final composition to compensate for loss by oxidation. The melt should be superheated 75°F. above the desired pouring temperature. Before pouring, the bronze should be skimmed and deoxidized with 2 or 3 ounces of 15 percent phosphor-copper per 100 pounds of metal and stirred well. The pouring range of this alloy is between 2,000°F. and 2,200°F., depending upon the section size of the casting, thin sections requiring hotter metal.

M Bronze. This alloy has a nominal composition of 88 percent copper, 6.5 percent tin, 1.5 percent lead, and 4 percent zinc. The melting procedure is the same as that for G bronze. Lead is added after the tin, and the melt should be stirred thoroughly by an acceptable stirring rod. The deoxidation practice and pouring range for this bronze are the same as for Composition G.

Manganese Bronze. The nominal chemical composition of this alloy is 58 percent copper, 1.0 percent aluminum, 0.5 percent manganese, 1.0 percent iron, 0.50 percent tin, and remainder zinc. The manganese, iron, and aluminum are most easily added as master alloys. When the copper is at approximately 2,000°F., the copper-manganese, iron, and aluminum master alloys are added. The metal should then be allowed to cool sufficiently to dissolve the zinc without flaring. The zinc loss is about 1 percent. The metal should be poured between 1,900° and 1,975°F., depending upon the size of the casting.

TEMPERATURE CONTROL

Temperatures should be measured with immersion-type pyrometers. The instruments should be maintained in proper operating condition. The immersion end of the pyrometer should be cleaned of any adhering metal or dross before taking readings. The power should preferably be shut off when taking a temperature reading in an induction furnace. Otherwise, a faulty reading may be obtained.

As an emergency measure, the power input for the electrical furnaces may be used for estimating temperatures of the melt if adequate records have been kept on previous heats. Temperatures obtained with the pyrometer and the power input of the furnace should be recorded for various heats and used as a reference in estimating temperatures. This should not be made a general practice but should be used only in an emergency when no pyrometer is available.

POURING

Copper-base alloys require the same precautions in pouring as do any of the other alloys handled aboard repair ships. The high-zinc alloys, aluminum bronze, and silicon bronze in particular, require added attention to pouring techniques. Any agitation of these alloys will result in poor castings. High-zinc alloys will lose zinc in the form of zinc vapor if agitation occurs during pouring. Agitation of aluminum bronze during pouring produces dross, which is trapped to produce defective castings. Silicon bronzes form a skin that if agitated during pouring, will act similar to dross and result in a defective casting.

In the pouring of copper-base castings, extreme care should be taken to prevent agitation of the molten metal and to insure a quiet stream of metal entering the casting. Refer to Chapter 9, "Pouring Castings," for information on proper pouring techniques.

Typical pouring temperatures for the various alloys are listed in table 25. It must be remembered that these temperatures are suggested pouring temperatures, and actual experience aboard ship may indicate that pouring temperatures different from those listed are more satisfactory. This is one of the reasons why it is helpful to keep records of castings made aboard ship and to record information such as the pouring temperatures used.

TABLE 25. TYPICAL POURING TEMPERATURES FOR COPPER-BASE ALLOYS

Alloy	Average Section Size of Casting		
	Light, less than 1/2 inch	Medium, 1/2 to 1-1/2 inches	Heavy, over 1-1/2 inches
Composition G Composition M Hydraulic bronze	2200°F.	2150°F.	2050°F.
Yellow brass Naval brass Commercial brass Manganese bronze	2000°F.	1900°F.	1850°F.
Aluminum bronze	2300°F.	2200°F.	2100°F.
Silicon bronze	2200°F.	2100°F.	2050°F.
Cupro-Nickel Nickel Silver	2800°F. 2500°F.	2700°F. 2400°F.	2650°F. 2350°F.

CLEANING

Copper-base alloy castings do not present any problems in cleaning. Any sand adhering to the casting can be easily removed with a wire brush or by grit or sand blasting.

CAUSES AND CURES FOR COMMON CASTING DEFECTS IN COPPER-BASE CASTINGS

Copper-base alloy castings develop the same defects as other types of castings. For descriptions of these defects and their cures, see Chapter 11, "Causes and Cures For Casting Defects."

METAL COMPOSITION

There are various elements that, either by their presence alone or because of an excess, are detrimental to the physical properties of copper-base alloy castings.

Iron can be tolerated up to 0.25 percent in tin bronze and red brass. A higher iron content causes harder and more brittle alloys and hard spots. Large amounts of iron in manganese bronze reduce corrosion resistance.

Sulfur in small quantities has little effect on the strength of red brass or tin bronze. Too much sulfur decreases fluidity, produces excessive dross, and may cause dirty castings.

Phosphorus in excess increases the fluidity and may result in severe metal penetration into the sand. In aluminum bronze, it produces embrittlement.

Antimony in excess of specification requirements usually results in a weakened alloy. In excess of 0.1 percent in yellow brass it causes hot shortness.

Zinc in excess of specification requirements produces hardness and brittleness in manganese bronze.

Aluminum is detrimental to red brass and sometimes causes lead sweat. Its presence should be avoided for pressure castings. In combination with lead, it weakens tin bronze.

Silicon in excess of 0.05 percent causes embrittlement in aluminum bronze. In combination with lead it weakens tin bronze.

In short, use the specified compositions. Deviations from them are an invitation to trouble. Remember that the specifications have been worked out over many years.

POURING

Metal that has been poured too hot may produce cracks in the side walls of castings. This crack can be orange, yellow, or golden red in color and will leak under pressure. Such a crack is a good indication that the metal was poured approximately 200°F. above the proper pouring temperature.

Improperly dried ladles or furnace linings cause their own type of porosity in copper-base castings. Holes caused by improperly dried ladles will have a color ranging from orange to yellow and will be associated with a gray to yellow crystalline fracture of the casting in the vicinity of the holes. The casting will generally leak under pressure. The cure for this type of defect is to dry the ladles, crucibles, and linings thoroughly before use.

Dross inclusions in copper-base castings have the appearance of fins when occurring in side walls and have a red and green color when fractured. Proper skimming of the ladle followed by complete filling of the sprue with a steady uninterrupted stream of metal are the cures for this defect.

MELTING PRACTICE

Gas holes in castings are usually caused by a reducing atmosphere instead of a correct oxidizing atmosphere. Hydrogen is dissolved by the metal under reducing conditions and produces the gas holes during the solidification process. The defects may occur as rounded gas holes or may be present as microporosity. Melts that are gassy will not shrink normally in a riser or sprue. This is illustrated in figure 238. Note that the gassy metal has a dome shaped surface (sample on left), while the gas-free metal showed a conventional pipe. If the metal isn't fed from the riser, there must be holes or porosity in the casting. Remember that all of these alloys shrink when they solidify. Gas holes in copper-base castings will be colored brown, red, orange, or yellow gold.

An evenly distributed porous structure is caused by overheating and soaking the metal for too long a time. It is a type of gas defect that is usually associated with correct shrinkage in the sprue and riser, followed by the ejection of a small globule of metal. The correct pouring temperature and proper temperature control are the cures for this defect.

MISCELLANEOUS

Veining in copper-base castings is usually caused when the lower melting constituent of the alloy penetrates into cracks that have occurred in weak cores. The casting may show a porous structure in the vicinity of the veining. This defect is usually caused by a core mix with a low hot strength and can be cured by using a core mix with a higher hot strength. This usually can be obtained with additions of clay, silica flour, or iron oxide.

Tin sweat is a defect that is found in high-tin copper-base alloys. A high gas content in the melt causes pressure that forces the low-melting-point tin-rich part of the alloy to the outside of the casting through interdendritic spaces. The "sweated" metal occurs as small droplets on the surface of the casting. Correct melting practice and proper degassing procedures are the cures for this defect.

WELDING AND BRAZING

Copper-base alloys have a very high heat conductivity. As a result, brazing of these alloys is difficult and should be done only by trained personnel. When repairs by welding or brazing are required, refer to the "General Specifications for Ships of The United States Navy," Section S9-1, "Welding," for general guidance.

SUMMARY

Copper-base castings make up a major part of the foundry work that is done aboard repair ships. It is necessary for molders to become familiar and proficient with castings made from these alloys. Various castings made from these alloys are very often repeaters and the maintaining of records on sand and core mixes used, gating and risering arrangements, and any measures taken to correct specific defects will prove helpful in reducing the time required to produce a good casting.

The important points to consider in melting copper-base alloys are as follows:

1. Use clean uncontaminated crucibles.

2. Use clean uncontaminated melting stock.

3. Melt under oxidizing conditions.

4. Melt rapidly.

5. Do not use excessive superheat. Heat only as hot as necessary.

6. Do not hold the metal at high temperatures.

7. Pour the casting as soon as possible after the metal is melted.

8. Skim carefully and avoid agitation.

9. Allow metal to cool to pouring temperatures in the open air. Do not use cold metal additions to reduce the temperature.

10. Use a properly maintained and calibrated pyrometer.

11. Use deoxidizers only in recommended amounts.

12. Do not agitate or stir the melt immediately before pouring.

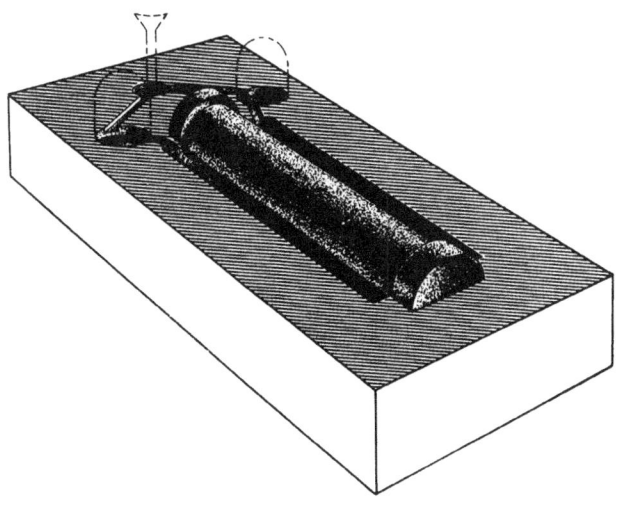

Figure 224. Horizontal holding of a bushing.

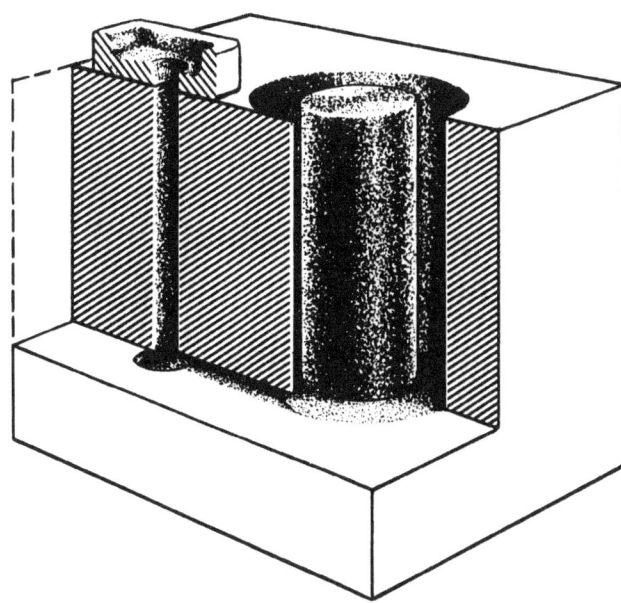

Figure 225. Vertical molding of a bushing.

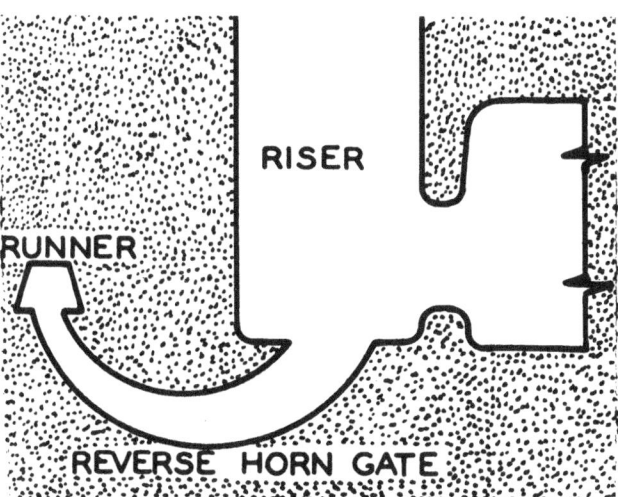

Figure 226. Gating a manganese bronze casting.

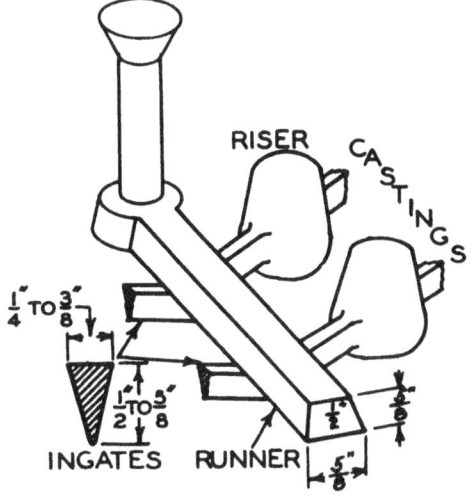

Figure 227. Gating a number of small castings in manganese bronze or red brass.

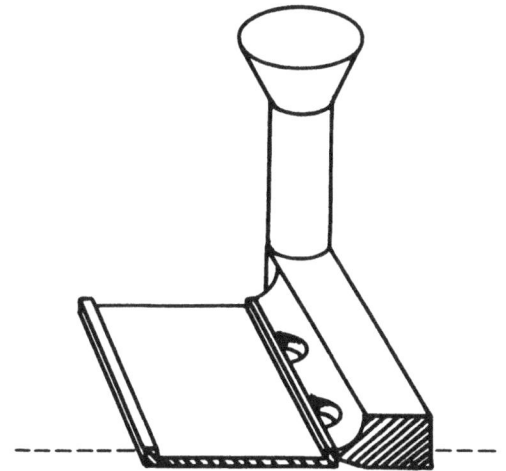

Figure 228. Gating for a thin nickel-silver casting.

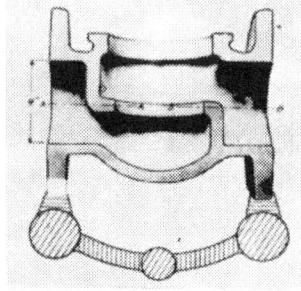

Figure 229. Poor gating system for a cupro-nickel check valve.

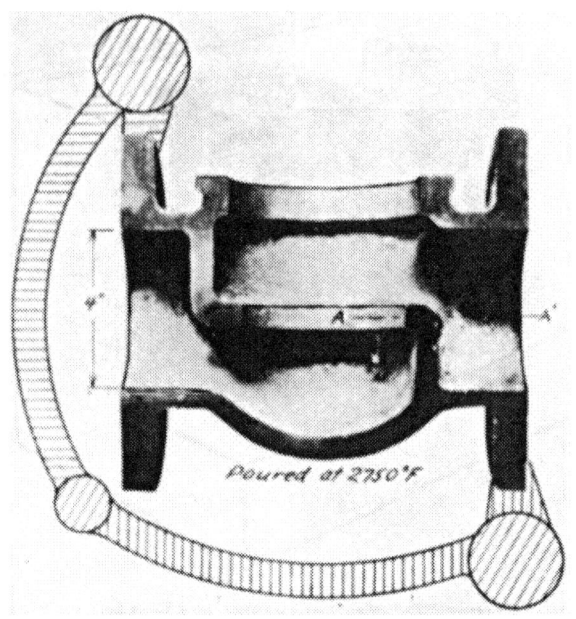

Figure 230. Improved gating that produced a pressure-tight casting.

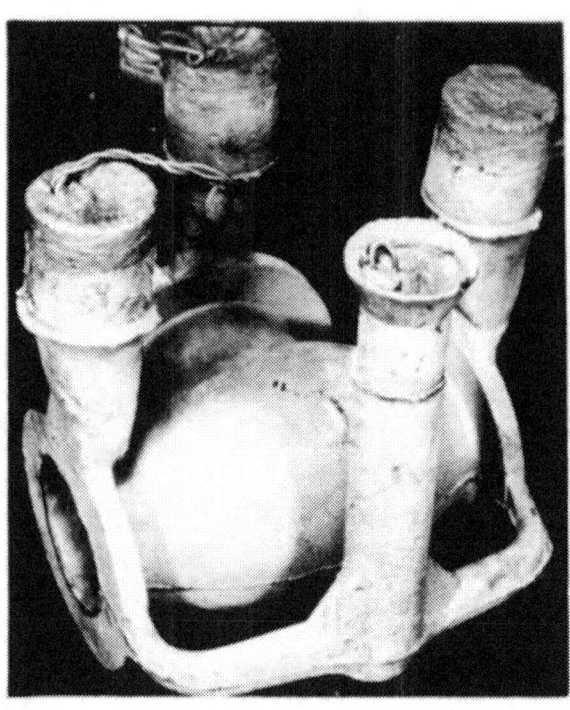

Figure 232. Globe valve - improved risering practice

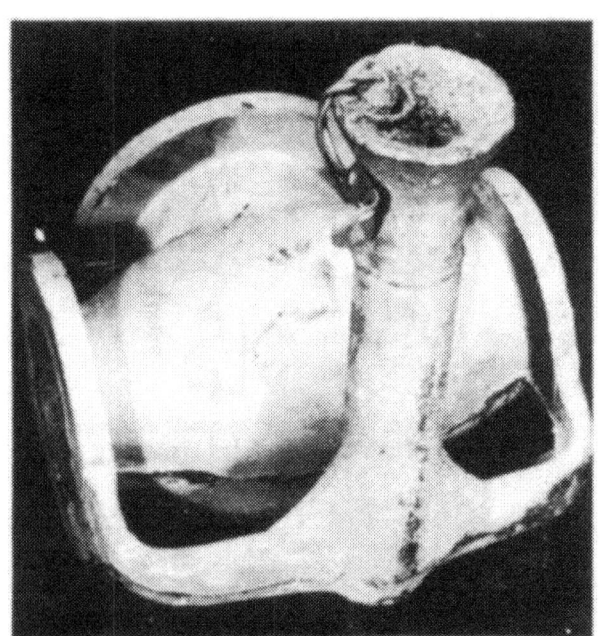

Figure 231. Globe valve - poor risering practice.

Figure 233. High pressure elbow - poor risering practice.

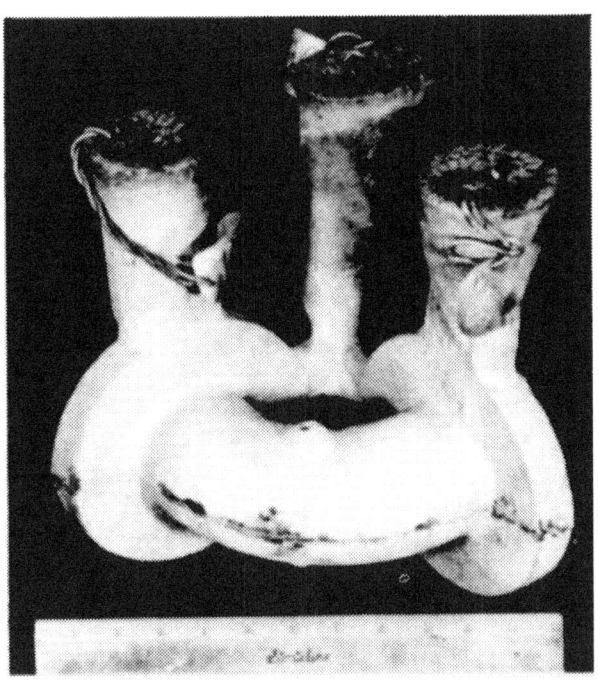

Figure 234. High pressure elbow - improved risering practice.

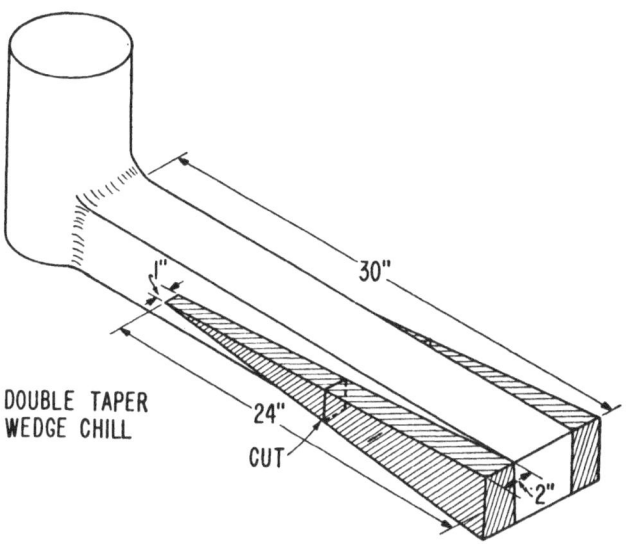

Figure 236. Tapered chills on a flat G metal casting.

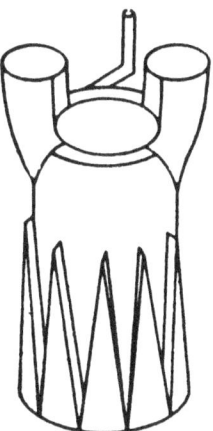

Figure 237. Tapered chills on a G metal bushing.

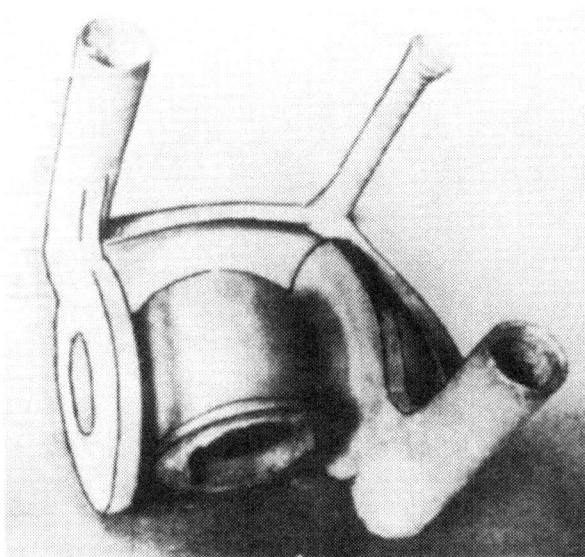

Figure 235. Risers for a cupro-nickel valve body.

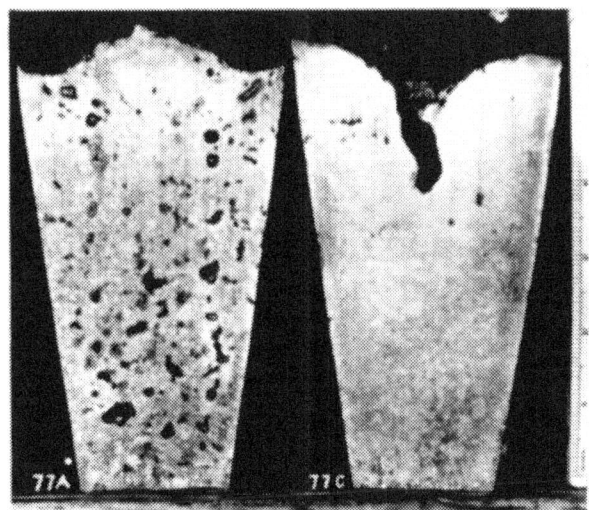

Figure 238. Examples of gassy and gas-free metal.

Chapter XV
ALUMINUM-BASE ALLOYS

The two principal aluminum alloys; used for making castings for shipboard use are the aluminum-silicon alloys and the aluminum-copper alloys. An aluminum-zinc alloy is also available. Refer to table 2, Chapter 13, "Compositions of Castings."

Aluminum-base alloys are used because of their light weight, good corrosion resistance, good machinability, and each castability. Their particular characteristics that affect foundry practice are:

1. Light weight and high shrinkage call for well-vented and permeable molds that are rammed fairly soft to vent gases and give little resistance to contraction of the casting.

2. Ease of absorption of hydrogen requires proper melting practice (oxidizing atmosphere) and close control of moisture in all forms.

3. Ease of reaction with oxygen promotes a dross or scum that must be properly removed and must be minimized by proper gating and pouring practice.

4. Gas absorption calls for pouring at the lowest possible temperature.

ALUMINUM SILICON

The aluminum-silicon alloys have very good casting properties. The fluidity (castability) increases as the silicon is increased. The tendency to hot tearing (which is characteristic of many aluminum alloys) is reduced by the silicon additions. These alloys are well suited for castings which must be pressure tight. Corrosion resistance is very good. The aluminum-silicon alloys do not machine as easily as some of the other aluminum alloys; carbide tools are recommended for the alloys with high silicon content. These alloys have been generally replaced with aluminum-silicon-magnesium alloys or aluminum-silicon-copper-magnesium alloys.

ALUMINUM COPPER

The addition of copper to aluminum increases its strength and hardness. These are the main advantages of the aluminum-copper alloys. The disadvantages of these alloys is that they are more susceptible to hot tearing and that they have a relatively low resistance to corrosion. Machinability of the aluminum-copper alloys is good. These alloys have been generally replaced by the aluminum-silicon-copper alloys.

ALUMINUM ZINC

This alloy has high strength and ductility without heat treatment. However, these properties can be improved by room-temperature aging or by heat treatment. Its machining properties are good. It is somewhat more difficult to cast than are the aluminum-silicon or the aluminum-copper alloys.

ALUMINUM MAGNESIUM

This alloy has the best strength and corrosion resistance of any aluminum alloy when properly cast and heat treated. Heat treatment of the 10 percent magnesium alloy is mandatory but the 7 percent magnesium alloy is used as cast. These alloys are difficult to cast properly.

HOW ALUMINUM SOLIDIFIES

Aluminum solidifies in the manner described in Chapter 1, "How Metals Solidify." Its solidification shrinkage is high. Solidification starts at the mold and core surfaces and proceeds inward. Aluminum, however, is an excellent conductor of heat. Within a short time after a casting is poured, all of the molten metal has cooled to near the solidification temperature. Nucleation and growth of crystals then start throughout the melt and solidification is rapid. This makes it difficult to feed a casting. Large risers and strong directional solidification are needed. The addition of copper and silicon to aluminum (as in the alloys mentioned) increases the fluidity and makes feeding easier. The aluminum-zinc alloy has a greater shrinkage than the aluminum-silicon or aluminum-copper alloys. For easiest castability, use the aluminum-silicon alloys, if they are otherwise suitable for the purpose intended.

PATTERNS

Patterns used in making molds for aluminum castings are the same as for other metals. Patterns should be checked to make sure that any section junctions are properly blended as shown in figure 17, Chapter 2, "Designing a

Casting." Blending of junctions is especially necessary in aluminum-alloy castings because they are highly susceptible to hot tearing.

MOLDING AND COREMAKING

Molding and coremaking procedures are generally the same as for other metals. See Chapter 5, "Making Molds," and Chapter 6, "Making Cores."

SAND MIXES

Aluminum-alloy castings can be made in either a natural sand (such as Albany sand) or in a synthetic sand. Synthetic sand will usually produce a sounder casting, but the natural sand will usually give a smoother casting surface.

Properties of a typical natural sand for aluminum are as follows:

Green Compressive Strength, p.s.i.	6-8
Permeability, AFS units	5-15
Clay Content, percent	15-22
Grain Fineness Number	210-260
Moisture Content, percent	6-7

A sand of this type must be properly reconditioned and the moisture content closely controlled. Negligence will lead to blows and gas porosity. Although one advantage of a natural sand is that it can be reconditioned by shoveling, the use of a sand muller will give more uniform mixing and extend the quality and usefull life of the sand.

Synthetic sands used for aluminum-alloy castings should have properties in the ranges listed as follows:

Green Compressive Strength, p.s.i.	6-10
Permeability, AFS units	25-120
Clay Content, percent	4-10
Grain Fineness Number	70-160
Mositure Content, percent	3-5

This type of synthetic sand *must* be reconditioned with a sand muller in order to obtain the best properties. Manual conditioning of synthetic sand is not recommended.

PROCEDURES

Coremaking. Coremaking for aluminum castings follows the same techniques as used for other metals except that a minimum amount of binder should be used to give the weakest core that can be handled. Binders should also be maintained at the proper amount to prevent any excessive gas generation. Aluminum and its alloys are sensitive to gas pickup. The cores, therefore, should be thoroughly dried and well vented to prevent any defects from moisture.

Molding. Because aluminum has a density approximately one-third that of brass, cast iron, or steel, it creates a much lower pressure than the other metals against the mold wall. Because of this, the mold can be rammed lighter. Light ramming is particularly important when casting the aluminum alloys that are hot-short or susceptible to hot tearing. Light ramming permits the casting to contract more easily and avoids hot tears. The placing of gates and risers must also be planned so that they do not hinder the contraction of the casting while it is cooling.

Washes. Refractory washes and sprays may be used on mold and core surfaces, but any organic materials should be avoided because they are sources of gas. Commercial washes of alcohol and clay are generally used on cores. Even though these washes contain alcohol as the liquid, the cores should preferably be dried in the core oven after they have been coated.

GATING

Aluminum and its alloys have the disadvantage that they form dross (metal oxides or scum) if they are agitated when molten. To eliminate or minimize drossing, the gating system must be designed to conduct the molten metal to the mold cavity as quietly as possible and with as little agitation as possible.

For small castings, the casting should be gated with a singe gate on one side. Larger, simple shaped castings and castings of complicated shapes will have to be gated from more than one side with multiple gating systems. However, it must be remembered that the metal must be poured hotter with the multiple gating system to prevent cold shuts and laps in the larger castings.

Studies in recent years have shown that if the sprue and gating system can be filled quickly, sucking in of air and the resulting formation of dross can be minimized or even eliminated. It is also necessary to avoid abrupt changes in the direction or cross-sectional area of the gating system. This principle seems to be violated at the junction of the sprue and runner, but this is necessary in order to distribute the metal and to reduce its velocity.

The best results can generally be obtained from a sprue that has a reduction in cross-sectional area of 3 to 1. In other words, the top of the sprue has an area three times larger than the bottom of the sprue. Sometimes a square or rectangular sprue may reduce turbulence more than a round sprue, but this must be determined by experiment. Usually a reduction of turbulence at the start of the pouring operation is obtained by the choking action at the base of the sprue and rapid filling of the gating system. Two sprue-base designs have been developed to reduce turbulence and aspiration of air caused by the abrupt change in direction at the base of the sprue. The first is known as the enlargement type of sprue-base design, and the second as the well type.

The enlargement type of sprue base is shown in figure 239. The runner is enlarged below the base of the sprue. This arrangement reduces the velocity of the molten metal as it passes from the sprue to the runner. The diameter of the enlargement should be approximately 2-1/2 times the width of the runner. This type of sprue base is most effective with a narrow deep runner.

The well-type sprue base is shown in figure 240. The best results with this design were obtained when the area of the well was approximately five times the area of the sprue base and its depth was about two times the depth of the runner. This type of sprue base is most effective with wide shallow runners and with square runners.

In the runners and gates, turbulence can be avoided by rounding the corners with as large a radius as possible. As each gate is passed, the cross-sectional area of the runner should be reduced by the area of the gate passed. For example, if the runner ahead of the first gate is 1/2 x 1 inch (0.5 square inch) and the first gate is 1/2 x 1/4 inch (0.125 square inch), the runner past the first gate should be 0.5 minus 0.125 or 0.375 square inch (say, 3/8 x 1 inch). This type of gating will keep the runner system full of molten metal and distribute it uniformly to the mold. If the runner system is not reduced in area as described, the gates farthest from the sprue will carry most of the metal. As general practice, or as a starting point for planning the gating system, the total area of the gates should be equal to (or slightly larger than) the cross-sectional area of the runner between the first gate and the sprue. The cross-sectional area of the runner between the sprue and the first gate should be two to three times greater than the cross-sectional area of the small end of the sprue.

The relationship between the cross-sectional areas of the gating system is called the gating ratio, and is expressed as three numbers. As an example, a gating ratio of 1:4:4 means that the area of the gates is the same as the area of the runner, and the area of all the runners (or all the gates) is four times as large as the cross-sectional area of the small end of the sprues.

The size and location of gates is determined by the size of the casting, wall thickness, and weather it is a flat plate or chunky. For general usage, the following may be used in locating gates: (1) the width of the gate should be approximately three times the thickness of the gate, (2) the thickness of the gate at the mold cavity should be slightly smaller than the thickness of the casting at that point, (3) the gate should be slightly longer than it is wide, and (4) the spaces between the gates should be approximately twice the width of the gates. This information is given to show good gating procedure. The rules are not hard and fast, but the molder should have a good reason for changing them.

Large castings should be gated at the bottom to insure a minimum of turbulence. In small castings where the drop from the parting line to the drag side of the mold is four inches or less, parting-line gates may be used.

The runner should be in the drag and gates in the cope. The gating system will fill with metal before flowing into the mold cavity, trapping the dross against the cope surface, and result in more uniform distribution of metal. Wide flat runners and ingates provide more cope surface for the trapping of dross. This is the major difference between gates for light and heavy metals.

Risers. Aluminum alloys have a high solidification shrinkage and must be risered properly to prevent any defects due to shrinkage. Risers should be used to obtain directional solidification as much as possible. Gating through risers so as to have the last and hottest metal in the riser should be used wherever possible. Refer to Chapter 7, "Gates, Risers, and Chills."

Chills. Strong directional solidification is difficult to obtain in aluminum alloys without the use of chills. The tendency for solidification to start throughout the metal makes proper feeding difficult. Chills must often be used to obtain satisfactory directional solidification.

External chills for aluminum castings can be made from cast iron, bronze, copper, or steel. They should be clean. Chills are occasionally coated with plumbago, lampblack, red oxide, or other compounds to prevent the cast metals from sticking to them, but this procedure is generally not necessary. Organic coatings should never be used on chills for aluminum alloys. The chills should be absolutely dry (and preferably warm) before being set in the mold.

Vents. Aluminum is approximately one-third the weight of cast iron, steel, or bronze. This low density makes it more difficult for aluminum to drive mold gases or air from the mold cavity. The mold, therefore, must be thoroughly vented to permit easier escape of the gases. This is the reason that molding sands for aluminum must have a high permeability.

MELTING

TYPE OF FURNACE

Aluminum and its alloys can be melted only in the oil-fired crucible furnace, or high-frequency induction furnace. The oil-fired furnace has the disadvantage of exposing the molten aluminum to the products of combustion in the furnace atmosphere. Close control and constant attention are required during melting in these units.

PROCEDURE

Charges for aluminum heats should be made of ingot material and foundry remelt. Machine turnings and borings should not be used. It is difficult to clean turnings and borings so as to remove any oil, and the large surface area of the chips causes high oxidation losses. If turnings and borings must be used, they should be thoroughly cleaned, melted down, and poured into ingots, and then the ingots used as charge material. Such a procedure will reduce the gas content of the material and eliminate the large amount of dross which would result from direct use of the turnings.

Oil-Fired Crucible Furnace. Aluminum alloys can be melted in graphite or silicon carbide crucibles. The crucibles must be kept clean to avoid contamination. Sometimes cast iron pots are used, but they require a refractory wash to prevent pickup of iron by the aluminum. The pots must be thoroughly cleaned before the protective coating is applied. A coating for iron pots can be made from seven pounds of whiting mixed with one gallon of water. A small amount of sodium silicate may be added to provide a better bond. The pot should be heated to a temperature slightly above the boiling point of water and the wash applied. Care must be taken in charging these pots in order to avoid chipping of the coating, which will expose the iron.

The metal charge should not extend above the top of the crucible. Such charging practice will result in high oxidation losses and severe drossing. It is better to melt down a partially filled crucible and then charge the remainder of the cold metal.

The furnace atmosphere should be slightly oxidizing to prevent excessive absorption of gases by the melt. Hydrogen is the gas that is most harmful to aluminum. Hydrogen is dissolved by the aluminum and produces gas defects if it is not removed or permitted to escape. Oxygen combines with aluminum to form the familiar dross which is easily removed.

High-Frequency Induction Furnace. Melting practice in the high-frequency induction furnace is essentially the same as that described for the oil-fired furnace. Temperature control is much easier with an induction furnace because the temperature will rise only slightly after the power has been shut off.

TEMPERATURE CONTROL

The temperature of molten aluminum cannot be determined by visual observation, as is sometimes done with iron and steel. It is necessary to use an immersion pyrometer for temperature readings. Without a pyrometer, overheating would probably result and produce a gassy heat or a casting having a very coarse grain structure. Remember that a melt of aluminum alloy that is overheated is usually permanently damaged. Merely cooling the metal back to the proper temperature will not correct the damage.

An immersion Chromel-Alumel pyrometer is satisfactory for temperature measurements in molten aluminum. The thermocouple should be protected with a cast iron tube coated with the same wash as described for iron melting pots. If temperatures are to be taken in the ladle, an open-end thermocouple should be made from 8-gage asbestos-covered Chromel-Alumel wires. While the thermocouple is in the melt, it should be moved with a slow circular motion.

DEGASSING

Because aluminum and its alloys absorb gases so readily, the proper removal of gases is an important step in preventing defective castings. Degassing of aluminum can be accomplished with gaseous fluxes.

Aluminum should be degassed with dry nitrogen. Chlorine, either solid or gaseous, should not be used in shipboard foundries. A carbon or graphite pipe is connected to the tank of compressed gas with suitable rubber hoses. When the metal temperature reaches about 1,250°F., the gas should be turned on and the preheated tube inserted in the metal to the bottom of the crucible. The flow of gas should be adjusted to produce a gentle roll on the surface of the metal. Fluxing times should be from 10 to 15 minutes for a 100-pound heat. (Differences in size of heat will not change the fluxing time appreciably.) Temperature control will generally require the fuel to be shut off as soon as the charge has melted. The metal temperature should never be allowed to go over 1,400°F. After the fluxing operation, the surface of the melt should be skimmed and the metal poured.

POURING

The pouring temperature and method of pouring determine whether a properly melted heat and a properly made mold will produce a good casting. Aluminum and its alloys should be poured at as low a temperature as possible without causing misruns. For any given alloy, the pouring temperature will determine whether a casting will have a fine grain structure and good properties or a coarse grain structure and lower properties. A high pouring temperature will tend to give a large grain size, and a low pouring temperature will tend to give a small grain size. The pouring temperature will vary between 1,240°F. and 1,400°F., depending on the alloy and section size of the casting. If a casting poured at 1,400°F., misruns, the gating should be revised to allow faster pouring.

Because aluminum absorbs gases easily, pouring should be done with the lip of the ladle as close as possible to the sprue of the mold. The stream of molten metal should be kept as large as possible (as large a stream as the sprue will handle). A thin stream or trickle of molten metal from a ladle that is held high above the mold will cause a gas pickup and unnecessary agitation of the metal.

CLEANING

Aluminum-alloy castings are easy to clean. Burn-in is absent and any loosely adhering sand can be easily removed with a wire brush. Gates and risers can be removed with hack saws, band saws, or by chipping hammers. When using chipping hammers, the cutting should be done slightly away from the surface of the casting to prevent breaking into the casting. The remainder of the metal can then be removed by grinding. After any cutting operation for the removal of gates and risers, grinding is used for a rough finish. Final finishing can be done with a single deep-cut coarse curved-tooth file.

Grit or sand blasting is useful for cleaning up the casting surface and gives it a pleasing appearance. Unsoundness that may be present just below the surface of the casting is usually revealed by a blasting operation. Grit or sand blasting is also useful for removing minor surface roughness and burrs from a casting.

CAUSES AND CURES FOR COMMON DEFECTS IN ALUMINUM CASTINGS

Aluminum castings are subject to the defects described in Chapter 11, "Causes and Cures for Common Casting Defects."

METAL COMPOSITION

An embrittled and coarse grain structure is produced in aluminum alloys by iron contamination. A structure of this type can be seen in the fracture of a casting shown in figure 241. An increase in iron content is usually due to improperly coated melting tools and iron melting pots. This defect can be remedied only by proper coating of the molding tools and melting pots with a protective wash.

Aluminum-copper alloys cannot tolerate magnesium contamination, aluminum-silicon alloys cannot tolerate iron or magnesium and aluminum-magnesium alloys cannot tolerate copper, iron, or silicon. The contaminating elements result in embrittlement and lowering of physical properties. Aboard ship, where chemical analyses are not available, proper scrap segregation is the only way of preventing contamination from harmful elements.

POURING

Dross inclusions are caused by poor pouring practice or poor gating practice. Such a defect is shown in figure 205, Chapter 11, "Causes and Cures for Common Casting Defects." This type of inclusion can be eliminated by using a proper pouring and gating technique as described under the section, "Pouring," in this chapter. Thorough skimming of the melt must also be a part of the pouring practice. Gas trapped in the stream of molten metal during pouring will cause porosity and can be corrected only by proper pouring technique.

MELTING

Pinhole porosity in aluminum castings is caused by poor melting practice. This type of defect is common and shows up as very small gas holes that are scattered through the casting. They may or may not show up on the casting surface. Porosity of this nature can be cured only by correct melting practice. In the case of an oil-fired furnace, a slightly oxidizing atmosphere must be used. Melting tools must be clean and dry to prevent any pickup of moisture by the melt. <u>Degassing procedures must be used to remove any gases</u> which are dissolved in the melt.

MISCELLANEOUS

Excessive moisture in the molding sand will cause porosity in aluminum castings. This defect can be easily identified because it occurs just below the surface of the casting and on all surfaces. Such porosity is shown in figure 242. The cure for this defect is to use the correct moisture content in the molding sand. This can be done only through proper testing procedures.

The causes of hot cracks in aluminum alloys are described in Chapter 11, "Causes and Cures for Casting Defects." Because some of the aluminum alloys are so likely to hot tear and hot crack, special precautions must be taken to reduce the resistance of the mold so as to permit free contraction of the casting.

WELDING AND BRAZING

When used with the proper welding rods and fluxes, oxyacetylene, oxyhydrogen, carbon-arc, and metallic-arc welding can be done on aluminum-alloy castings. When repairs or other welding are required, refer to the "General Specifications for Ships of the United States Navy," section S9-1, "Welding," for general guidance.

SUMMARY

The light-weight and good corrosion resistance of aluminum alloys make them ideal for certain applications aboard ships. The production of aluminum castings is not easy, because dross forms and gases are absorbed readily by molten aluminum. Careful attention to the melting and molding procedures are necessary for the production of good castings. Variations in sand properties, superheating, and pouring temperatures must be maintained more closely than for the other metals cast aboard ship.

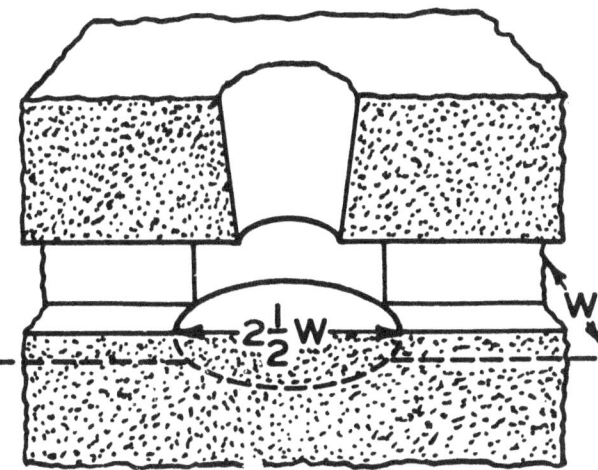

Figure 239. Enlargement-type sprue base.

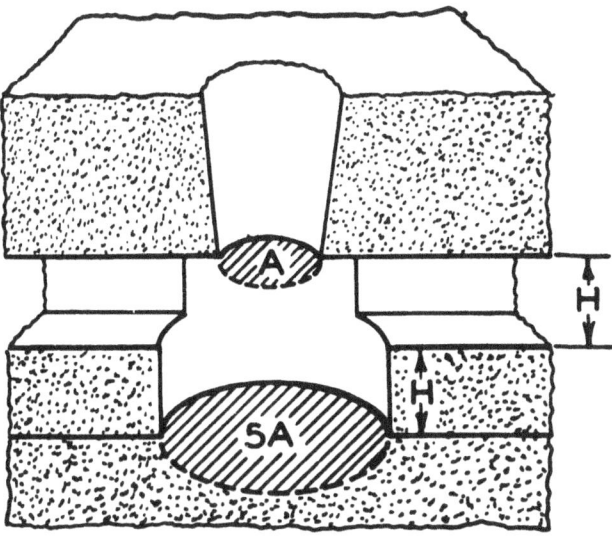

Figure 240. Well-type sprue base.

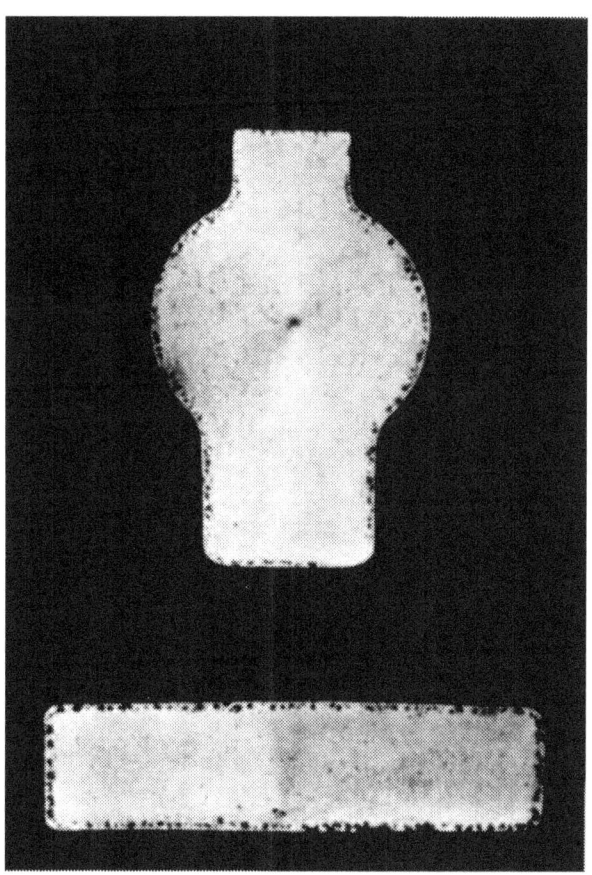

Figure 242. Porosity. (Caused by excessive moisture in the sand)

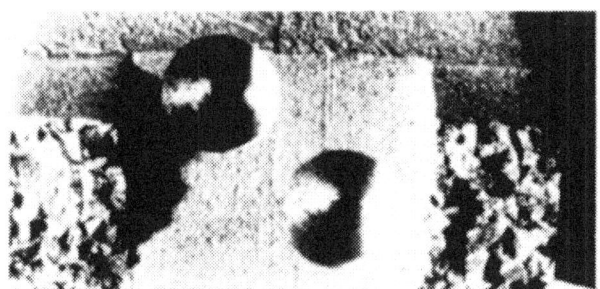

Figure 241. Coarse-grained structure. (Caused by iron contamination)

Chapter XVI
NICKEL-BASE ALLOYS

SELECTION OF ALLOYS

The nickel-base alloys available for use aboard ship are Monel and modified S-Monel. Nickel-base alloys are used mainly because of their excellent corrosion resistance and strength, even at high temperatures. They have a pleasing appearance and are heavy.

Peculiarities of nickel-base alloys which influence foundry practice are:

1. High melting temperatures require hard well-rammed molds made on sturdy patterns.

2. Nickel is poisoned by sulfur.

3. Gas is absorbed during poor melting practice or by overheating during melting and causes defective castings.

4. The molten metal is hotter than it looks.

5. Risers develop a deep pipe, so they must be large.

MONEL

Castings requiring a medium strength, high elongation, and extra-high corrosion resistance are made from Monel. High-pressure valves, pump impellers, pumps, bushings, and fittings are typical parts cast in Monel.

MODIFIED S-MONEL

This alloy has a high hardness that is retained up to 1,000°F. It has excellent resistance to galling and seizing. Modified S-Monel is useful under conditions where there is very poor lubrication. Refer to the table in chapter 13, for the actual alloy compositions.

HOW NICKEL-BASE ALLOYS SOLIDIFY

Nickel-base alloys solidify with a high shrinkage and have a narrow solidification range. Risers show the deep pipe which is characteristic of this type of solidification. Large risers are usually required to supply the molten metal necessary to make the casting sound.

PATTERNS

Patterns for nickel-base-alloy castings require rugged construction. The molds are rammed hard (similar to molds for steel castings). Solidly built patterns are required to insure castings which are true to pattern.

Shrinkage allowances vary for the type of alloy cast and the casting design. Monel and Modified S-Monel have a shrinkage of 1/4 inch per foot when unrestricted. The shrinkage may be as little as 1/8 inch per foot in heavily cored castings or in castings of intricate design.

MOLDING AND COREMAKING

Generally, the sand and core practices for nickel-base alloys are similar to those for steel castings.

SAND MIXES

Sands for Monel and modified S-Monel should have a high permeability and low clay content. Organic binders should be kept to a minimum. The all-purpose sand described in Chapter 4, "Sands for Molds and Cores" is satisfactory for use with these nickel-base alloys.

PROCEDURES

Molding. Molding practices for nickel-base alloys are the same as for steel. The molds should be rammed hard to provide a surface that will resist the erosive action of the molten metal at the high pouring temperatures.

Dried, oil-bonded sand molds should be used for Monel and modified S-Monel castings weighing more than 15 pounds. Smaller castings are made in skin-dried or green sand molds.

Coremaking. Pure silica sand with linseed oil binders should be used for cores. The cores should be well vented and have good collapsibility. Core practice for nickel-base castings should follow the practices used for copper-base alloy castings.

Washes. Either graphite or silica-flour washes can be used for cores.

GATING

Gating for the nickel-base alloys should permit rapid filling of the mold without erosion of the mold surface or exposure of the mold to radiated heat from the molten metal for longer than absolutely needed. Gating practices similar to those used for steel castings are best for nickel-base castings.

RISERING

Risers for nickel-base alloys must be large in order to provide enough molten metal to feed heavy sections and to compensate for the high solidification shrinkage. Risers must be located properly to obtain complete feeding of heavy sections. The gating and risering arrangement shown in figure 243 resulted in shrinkage porosity in the section marked A-A. A change in the gating as shown in figure 244 permitted proper feeding of all the heavy sections from the risers and produced a sound casting. Notice the deep piping in the risers in figure 244. Refer to Chapter 7, "Gates, Risers, and Chills," for details of proper gating and risering practice.

CHILLS

Chills should be used as needed to obtain directional solidification and to insure soundness in heavy sections. See Chapter 7, "Gates, Risers, and Chills."

VENTING

Because high pouring temperatures are required for nickel-base alloys, liberal venting of molds and cores is necessary. Cores in particular should be well vented because some of the nickel-base alloys are sensitive to the gases generated by organic core binders (linseed oil).

MELTING

TYPE OF FURNACE

Nickel-base alloys can be melted in oil-fired crucible furnaces, indirect-arc furnaces, resistance furnaces, and induction furnaces. Fuel in oil-fired crucible furnaces may damage the alloys because of the sulfur content or the possibility of producing a reducing atmosphere. See Chapter 8," Description and Operation of Melting Furnaces," and manufacturer's literature for operating procedures for the various furnaces.

PROCEDURE

The charge for Monel or modified S-Monel is made up from Monel block and remelt scrap in the form of gates and risers. The Monel block is 2 inches x 2 inches x 4 inches and weighs about 6 to 8 pounds. Remelt scrap should be free of all sand and foreign materials. Sandblasting should be used to make sure it is cleaned properly. Machine-shop borings and turnings should not be used at all. Sulfurized oils and lead oxides used for machine lubrication cannot be removed satisfactorily and are likely to cause contamination of the melt. Up to 40 percent remelt scrap is used in normal operation, but as much as 50 percent can be used. Electrolytic nickel and nickel shot are available for adjustments of nickel content. Manganese is added as 80 percent ferromanganese and silicon as 95 percent metallic silicon.

The procedures for all types of equipment are the same as far as the actual meltdown is concerned. In oil-fired crucible furnaces, clay-graphite crucibles should be used and an oxidizing atmosphere maintained in the furnace. If the crucible is closed with a clean cover, no slag is needed. A lead-free glass slag may be used in the crucible instead of a cover.

The charge of Monel block and scrap is melted down and brought to a temperature about 50° to 75°F. above the desired pouring temperature. The major additions of manganese and silicon are made as part of the deoxidation practice.

TEMPERATURE CONTROL

The nickel-base alloys are sensitive to the proper pouring temperature. The surface appearance of the melt is deceptive when trying to judge temperature. Usually the metal is much hotter than it appears. Any unnecessary superheating of these alloys results in gas absorption.

Temperature control should be maintained with the use of properly operating pyrometers. The surface of the melt should be clear of slag before temperature readings are taken with an optical or immersion pyrometer.

DEOXIDATION

Three to 5 minutes before the heat is ready to pour, manganese is added as 80 percent ferromanganese and is followed by the silicon addition of 95 percent metallic silicon. Final deoxidation is accomplished with 0.1 percent of magnesium, which is plunged below the surface of the melt with a pair of tongs, or with a specially made rod. This causes a vigorous reaction and proper safety precautions should be observed. If the magnesium burns on the surface, it is ineffective as a deoxidizer.

POURING

Nickel-base alloys should be poured as rapidly as possible in order to fill the mold quickly and to prevent chilling of the molten metal. A pouring basin should be used and kept full at all times during the pouring of the casting.

Pouring temperatures for Monel are between 2,700°F. and 2,850°F., depending on the size and section thickness of the casting. Modified S-Monel is poured between 2,650°F. and 2,800°F.

Haphazard pouring may cause slag or sand to flow into the mold with the stream of molten metal. The ladle should be skimmed of all slag, or a skimming rod should be used to keep the slag from entering the sprue. The mold should be rammed hard around the sprue to prevent sand erosion.

CLEANING

Nickel-base alloy castings usually come free from the mold easily and wire brushing is the only cleaning necessary. Grit or sand blasting can be used to give a better surface.

CAUSES AND CURES FOR COMMON CASTING DEFECTS IN NICKEL-BASE ALLOY CASTINGS

Nickel-base alloy castings are susceptible to all of the various defects described in Chapter 11, "Causes and Cures for Common Casting Defects." There are, however, some defects that are particularly apt to occur in nickel-base alloys, such as gas contamination.

METAL COMPOSITION

The carbon content of Monel and modified S-Monel must be kept low. Carbon in excess of the very small amounts that these alloys can tolerate will be precipitated as free graphite and cause intercrystalline brittleness. Sulfur causes hot shortness in nickel-base alloys and makes them susceptible to hot tearing. Modified S-Monel cannot tolerate any lead. Lead in the presence of silicon (which is an alloying element in modified S-Monel) causes a coarse grain structure and cracking.

MELTING PRACTICE

Nickel-base alloys are especially sensitive to gas absorption and will develop porosity and gas holes if proper attention is not paid to melting practice. Proper temperature control should be maintained to prevent the possibility of excessive superheating which will increase the susceptibility to pick up of gas. Pouring ladles and any other pouring equipment should be thoroughly dried to prevent moisture pick up from these sources.

WELDING AND BRAZING

Nickel-base alloys may be welded by metallic arc, electric resistance, oxyacetylene, and atomic hydrogen processes. When repairs or other welding are required, refer to the general "Specifications for Ships of the United States Navy," Section S9-1, "Welding," for general guidance.

SUMMARY

The high melting point of nickel-base alloys, combined with their sensitivity to absorption of gases, makes the proper control of all foundry procedures mandatory for the production of good castings. Proper temperature control by the use of pyrometers cannot be stressed too strongly. Determination of metal temperature by visual observation is at best a gamble <u>and should never be used.</u>

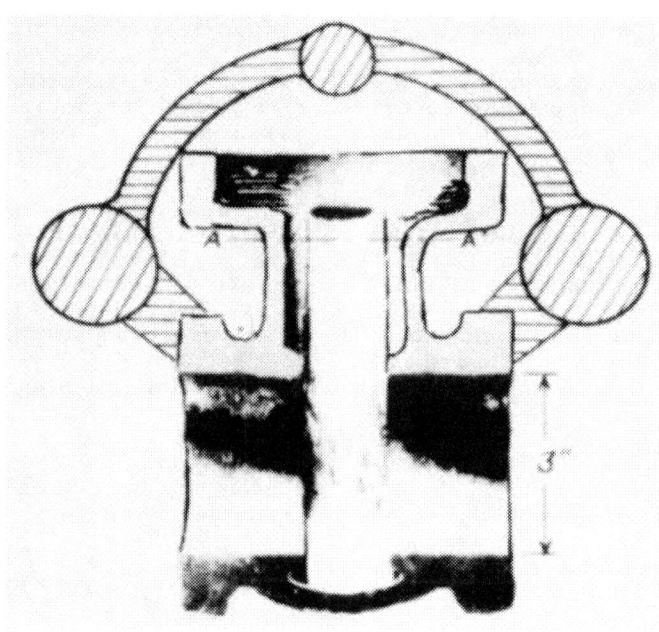

Figure 243. Poor gating and risering practice for a nickel-base alloy casting.

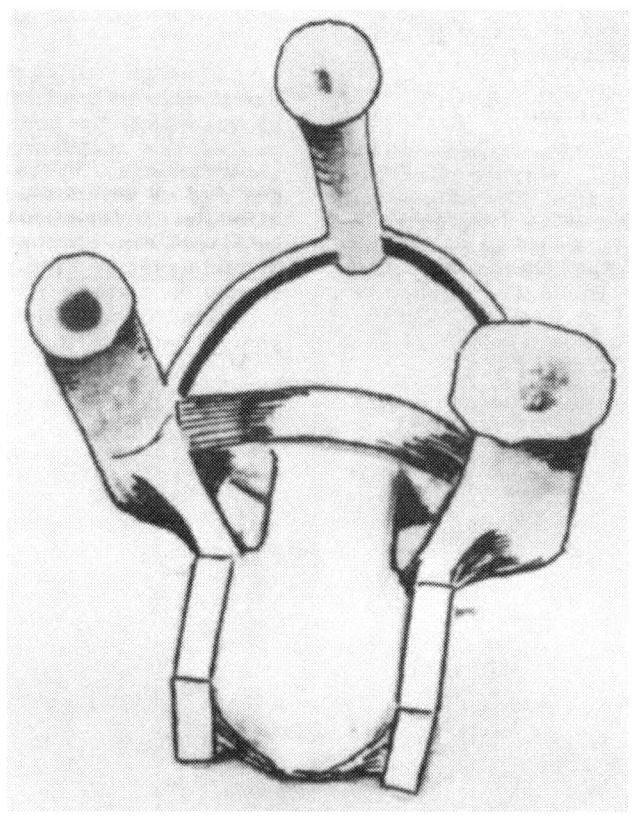

Figure 244. Improved gating and risering for nickel-base alloy casting.

Chapter XVII
CAST IRON

Cast iron has many properties that make it a good alloy for castings. Gray cast iron has: (1) excellent castability, (2) good machinability, (3) good water resistance, (4) high damping capacity, (5) high compressive strength, (6) good tensile strength, and (7) good yield strength. Its two major disadvantages are: (1) low impact strength, and (2) low ductility.

The unique feature of cast iron is its high carbon content. In gray iron, the casting has been cooled in the mold at a rate which allows graphite flakes to precipitate. These graphite flakes act as built-in "cushions" in the metal and as lubricants during machining or when the casting is subjected to wear. The metal is called "gray" iron because the graphite flakes impart a gray color to a fractured surface. Although the graphite flakes impart many desirable features to gray iron, they also make it brittle and reduce its strength.

In white cast iron, the casting has been cooled so that no graphite flakes are formed. The high carbon content shows up in hard massive particles of chilled iron (cementite), which make the iron act much as high-speed steel.

Excellent castability enables cast iron to be cast into very thin sections without any particular difficulty. High machinability permits fast and easy machining without the difficulties encountered with many other metals. Wear resistance permits the use of cast iron in moving and rubbing parts without any special treatment. A high damping capacity means that cast iron is capable of absorbing vibrations. This makes it useful for machine bases and tool holders. High compressive strength permits it to withstand heavy loads. Good tensile and yield strengths enable cast iron to withstand normal stresses required of engineering materials. However, cast iron cannot be subjected to sudden blows and it will not stretch like steel when an overload is applied.

White cast iron, on the other hand, is almost impossible to machine or saw because of its very high hardness. It is highly useful where a very hard material is needed to resist wear, but is more brittle than gray cast iron.

SELECTION OF ALLOYS

Cast iron is a general name given to a group of alloys rather than a name for one particular alloy. It is primarily an alloy of iron with carbon and silicon. By changing the amounts of carbon, silicon, and other alloying elements, it is possible to produce a series of alloys with a wide range of properties.

Carbon is the most important alloy in cast iron. The total amount of carbon and its condition are the major factors that determine the properties of the iron. The presence of other elements (such as silicon and phosphorus) affect the solubility of carbon in iron. The eutectic composition of carbon and iron (the mixture with the lowest melting point) has 4.3 percent carbon. Silicon and phosphorus each reduce the amount of carbon required to form the eutectic mixture by approximately one-third of one percent of carbon for each one percent of silicon or phosphorus in the iron. From this has developed the term "carbon equivalent" that is used to express the composition of the cast iron with respect to its carbon content.

The carbon equivalent is determined by using the following equation: carbon equivalent, % = total carbon, % + 1/3 (silicon, % + phosphorus, %).

Cast irons that have a carbon equivalent greater than 4.3 percent may have a coarse open-grained structure with large amounts of "kish" graphite (graphite that floats free of the metal during pouring and solidification). Cast irons that have a carbon equivalent less than the eutectic amount will have finer graphite flakes and a more dense grain structure.

Cast irons are more sensitive to section size than most other casting alloys. This is caused by the great effect that cooling rates have on the formation of graphite flakes.

It is important to realize that the properties and strength of cast iron can be affected greatly by cooling rate. For example, if an iron is poured into a casting with variable wall thickness, a high-strength iron might be obtained in 2-inch walls and brittle unmachinable white iron in 1/4 inch walls and at edges where the cooling rate is high.

REGULAR GRAY IRON

Regular or ordinary gray iron is often used for space-filling castings where high strength or ductility are not needed. It also finds considerable use in castings that serve as replaceable wearing parts. Typical examples are cylinder blocks, heads, piston rings, and cylinder liners for internal-combustion engines.

Ordinary cast irons have a tensile strength varying from 18,000 p. s. i. to 24,000 p. s. i. Typical compositions and mechanical properties obtained from various section sizes are listed in table 26.

TABLE 26. TYPICAL COMPOSITIONS OF ORDINARY GRAY CAST IRON

Composition, percent					Wall Thickness, inch	Tensile Strength, p.s.i.	Brinell Hardness
T.C.	Si	P	S	Mn			
3.50 to 3.80	2.40 to 2.60	0.20 max	0.15 max	0.50 to 0.70	Up to 1/2	22,000 to 26,000	160 to 190
3.40 to 3.60	2.30 to 2.50	0.20 max	0.15 max	0.50 to 0.80	1/2 to 1	18,000 to 24,000	160 to 180
3.10 to 3.30	2.20 to 2.40	0.20 max	0.15 max	0.50 to 0.80	Over 1	18,000 to 22,000	130 to 170

HIGH-STRENGTH GRAY IRON

Castings requiring a higher strength than ordinary gray cast iron but no other improvements in properties can be made from an iron having the following typical analysis:

Total carbon   3.30 - 3.35 percent
Manganese   0.80 percent
Silicon   2.00 percent
Phosphorus   0.20 percent
Sulfur   0.12 percent

An iron made with this analysis would have an average tensile strength of 34,000 to 40,000 p. s. i. in a 1.2-inch-diameter bar. Typical compositions of high strength gray cast iron are listed in table 27.

TABLE 27. TYPICAL COMPOSITIONS OF HIGH-STRENGTH GRAY CAST IRON

Composition, percent					Wall Thickness, inch	Tensile Strength, p.s.i.	Brinell Hardness
T.C.	Si	P	S	Mn			
3.10 to 3.30	2.00 to 2.20	0.20 max	0.12 max	0.45 to 0.70	Up to 1/2	36,000 to 40,000	180 to 230
3.00 to 3.25	1.80 to 2.10	0.20 max	0.12 max	0.45 to 0.70	1/2 to 1	35,000 to 39,000	205 to 230
2.80 to 3.10	1.60 to 2.00	0.20 max	0.12 max	0.45 to 0.70	Over 1	35,000 to 38,000	180 to 220

As a general rule, the higher the tensile strength of a gray iron, the greater will be the shrinkage during solidification and the poorer the machinability. Tensile strengths of at least 50,000 p. s. i. are readily obtainable in gray iron by proper selection of the carbon and silicon contents and by alloying. As a general rule, for tensile strength over 45,000 p. s. i., the carbon content of the iron should not exceed 3.10 percent and the silicon content should not exceed 2.00 percent. Useful combinations of alloying additions to increase the strength of such irons are: (1) 1 percent of nickel and (2) 1 percent of nickel with 0.50 percent of molybdenum or chromium.

SCALE AND CORROSION-RESISTANT GRAY IRON

Ordinary and high-strength gray cast irons sometimes are not satisfactory where high temperatures will be encountered continuously. When heated, these irons have a tendency to scale or warp and to "grow." In simple words, "casting growth" means that when unalloyed cast irons are heated to high temperatures for a long time, they increase in size. To provide a casting material that will have the properties of gray iron and be more useful at high temperatures, the alloyed cast irons known as scale and corrosion-resistant gray cast irons have been developed.

Two analyses for these irons are given in the Navy Specifications:

	MIL-G-858	
	Class 1	Class 2
Total carbon	2.60- 3.00	2.60- 3.00
Manganese	1.0 - 1.5	0.8 - 1.3
Silicon	1.25- 2.20	1.25- 2.20
Copper	5.5 - 7.5	0.50
Nickel	13.0 -17.5	18.0 -22.0
Chromium	1.8 - 3.5	1.75- 3.50

In addition to scale resistance at high temperatures, these irons are resistant to corrosion by acid, caustic, and salt solutions. They are, however, low-strength irons having a tensile strength of only about 25,000 p. s. i.

WHITE CAST IRON

Cast irons in this classification are not provided for in Navy Specifications. However, information is supplied as background material in case the occasion should ever arise for their use.

A white cast iron gets its name from the fact that a newly fractured surface has a white metallic appearance. A white cast iron is obtained by: (1) severe chilling of a low-carbon low-silicon iron, (2) addition of elements (especially chromium) that will prevent the formation of free graphite, or (3) by a composition that will produce a white iron when poured into a sand mold. Malleable iron in the as-cast condition is an example of the last method.

HOW GRAY CAST IRONS SOLIDIFY

Gray cast irons solidify generally the same as other eutectic alloys. Solidification starts with the growth of dendrites as described in Chapter 1, "How Metals Solidify." The first dendrites to form in gray cast iron are austenite because austenite is the constituent having the highest melting point and at this temperature ferrite cannot exist. As the austenite dendrites grow, they contain some dissolved carbon and the melt surrounding them changes in composition until it reaches the eutectic composition. In turn, the remaining liquid starts to solidify. Up to this point, the solidification is the same as for a normal eutectic alloy.

As the eutectic starts to solidify, however, another process begins. This is the nucleation and growth of the graphite flakes. The growth of graphite flakes starts at nuclei different from those that initiate the growth of the dendrites. The graphite flakes continue to grow so long as there is liquid eutectic around the flakes. This shows the importance of cooling rate to gray cast iron. The slower the cooling, the longer it takes for solidification, and graphite flakes can become quite large. A faster cooling rate means a shorter solidification time and smaller graphite flakes. Once the eutectic mixture has solidified, the structure of the graphite flakes is established. No further change in the pattern of the graphite flakes occurs as the iron cools to room temperature.

PATTERNS

Section thickness plays an important part in gray iron castings. There are limiting wall thicknesses below which a chilled structure (white iron) will be produced. Chill can be removed by annealing, but other physical properties are also reduced. The limiting wall thickness for ordinary gray iron is 1/8 inch, and for high-strength gray iron 3/8 inch. Patterns should be checked to determine that sections are not thinner than these limits.

As will all cast metals, gray iron is also sensitive to abrupt changes in section size and to sharp corners. Patterns should be checked for proper blending of unequal sections and for proper filleting of junctions. Refer to Chapter 2, "Designing a Casting."

The shrinkage allowance for gray cast iron is 1/8 inch per foot and pattern draft is 1/16 inch per foot. Machining allowances and approximate tolerances in the as-cast condition for various sizes of castings are listed in table 28.

TABLE 28. MACHINING ALLOWANCES AND AS-CAST
TOLERANCES FOR GRAY IRON

Approximate Length of Casting, inches	Machining Allowance, inch	Approximate As-Cast Tolerance, inch
Up to 8	1/8	1/16
Up to 14	5/32	3/32
Up to 18	3/16	1/8
Up to 24	1/4	5/32
Up to 30	5/16	3/16
Over 30	3/8 to 1/2	1/4

This table is to serve only as a guide. Information of this type is best determined by experience. Refer to Chapter 3, "Patternmaking," for information on repair and storage of patterns.

MOLDING AND COREMAKING

Molding and coremaking procedures for gray iron are the same as for other metals as described in Chapter 5, "Making Molds," and Chapter 6, "Making Cores."

SAND MIXES

Various sand mixes that may be used for gray iron castings are listed in table 9, Chapter 4, "Sands for Molds and Cores." Core mixes are listed in table 14 of the same chapter. Typical properties of gray iron sands and the types of castings poured in these sands are listed in table 29.

TABLE 29. TYPICAL PROPERTIES AND USES OF GRAY IRON SANDS

Moisture, percent	Green Compressive Strength, p.s.i.	Permeability, AFS units	Type of Casting
4.5	9	28	Piston rings
4.5-5.5	10-12	20-25	Small
5-6	9-12	50-60	Medium
5.7	9	55	Machinery parts up to 300 pounds
7	9	50	20 to 50 pounds
9	8	63	High-pressure pipe fittings
9.6	9	20	200 to 500 pounds

PROCEDURES

Coremaking and Molding. Cores and molds for cast iron present no unusual problems. Cores and molds must, however, be rammed firmly and uniformly to resist the high temperatures at which cast iron is poured. The quality of the surface finish is determined mainly by the fineness of the sand. The finer the sand, the better will be the surface finish of the casting.

Cores for cast iron should be well vented, and core vents should connect with vents in the mold. Because of the high heat content of the molten iron, it is important that cores be well baked to avoid core blows in the casting from gas generated in the core.

Green-sand molds are used for iron castings. Skin-dried or completely dried molds are rarely required. Sea coal (ground bituminous coal) in amounts of 1 to 6 percent of the weight of the sand, is almost always added to molding sands for gray iron. The sea coal generates a controlled amount of gas when the casting is poured. This gas forms a film over the surface of the mold and helps to give a better surface finish of the casting. It also has a cushioning effect which helps to prevent sand expansion defects.

Washes. Iron castings do not usually require that molds or cores be protected with a wash if the sand is fine enough. It a coarse sand must be used, a silica wash can be applied to the mold as described in chapter 4. Another variation that gives an excellent surface finish is to rub or brush dry powdered graphite onto the surface of the mold or core.

Gating. The good castability and fluidity of gray cast irons permit the use of gating systems that seem small in comparison with the gating of other metals. Gating systems for gray iron castings can use the same ideas as described in Chapter 7, "Gates, Risers, and Chills," except that they can be made quite small in cross section. Experience and the use of properly kept records are the best sources of information for selecting a gating system for iron castings.

There are two ingates that have proven useful in the casting of flat castings in gray cast iron. These ingates are the knife gate shown in figure 245 and the lap gate illustrated in figure 246. Both gates permit the molten metal to enter the casting through a thin gate. The cross-sectional area of gates of this type is the same as for conventional ingates, but they have the advantages of filling the casting more uniformly and are easily removed by simply breaking them off. As part of the gating system for gray cast iron, it is good practice to include a strainer core or whirl gate in the system to remove dirt and slag.

RISERING

As for other metals, the size of riser for a particular casting depends on the thickness of the section that must be fed and the type of iron being poured. Generally the same practices as described in Chapter 7, "Gates, Risers, and Chills," may be followed. A rule-of-thumb for gray iron is that the cross sectional area of the riser should be about 80 percent of the cross-sectional area of section that must be fed.

The ordinary gray cast irons with a tensile strength of less than 25,000 p.s.i. can often be cast without risers. In such irons, the high carbon content (over 3.40 percent) often produces sufficient graphite flakes to offset most of the normal solidification shrinkage of iron. Indeed, if the gating system is selected carefully, sounder castings of ordinary gray iron can often be made without risers than with risers. However, this practice is not recommended. As the carbon content and silicon content of gray iron are decreased (tensile strength increased), larger risers become necessary; but even in the highest-strength gray irons, smaller risers are required than for almost any other metal. White cast iron, on the other hand, does not have the offsetting advantage of graphite formation and requires the large risers typical of steel practice.

A gating and risering system that is useful in industry uses a shrink-bob riser and a whirl gate. It is a modification of the simple but good system of gating through the riser.

Figure 247 illustrates a good type of riser for a casting that is cast in both the cope and the drag. The risering of a casting that is molded in the drag is shown in figure 248. A cope casting is risered as shown in figure 249. The plan view of all three risers is the same and is shown in figure 250. Notice that for all three systems of risers, the top of the riser must be a minimum distance of four inches above the uppermost part of the casting.

The runner is usually rectangular in cross section but has a slight taper. It is slightly curved and enters the riser tangentially to provide a swirling motion of the flowing metal. In all three cases of casting location in the mold, the runner is in the drag. The ingates for all three casting locations are made to conform with the casting position. In other words, if the casting is in the cope, the ingate is in the cope and if the casting is in the drag, the ingate is in the drag. The cross section of the ingate can be either square or round.

The riser size is the same as previously mentioned. The cross-sectional area of the riser should be about 80 percent of the cross-sectional area to be fed. This type of riser must have what is called the drag portion of the shrink bob. This is as important as the other dimensions of the riser to obtain good feeding. The drag part of the shrink bob is important because the part of the riser next to the ingate must be kept molten to permit proper feeding. The drag part of the riser heats the sand in this area and keeps the metal in a molten condition for a longer period of time. If the drag part of the riser is omitted or is too shallow, the last metal to solidify will be well up in the cope section of the riser and will be unable to perform its function of feeding the casting.

Proper feeding of gray iron castings with this system can best be assured by using the sizes of risers, runners, and ingates listed in table 30. As mentioned before, the exact size of riser, runner, and ingate is determined by the casting size and the type of iron poured. The use of records that show successful gating and risering systems is invaluable in rigging new castings.

CHILLS

In the manufacture of gray iron castings, it is usually not necessary to use chills to promote directional solidification. The desired degree of directionality can usually be obtained by intelligent use of a gating and risering system. The use of chills can be dangerous.

TABLE 30. PRINCIPAL DIMENSIONS OF GRAY IRON GATING AND RISERING SYSTEM

Size of Casting, pounds	Riser Diameter, inches	Depth of Riser Below Neck A, inches*	Runner Size, inch	Ingate Size, inch
Lightest castings that require feeding - under 50	2	2-1/4	3/8 × 1/4 (1 wide)	1/2 to 3/4 square
	2-1/2	2-1/4		13/16 to 7/8 diameter round
50 to 400	3	2-3/4	1/2 × 3/8 (1 wide)	1 to 1-1/4 square
	3-1/2	2-3/4		1-3/16 to 1-7/16 diameter round
	4	3-1/4		
	4-1/2	3-1/4		

*Refer to figures 239, 240, and 241.

If chills are necessary, they must be used cautiously with gray iron castings. Chills must be located with the understanding that there is a possibility of producing chilled or white iron in that section. If the casting can be annealed to remove chill, this is not too important. If the casting cannot be annealed, this situation is very important and considerable thought must be given to the size and location of chills. The use of chills on thin sections is particularly risky. Thin sections solidify rapidly, even with only the sand mold surrounding them, and in some cases may result in chilled iron. If the chilled portions are in an unstressed location or where no machining is to be done, it is probable that the chills will cause little or no harm.

Chills are highly useful when it is desired to produce a cast iron part entirely or partly of white iron. Because chills speed the cooling rate, they promote the formation of white iron. For example, assume that a gray iron casting is desired with a white iron section to resist wear at some point. An iron or steel external chill with a thickness about twice that of the casting can be inserted in the mold (or rammed up against the pattern) at the place the white iron is desired. The iron should be cast directly against the chill, which should be free of all rust and dirt, warm, and coated with a thin mixture of plumbago and fire clay in water. The coating must be dried before the casting is poured.

VENTING

Cast iron is not particularly susceptible to damage by gases other than water vapor and is heavy enough to displace most mold gases if they are given a reasonable chance to escape. High spots on a casting should be vented by using a hacksaw blade passed through the mold. This is good practice with all metals.

MELTING

The electric indirect-arc furnace, electric-resistor furnace, and electric induction furnace are all capable of melting cast iron. In an emergency, an oil-fired crucible furnace can be used to melt the high-carbon, high-silicon cast irons. This practice, however, is slow and permits only one or two heats per crucible and results in reduced refractory life in the furnace. Because of its high pouring temperature, cast iron is difficult to melt in an oil-fired crucible furnace. Electric methods should be used whenever possible.

ELECTRIC INDIRECT-ARC FURNACE

Charging. After the furnace has been preheated in accordance with instructions in the manufacturer's manual, the shell should be rotated until it is 45° down either front or rear from the top center position. The charging position should be varied from time to time to prevent excessive wear on one section of the lining. The ideal method of charging is to have the furnace door in the top-center position. The carbons should be moved back until they are flush with the furnace wall to prevent damage during the charging period.

Foundry returns (gates and risers) should be charged first and should be free of sand. Sand causes a slag blanket to form on the surface of the molten metal during the melting cycle. This insulates the bath from the heat generated by the arc and makes it difficult to reach and to determine the desired tapping temperature. Usually, heavy pieces should be charged first. If cast iron or steel borings are used, they should be added next because they will filter down through the scrap, give a more compact charge, and will be free from direct contact with the arc. Additions of nickel, chromium, molybdenum, and vanadium may now be made. If charged on top (close to the arc), some loss of the finer alloys may occur because of "blowing-out" by the arc. Pig iron should be charged next and steel scrap on top (closest to the arc). Ferromanganese and ferrosilicon should not be added with the charge, but should be added to the molten bath before tapping.

Charging should be accomplished as quickly as possible to prevent excessive loss of heat from the lining. It is not good practice to exceed the rated capacity of the furnace. Lastly, the furnace door should be closed and clamped securely.

Working the Heat. The charge should be observed periodically during the melting cycle through glasses with a No. 12 lens. The angle of rock should be increased steadily as the pool of molten metal collects under the arc. As melting proceeds, the melted metal will wash over the rest of the charge until it is entirely melted.

Ferromanganese and ferrosilicon should be added through the spout in the order mentioned about 3 to 5 minutes before tapping. The bath should not be superheated from 2,700°F. to 2,800°F. to permit good separation of slag from the metal. The metal can be cooled in the furnace or tapped and cooled in the ladle to the desired pouring temperature. The temperature of the molten bath may be estimated rather closely from the kwhr input. However, sufficient data must be collected before this is possible.

A typical operating log for a few cast iron heats is shown in figure 251. Notice that the kwhr input decreases slightly with each consecutive use of a barrel because of the heat retained by the lining. Another rough method for determining the temperature of the bath is to work a 1/2 inch-diameter soft-iron rod in the molten bath for a period of 15 to 20 seconds. If the tip sparkles, the temperature is approximately 2,700°F. Should the tip melt in that period, the bath is 2,800°F. or higher and the iron should be tapped. The iron rod should be bent and care taken to avoid striking the electrodes. The behavior of the molten bath gives an indication of the proper tapping temperature. A bubbling action can be observed at approximately 2,800°F.

Prior to tapping, a chill-test specimen should be poured and fractured to determine if the heat has the desired characteristics. If the chill depth is too great, the chill characteristics can be adjusted by the addition of graphite, ferrosilicon, or other graphitizing inoculants to the bath or ladle. The effective use of this fracture test requires experience in order to judge the relationship between the depth of chill and the carbon equivalent of any heat as it applies to the controlling section thickness of the casting to be poured. The use of a chill test is described later in this chapter.

Slag is objectionable in this type of furnace because it serves as an insulating blanket between the arc and the surface of the bath and it reflects an abnormal amount of heat to the refractories above the slag. Under these conditions, the refractories start to melt, more slag is formed, and the entire lining can be melted out rather quickly. Melting of the lining is shown by a "runny" appearance of the refractory. When this occurs, the power should be shut off, the door removed, and clean dry sand spread over the slag to thicken it. The slag should then be pulled from the furnace and operations resumed.

Tapping. The furnace should be operated at reduced input (just sufficient to maintain the temperature of the bath) throughout the tapping period. The "Automatic Rock" switch should be placed in the "off" position and the furnace should be operated by the portable push-button station during pouring. If the entire heat is to be tapped into one ladle, ladle inoculations of ferrosilicon, graphite, ferronickel, or proprietary inoculants may be used. If the heat is to be shanked, ladle inoculants may cause nonuniformity of composition unless the weight of metal tapped can be weighed separately.

There should be no delay in tapping once the proper temperature has been reached. If a slight delay is unavoidable, the arc and the "Automatic Rock" switch should be shut off. The temperature of the bath will not fall or rise appreciably during the first few minutes. If a longer period of delay is necessary, the furnace should be operated intermittently at reduced input and at full rock in order to maintain the desired temperature. A well dried and preheated ladle should always be used for tapping.

ELECTRIC-RESISTOR FURNACE

Melting of cast iron in an electric-resistor furnace follows the same practice as described for the electric indirect-arc furnace.

ELECTRIC-INDUCTION FURNACE

The procedure for melting gray cast iron in the electric induction furnace is a simple operation. Refer to Chapter 8, "Description and Operation of Melting Furnaces," for the proper charging procedure and handling of the heat. The sequence of charging and alloy additions are the same as for the electric indirect-arc furnace.

TEMPERATURE CONTROL

The melting and tapping temperatures of cast irons are too high to permit the use of the type of immersion pyrometers used aboard repair ships. The optical pyrometer should be used to determine the temperature of the molten metal and readings must be taken on the clean metal surface. The surface of the metal bath should be as free of slag as possible. Otherwise, a false reading will be obtained.

Tapping temperatures may be estimated from the kwhr input and the time at the particular power input. Estimations of this type can be made only after considerable experience with the particular melting unit used and comparison of the power input with actual pyrometer temperatures. This type of procedure should be relief upon only as an emergency measure. Pyrometer readings are the only reliable indications of the true temperatures of the molten metal.

A common error with an optical pyrometer is to focus the instrument on the brightest portion of the metal. This is usually slag or an iron oxide and will give a temperature reading that is too high. Sighting the pyrometer into an enclosed chamber will also give a high reading. For consistent and dependable results, focus the instrument on metal in the open end of the furnace on the darkest portion of the metal.

LADLE TREATMENT

Inoculants such as ferrosilicon, graphite, or other commercial materials are added to the melt to obtain better properties in the base iron or to obtain the proper carbon equivalent. The chill-test as a melting control is discussed in the following section and described in Chapter 21, "Process Control."

The addition of the inoculant should be made in the stream of molten metal if a small ladle is used, or to the heat in the furnace if the entire heat is to be tapped at one time. The ladle addition to the stream of metal is a tricky process because the amount of metal tapped into a ladle can only be estimated. Small additions that are weighed out beforehand and added gradually to the molten metal stream are the easiest to make. The purpose of adding the inoculant to the metal stream is to assure good mixing with the molten metal. If the addition is placed in the bottom of the ladle before the iron is tapped, the addition is likely to fuse to the ladle lining and its effectiveness is reduced. Likewise, if the addition is made to the top of the metal bath after the ladle has been filled, intimate mixing is impossible and effectiveness of the inoculant is again reduced.

The inoculant should be thoroughly dry before adding it to the molten metal. Moisture in an inoculant causes severe boiling and spattering of the molten metal. The size of the inoculant is important in its effectiveness. If the material is too fine and powdery, it will be blown away during the ladle-filling operation and recovery is reduced. If the material is too coarse, it will not dissolve readily and will be carried into the casting as undissolved particles.

USE OF CHILL TEST AS A MELTING CONTROL

One of the major problems in selecting, melting, and controlling gray iron is to avoid the formation of hard white iron (chill) in thin sections and on edges and corners of castings. Fortunately, however, a rapid chill test can be made while the iron is still in the furnace.

Assume that it is desired to pour a certain iron into a 1/4 inch wall without danger of forming white iron. A logical test is to make a small test specimen 1/4 inch thick, pour it with the iron, cool it, and break it to examine the fracture. If white iron (chill) is present, it will be easy to see on the fractured face. Such a test can be made rapidly while the iron is still in the furnace and while there is still time to adjust the composition.

A test such as that above, however, would be restricted to one wall thickness (1/4 inch). A more versatile and universal test would be to pour a wedge (say, 4 inches long and with a triangular cross section tapering from 3/4 inch down to a knife edge). This wedge can then be broken through the middle. The triangular area of the fracture gives important information. The knife edge of the wedge will be white except for the very soft irons. The 3/4 inch base will be gray, except for the very strong irons. Most irons will have intermediate degrees of chill. By measuring <u>across</u> the wedge at the zone of chill depth, one has a direct measure of the thinnest casting that can be poured gray without chill.

Thus, the wedge test when fractured gives a measure of the chilling tendency of the iron. Ordinary low-strength irons will have low chill depths. High-strength irons will have high chill depths.

The wedge can be poured in green sand, but if used often can conveniently be poured in oil-sand cores made up ahead of time. The wedge does not have to be cooled slowly. As soon as it is solid, it can be shaken out of the mold, quenched in water, broken, and examined.

Intelligent and informative use of a chill test for iron depends upon experience gained in making a number of tests and examining the castings that result.

If chill depth of a chill test is too deep, the most rapid and satisfactory correction can be made by adding ferrosilicon to the melt. Graphite additions will serve the same purpose, but graphite is usually difficult to put in solution in iron.

If the chill depth is too low, the carbon and silicon content of the iron are probably too high. These elements cannot be readily removed from the iron, but they can be reduced by dilution if steel is added to the melt in the furnace. Another method to increase chill depth is to add up to 1 percent chromium or 1 percent molybdenum to the iron. If a totally white iron is desired, up to 4 percent chromium or nickel can be added to the iron in the furnace.

POURING

Pouring practices for gray cast iron are the same as for other metals as described in Chapter 9, "Pouring Castings." Proper skimming before and during the pouring operation are particularly important with cast iron. A small amount of slag passing into the mold usually results in a scrapped casting. It is best to use a pouring basin for cast iron. The high temperatures of molten cast iron tend to burn out the binder in pouring cups, cut into the top of the mold, and wash loose sand particles into the casting.

As with all metals, the castings should be poured at the lowest temperature that will permit complete filling of the mold. For thin and intricate castings, the pouring temperature may be as high as 2,600°F., but a good molding sand and good mold are required under these conditions to obtain a reasonably clean casting finish and freedom from trapped dirt. For fairly chunky castings with a 2 inch wall, the pouring temperature can be as low as 2,350°F. with excellent results. Because of its lower carbon and silicon contents, white iron castings must be poured slightly hotter (say, 50°F. to 100°F.) than gray iron. Even so, the white iron will not reproduce the pattern detail as well as gray iron.

CLEANING

Gray iron castings usually shake out of a mold easily and sand adherence is not a problem. Any sand sticking to the casting after shakeout can be removed with a wire brush or by sand blasting.

Gates on gray iron castings are normally small enough that the gates and risers can be snocked off easily with a hammer. If there is any doubt as to whether a gate or riser will break cleanly without damaging the casting, the gate or riser should be notched before attempting to knock it off. The remaining stub on the casting can be easily ground to the casting surface. A stand or portable grinder can be used, depending on the size of the casting.

CAUSES AND CURES FOR COMMON DEFECTS IN IRON CASTINGS

Gray iron castings can have the same defects as other types of cast metals. Refer to Chapter 11, "Causes and Cures for Common Casting Defects." There are some that are especially apt to occur in gray iron castings and are discussed in this section.

METAL COMPOSITION

The proper control of chemical analysis is important in the production of gray cast iron. Information given here can only serve as a rule-of-thumb method for recognizing the effects of the various elements.

The chill test is invaluable in determining if a particular heat is going to have the desired properties. A high carbon content, a high silicon content, or a combination of both will result in a weak iron. Small additions of some of the carbide-stabilizing elements (chromium or molybdenum) can be made to correct such a condition. A heat that reveals too strong a chill in the chill test can usually be corrected by additions of ferrosilicon.

Severe and unexplainable chilling of a test specimen or a casting may be traced to contamination by tellurium. There are tellurium mold washes that produce a severe chilling action. Their use should be avoided unless the molder is absolutely certain as to their use and results. If the charge is contaminated with tellurium, a faint odor of garlic may be detected when the charge becomes molten. Such a heat should be scrapped unless it can be used for castings that require chilled iron. If it is necessary to anneal an iron casting that contains chill or hard areas, a temperature of about 1,700°F. is required for at least several hours and the casting must be cooled slowly in the furnace to at least 1,200°F.

POURING

A particular defect known as a cold shot may be found in gray iron castings. This defect is caused by a small amount of molten metal entering the gating system or mold cavity and solidifying in the form of a small ball. This condition usually takes place if the man pouring the metal starts to pour, for some reason must stop momentarily, and then resumes pouring. The cold shot occurs because the small globule of metal is highly chilled and cannot dissolve or fuse with the molten metal entering the mold. The cure for this defect is to fill the gating system as rapidly as possible and to keep a steady uninterrupted stream of metal flowing into the mold.

MELTING PRACTICE

It is hard to damage gray iron by improper melting in the furnaces used aboard ship. The greatest source of trouble will be from overheating of the metal which will increase the loss of carbon, silicon, and manganese from the iron. Contamination of the charge or melt with phosphorus or sulfur in excess of specified maximums will cause defective castings.

Excessive carbon pickup by the iron can occur in an arc furnace if the arc is permitted to run smoky for a long time. This is bad practice with any metal. However, gray iron is less likely than any other metal to be damaged by such practices. It is almost certain that any melting practice that will damage gray iron even slightly will be impossible for any other metal.

MISCELLANEOUS

Cast iron is particularly susceptible to damage from water vapor released from damp linings or ladles. Damp linings will cause the iron to pick up gas which is rejected as blows or pinholes when the casting solidifies.

Hard cores will restrict contraction of the iron and may cause cracks. It is sometimes advisable to dig out cores before the casting is shaken out of the sand.

Because of its high pouring temperature, gray iron is rough on sand. For smooth castings, it is necessary to use a fine sand bonded with high-fusion clay or bentonite or to use a suitable mold wash.

WELDING AND BRAZING

Welding of iron castings is a difficult procedure and unless done properly usually results in a cracked casting. White iron castings are more likely to crack than gray iron when welded. Iron castings of any type should be preheated to at least 700°F. for welding or brazing. Under such conditions, excellent and dependable repairs can be made by brazing. If repairs by welding or brazing are required, refer to the "General Specifications for Ships of the United States Navy," Section S9-1, "Welding," for general guidance.

SUMMARY

Gray cast iron is a casting alloy that has a wide range of good properties that make it highly desirable as an engineering material. It is particularly adapted to structural castings of intricate design requiring thin sections. Gray iron is one of the easiest metals to cast.

Aboard repair ships, the determination of the composition of the iron is impossible. The use of the chill test and records of successful gating and risering systems are indispensable in producing good castings.

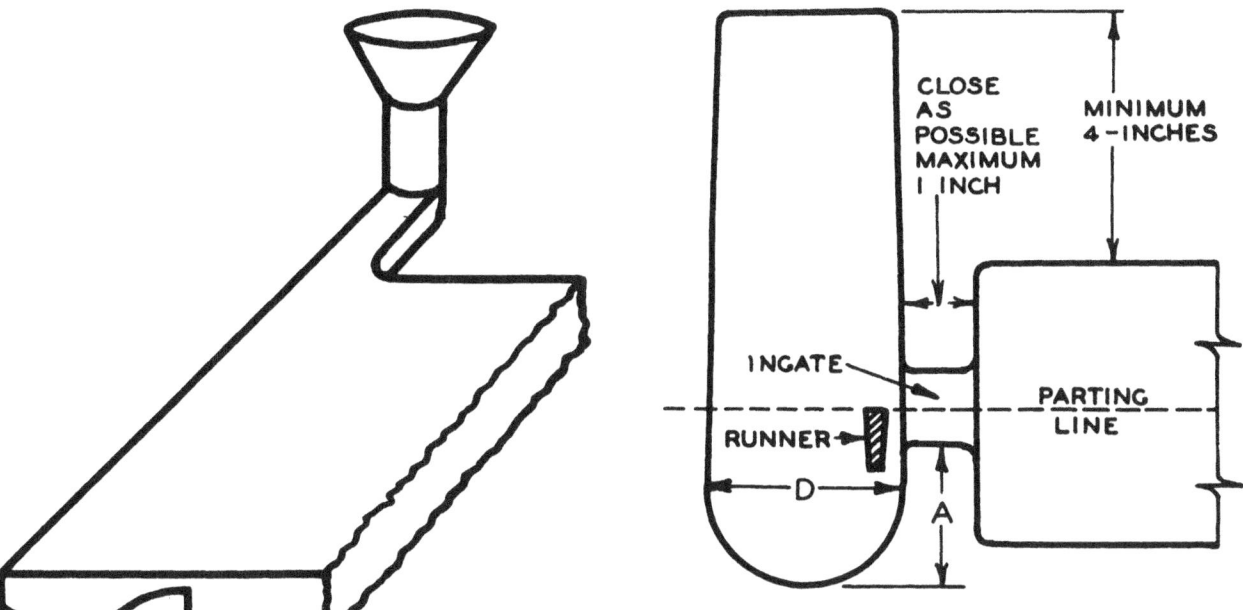

Figure 245. Knife gate.

Figure 247. Riser for a gray iron casting molded in the cope and drag.

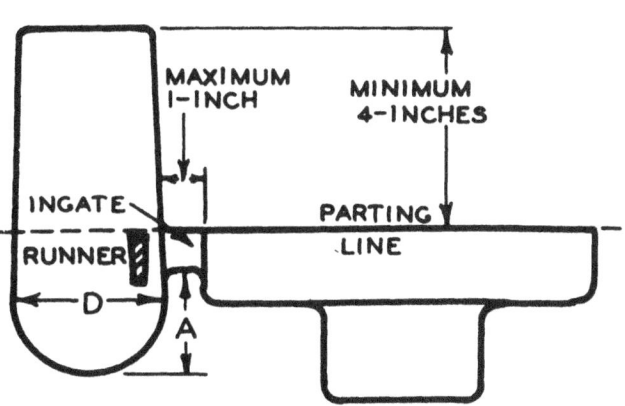

Figure 246. Lap gate.

Figure 248. Riser for a gray iron casting molded in the drag.

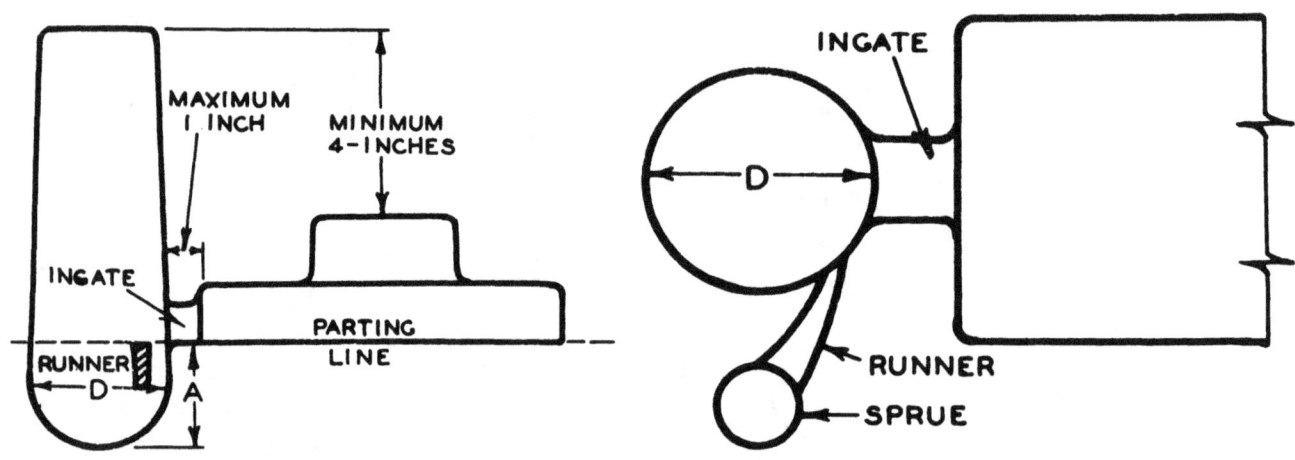

Figure 249. Riser for a gray iron casting molded in the cope.

Figure 250. Plan view of runner, riser, and ingate.

NYD - 348

DAILY OPERATION
DETROIT ROCKING ELECTRIC FURNACE

Furnace No. *El. 4* .... Operator *Turner* .... Plant *Navy Yard, N. Y.* .... Date *7-8-43*.

Day Heat No.	1	2	3	4	5	6	7	8	9	10	11	12	13	14	15	16	17	18
Lining Heat No.	46	47	48	49														
Time Start Charging	5.30	7.15	8.45	10.10														
Time Arc On	5.35	7.25	8.55	10.15														
Time Arc Off	6.55	8.35	10.05	11.20														
Kw. Hr. Start Rock	0	0	0	0														
Kw. Hrs. Reheat	–	–	–	–														
Total Kw. Hrs.	235	210	190	180														
Total Temperature °F	2900	2880	2900	2800														
Weight of Charge	800	800	800	800														
Composition and Analysis of Charge — #1 Pig	160#	160#	160#	160#														
Low Phos. "	240#	240#	240#	240#														
#4 "	80	80#	80#	80#														
Remelt	240#	240#	240#	240#														
Steel Sc.	80#	80#	80#	80#														
FeMn(B)	2½#	2½#	2½#	2½#														
FeSi(D)	4½#	4½#	4½#	4½#														
DELAYS: Minutes																		
Cause																		

REMARKS: 1. Ht. 46 — Preheating — 90 KWH

2. Ht. No.   Castings Poured
    46)
    47) Reel Heads, Arbors,
    48) Shells for Piston Rings
    49)

    4.45 Hours Operation
    4 Heats
    3200 Lbs. Total Melt
    815 Total Kw. Hrs.
    509.4 Kw. Hrs. per ton

3. 1% Fe Ni inoculation to shells

Figure 251. Operating log for cast iron heats.

Chapter XVIII
STEEL

Steel is an iron-base alloy fairly low in carbon content. All of the carbon in steel is dissolved in the iron so that no carbon is present as graphite.

The outstanding characteristics of steel castings are their high strength and toughness. With steel, it is possible to make castings stronger and tougher than with any other common casting alloy. Steel retains its high strength even up to fairly high temperatures but can become quite brittle as low temperatures.

The main disadvantages of steel as a casting alloy are: (1) it melts at a high temperature, (2) has a high shrinkage during solidification, and (3) is difficult to cast. Except for specially alloyed grades, steel will rust or corrode and is magnetic. As a general rule, low-carbon unalloyed steel is harder to cast than high-carbon or alloyed grades.

SELECTION OF ALLOYS

The proper grade of steel for a casting is usually designated on the blueprint which accompanies the work order. When the blueprints are not available or when the casting must be made from broken parts, the intended use of the casting is the best basis for determining the proper steel. Steels generally used aboard ship can be divided into three general classes: (1) unalloyed steels, (2) low-alloy steels, and (3) corrosion-resistant steels.

UNALLOYED STEELS

Unalloyed steels (plain carbon steels) are those which contain carbon and manganese as the principal alloying elements. Other elements are present in small amounts carried over from the customary steelmaking operations. The elements normally found in plain carbon-steel castings are as follows:

Element	Percent
Carbon	0.05 to about 0.90
Manganese	0.50 to 1.00
Silicon	0.20 to 0.75
Phosphorus	0.05 maximum
Sulfur	0.06 maximum

The unalloyed steels can be further subdivided into low, medium and high-carbon steels. Table 31 lists the ranges of composition and mechanical properties (after normalizing) of three classes of plain carbon steel.

TABLE 31. COMPOSITION AND PROPERTIES OF PLAIN CARBON STEELS

	Low Carbon	Medium Carbon	High Carbon
Composition			
Carbon, percent	0.09-0.20	0.20-0.40	0.40-0.90
Manganese, percent	0.50-1.00	0.50-1.00	0.50-1.00
Silicon, percent	0.20-0.75	0.20-0.75	0.20-0.75
Phosphorus, percent	0.05 max	0.05 max	0.05 max
Sulfur, percent	0.06 max	0.06 max	0.06 max
Physical Properties			
Tensile Strength, p.s.i.	42,000-70,000	60,000-80,000	70,000-120,000
Yield Strength, p.s.i.	20,000-38,000	30,000-40,000	35,000-70,000
Elongation in 2 Inches, percent	36-22	30-20	26-30

High-carbon unalloyed steel is used where parts are subject to high stresses and surface wear (such as in hawse pipes, chain pipes, and engine guides). Low-carbon unalloyed steel is used where strength is not of prime importance but where welding may be necessary.

LOW-ALLOY STEELS

Alloy steels contain either unusually large amounts of the common elements or fairly large amounts of special elements. These changes from the composition of plain carbon steel are made to obtain special properties (usually for higher strength). In the low-alloy steels, manganese may be used up to 3.00 percent and silicon up to 2.75 percent, with copper, nickel, molybdenum, and chromium in varying amounts. The total content of special materials added to low-alloy steels is usually less than 8 percent.

CORROSION-RESISTANT STEELS

Steels in this group are highly alloyed and are used where high resistance to chemical or salt-water corrosion or good strength at high temperatures is necessary. A typical corrosion-resistant steel has a low carbon content and additions of 18 percent chromium and 8 percent nickel. This steel is commonly called 18-8 stainless steel. The alloying elements are added to obtain particular properties, which vary with the element added, the amount of the element, and combinations of several elements.

HOW STEELS SOLIDIFY

Steels solidify by the nucleation of crystals at the mold wall and growth of crystals in the molten metal as described in Chapter 1, "How Metals Solidify." The amounts of carbon and other elements affect the method of solidification of the steel.

Figure 252 can be used to illustrate how a steel solidifies. The composition of a steel is represented by the vertical lines. Any steel above the temperature line marked "Liquidus" is entirely molten. Any steel below the line marked "Solidus" is entirely solid. At temperatures between the liquidus and solidus, the steel is mushy. Consider a steel containing 0.20 percent carbon that has been heated to 2,900°F. The intersection of the vertical line (0.20 percent carbon) and the horizontal line (2,900°F.) is above the liquidus line, so the steel is entirely molten. As the steel cools, its composition does not change, so this steel is always represented by the vertical line marked 0.20 percent carbon. When the steel cools to 2,775°F., it reaches the liquidus and the first crystals (dendrites) start to solidify. Solidification continues until the temperature reaches the solidus at 2,715°F. At this temperature, the steel is completely solid. Thus, a steel containing 0.20 percent carbon solidifies over a 60°F. range of temperature. Figure 252 shows that with 0.10 percent carbon, the solidification range would be about 70°F.; with 0.30 percent carbon, the range is about 90°F.; and with 0.60 percent carbon, the range is about 160°F.

Except for very low carbon contents, an increase in the carbon content of steel causes it to solidify at a lower temperature and over a wider range of temperature. Laboratory studies have shown that actual solidification for a 0.05 to 0.10 percent carbon steel takes approximately 40 minutes for a square bar, 8 inches by 8 inches. A similar casting of 0.25 to 0.30 carbon steel takes approximately 45 minutes to solidify, and a 0.55 to 0.60 percent carbon steel takes approximately 50 minutes. The longer solidification times of the medium and high-carbon steels permit dendrites to grow farther into the liquid metal and make proper feeding of these castings more difficult. The low-carbon steels, which have a short solidification time, do not have as much growth of dendrites and so are more easily fed to produce sound castings.

PATTERNS

Molds for steel castings must be rammed quite hard to withstand the erosive action of the stream of molten metal and to bear the weight of the casting. The patterns must be sturdy to stand the impact during molding. Any patterns which are to be used repeatedly should be made of aluminum or some other easily worked metal. Many times a wood pattern will be satisfactory if the areas of high impact and wear on a pattern are protected by using metal inserts. However, many accurate steel castings have been made with soft wood patterns.

If a pattern is designed so that hard ramming of the mold makes it difficult to draw, the draft (taper) of the pattern should be checked and increased if necessary. It must be remembered that the harder a mold is rammed, the more draft is required to permit proper drawing of the pattern.

MOLDING AND COREMAKING

SAND MIXES

Sands for molds and cores must be highly refractory in order to resist the intense heat of molten steel. The sand mixes must have good permeability. The binder must not burn out too readily but at the same time cannot have such a high hot strength that defective castings are produced. The synthetic all-purpose sand described in Chapter 4, "Sands for Molds and Cores," has the properties required for a good sand for steel castings and should be used aboard repair ships and at advanced bases for this purpose. Refer to chapter 4, table 8 for typical molding sand mixes, and table 13 for typical core-sand mixes for steel castings.

PROCEDURES

The procedures for molding and coremaking as applied to steel castings are generally the same as for other types of castings. For general procedures, refer to Chapter 5, "Making Molds," and Chapter 6, "Making Cores."

Coremaking for steel castings requires a slight change in practice as compared to the coremaking practices for other metals. Cores must be rammed harder than usual to provide the strength that is required to resist the weight of steel. Hard ramming is also necessary to make the surface resistant to the eroding action of molten steel. Erosion of sand from cores is more of a problem in steel castings than in other metals because the molten steel burns out the binder quickly and sand is easily washed away.

All cores require adequate internal support but cores for steel castings require increased internal support for additional strength because of the high temperatures involved. Internal support for smaller cores is usually by means of reinforcing wires. Larger cores require arbors.

Steel castings hot tear easily if the cores do not have good collapsibility. To avoid this cause of hot tearing, large cores are often hollowed out or the center parts are filled with some soft material that will let the core collapse as the metal contracts around it.

Molding requirements for steel castings are similar to those for core making. Ramming of the mold must be harder than for other types of metals to withstand the weight of steel and provide a surface that will resist the eroding action of molten steel. Steel castings are often so large that the sand must have high strength or be reinforced to support its own weight and maintain dimensions of the mold.

Steel castings can be made in green-sand molds or in dry-sand molds, or in intermediate types such as skin-dried or air-dried molds. The type of casting determines the type of mold to use. Green-sand molds offer little resistance to the contraction of the casting but are relatively weak. However, they are used successfully for many types of steel castings. Dry-sand molds are stronger, and are used in special cases. Molds which have had the surface dried about 1/2 inch deep with a torch are excellent for most general-purpose work.

If a skin-dried or air-dried mold is used for steel and the casting cracks, try a green-sand mold. If the mold deforms or does not hold the steel, try a dry-sand mold.

Washes are extremely useful on molds and cores for steel castings. Without a wash, the molten steel will often penetrate the sand and cause rough "burned-on" surfaces. Washes on a mold or core should be thin and uniform. If a wash is put on a hot core after removal from the core oven, the wash will usually dry properly. A core that is coated with a wash after it has cooled should be returned to the core oven for drying.

When applying a wash to the mold, particular attention should be paid to the coating of deep pockets. Many times a careless application of a wash in a pocket will result in a heavy coating in that area or a thick coating of wash at the bottom of the pocket. Washes applied in this way result in areas of high moisture concentration, which in turn are likely to cause defects. Washes should be applied as uniformly and as thinly as possible and must be dried before use.

GATING

Steel castings are gated according to the principles discussed in Chapter 7, "Gates, Risers, and Chills." The gating of steel castings varies a great deal and it is difficult to provide a basic gating system that can apply to steel castings in general. A few suggested gating sizes are given, but they should be taken only as a starting point for determining the proper gating system. Records of successful steel casting gating systems should be used to determine gating systems for new or unfamiliar castings.

The use of a single ingate is to be avoided because it produces a jet effect in the molten metal stream entering the mold; mold erosion results. Likewise, the gating system should not produce an appreciable drop of the molten steel into the mold cavity. This also produces severe mold erosion.

The use of splash cores at the base of the sprues to prevent erosion is advisable also. If the metal must fall far in the sprue, the entire gating system may be made profitably out of cores that are rammed in the mold. Mold washes on the runners and ingates may also be used to reduce the erosive action of molten steel.

RISERING

The risering of steel castings has been based primarily on experience. However, a great deal of information has been developed recently through research and proved through application to production castings. Much of this information is beyond the scope of this manual.

CHILLS

Because of its high shrinkage during solidification, steel must freeze progressively toward a riser. This is called directional solidification. One way to encourage directional solidification is to pour the metal through the riser and use risers that are large enough to stay fluid while the casting is solidifying. Sometimes, however, the part of the casting farthest from the riser still cools too slowly. When this happens, external chills can be used to speed up the freezing of the metal farthest from the riser. The use of chills for this purpose was discussed in detail in Chapter 7, "Gates, Risers, and Chills." Chills are used more extensively with steel castings than with any other common casting alloy.

Whenever chills are used, they must be clean and dry. External chills are usually coated with plumbago or clay and then thoroughly dried before use. Refer to chapter 7 for special precautions in the use of chills.

External chills used with steel castings should have tapered edges as shown in figure 253. A straight chill causes a sharp change in solidification characteristics at the edge of the chill and often causes a hot tear.

VENTING

Because steel is poured at higher temperatures than other metals, a greater volume of gas must be removed from the mold. Thorough venting of molds for steel is necessary to prevent defects caused by trapped air or gas.

MELTING

TYPES OF FURNACES

Steel can be melted in either an indirect-arc furnace, resistor furnace or induction furnace. These three units are particularly adapted to the melting of steel because of their ability to melt rapidly and to reach high temperatures. For details on furnace operation see Chapter 8, "Description and Operation of Melting Furnaces." Oil-fired crucible furnaces are not satisfactory for melting steel because they will not reach the melting temperature of steel.

PROCEDURE

Indirect-Arc Furnace. The indirect-arc furnace is somewhat difficult to use for melting steel and requires the constant attention of the operator throughout the melting operation. Operation and control of the furnace are described in Chapter 8, "Description and Operation of Melting Furnaces."

Charging. The charging precautions described in chapter 8 should be followed in the charging of steel. After the furnace has been properly preheated, the shell should be rotated until it is 45° down either front or rear from the top center position. The charging position should be varied from time to time to prevent excessive wear on one section of the lining. The ideal method of charging is to have the furnace door in the top center position. The electrodes should be moved back until they are flush with the furnace wall to prevent damage during the charging period.

Heavy pieces should be charged first but should be avoided when possible because they prolong the melting time. Foundry returns (gates and risers) should be charged next and should be as free as possible from sand. Excessive sand causes a slag blanket to form on the surface of the molten metal during the melting cycle. This insulates the bath from the heat generated by the arc and makes it difficult to reach or determine the desired tapping temperature. Steel borings should be added after the returns. They will filter down through the scrap, make the charge more compact, and be away from direct contact with the arc. Structural steel scrap should be charged last (on top). Alloys of nickel and molybdenum may be added to the cold charge. (Alloys that oxidize rapidly, such as chromium, manganese, and silicon, are added just before tapping.) Any alloys of fine size should be charged so that they are not near the arc. This will prevent losses due to "blowing-out" by the arc.

Working the Heat. There are two distinct methods for making steel in the indirect-arc furnace. One is the dead-melting method and the other is the boiling method. The boiling method more nearly approaches the methods developed for commercial steel-melting furnaces and is the recommended method. Up to the time the charge is completely melted, the procedure is the same for both methods.

The metal should be melted as fast as possible and the furnace brought to full rock quickly. In the dead-melting method, additions of ferromanganese and ferrosilicon should be made as soon as the last piece of steel melts. The arc should not be broken and the rocking motion of the barrel should not be changed when making the additions. The alloys should be sized so that they can be charged without difficulty. After the additions have been made, the heat will be almost at the tapping temperature and provisions should be made for tapping shortly thereafter (approximately 5 minutes). The bath should be watched carefully during this stage because the temperature of the molten metal is approaching the fusion point of the furnace lining. If the lining shows any indication of "running," or if a thick slag blanket forms on the surface of the bath, the heat should be tapped immediately.

The dead-melting method described in the preceding paragraph frequently results in porosity as a casting defect. The porosity is produced by gases present in the charge and absorbed during melting.

When using the boiling method (the recommended method), as soon as the charge is completely molten, 2 percent of iron ore should be added to the charge, which has previously been calculated to melt down at 0.25 percent carbon. As soon as the iron ore melts, an immediate reaction takes place between the carbon in the steel and the iron oxide in the ore. This reaction produces carbon monoxide gas. Hydrogen and nitrogen pass into the bubbles of carbon monoxide and are removed from the steel. The carbon monoxide formed in the steel burns to carbon dioxide at the furnace door. As soon as the reaction subsides (which can be determined by the force and brilliance of the flame at the furnace door), the furnace should be rolled so that all excess slag is drained off. At the time the ore is added, the power is reduced approximately one-third and is kept reduced during the balance of the heat. After all the excess slag has been removed, the addition of ferrosilicon and ferromanganese should be made at once followed by any ferrochrome required and the heat tapped within two or three minutes.

The ore addition may be reduced to 1 percent if very rusty scrap is used. This practice of using rusty scrap should be avoided except in an emergency. Larger additions of ferromanganese will have to be made because of the oxidized condition of the bath if rusty scrap is used.

After making several heats of steel by the boiling method, an experienced operator should have no difficulty in controlling composition. Making steel in this manner results in sound castings, but the handling of the slag is a greater problem than with the dead-melting method.

Temperature Control. Proper temperature control for the production of good castings cannot be overemphasized. The proper pouring temperature is determined by the composition of the metal being poured and the size and nature of the casting. A heavily cored mold or one having thin sections will require a higher pouring temperature than other types of castings. Castings with heavy sections require lower pouring temperatures to minimize the attack of the steel on the sand molds and cores.

When melting steel, the optical pyrometer is the best instrument available for shipboard use to determine the temperature of the molten metal. It must be remembered when using the optical pyrometer that the surface of the bath must be free of any slag and the reading must be taken on the surface of the metal. If there is any smoke or fumes in the furnace, an erroneous reading will result. For proper operation of the optical pyrometer, see Chapter 9, "Pouring Castings."

Melting records are useful for estimating the temperature of the bath should an emergency arise because of lack of an optical pyrometer. If accurate melting records are kept to show the power input and optical-pyrometer readings once the charge has become molten, the temperature of the bath in later heats can be estimated from the power input. This procedure is strictly for emergency use and should not be used as standard practice.

Another method useful for determining the proper tapping temperature is to use steel rods. A 5/16-inch diameter rod is cut off to make a blunt end. The rod is then immersed in the steel bath for a short period of time. If the bath is at the proper temperature, the rod will have a rounded end when it is drawn out. The appearance of the rod before and after immersion is shown in figure 254. Considerable skill is required for this test.

De-oxidation and Tapping. The tapping temperature for steel is preferably between 3,000° and 3,100°F. The molten metal should not be held in the furnace any longer than necessary. Holding of the heat in the furnace makes possible the absorption of gases by the melt.

The furnace should be tapped into a well-dried and heated ladle. The ladle lining should be red hot before it is used. When the ladle is half full, the necessary de-oxidizers should be added to "kill" the heat. Additions of calcium-silicon-manganese (Ca-Si-Mn) and aluminum give satisfactory results.

Care must be taken to make certain that the additions are effective. De-oxidizers which are not completely dissolved in the metal or which become trapped in slag are ineffective and result in an incompletely de-oxidized or "wild" heat.

PROCEDURE FOR RESISTOR FURNACE

The mechanical operation of the resistor furnace is the same as for the indirect-arc furnace. Charging is carried out with the same precautions as for the indirect-arc furnace described in Chapter 8, "Description and Operation of Melting Furnaces." For a detailed description of the operation of the resistor furnace, also see chapter 8.

The dead-melting method of making steel is used in the resistor-type furnace. The working, melting, and tapping of the heat are the same as for the indirect-arc furnace. The power input should be maintained constant during the melting cycle and tapping. For a 500-pound capacity furnace, this is 150 k.w.

PROCEDURE FOR CORELESS INDUCTION FURNACE

The dead-melting method may be used with this furnace but does not consistently produce sound castings. It is recommended that the following practice be used for melting all grades of carbon and low-alloy steel.

Charging. The initial charge is made up of gates, risers, and steel scrap. A mixture of 50 percent gates and risers and 50 percent steel scrap makes a satisfactory initial charge. Old castings, if used, should be considered as gates and risers.

The charge is made by placing about 10 percent of steel scrap on the furnace bottom. On top of this, 3 or 4 percent of iron ore is added and then the remainder of the plate scrap. Gates and risers are charged last. Careful handling of the charge material should be followed as standard practice to prevent any damage to the furnace lining.

Working the Heat. As the initial charge melts down, any remaining charge material can be added to the molten metal. Any material added to molten metal should be completely dry. As soon as melting is completed and a temperature of about 3,000°F. is reached, 1-1/2 to 2 percent of pig iron is added. Before being added, the pig iron should be placed on top of the furnace for four or five minutes to warm and to dry. When added, the pig iron should be held under the surface of the bath with a steel rod. The addition of carbon from the pig iron will react with the oxides in the molten metal to produce a boil that will serve to flush out undesirable gases.

After the boil has subsided, ferro-manganese and ferro-silicon should be added. When they are completely dissolved, the heat should be tapped. If the entire heat is to be tapped into one ladle, the power should be shut off and should remain off. If the heat is to be tapped into several ladles, the power should be shut off as soon as the ferro-manganese and ferro-silicon have dissolved. The slag will come to the surface where it can be skimmed off. As soon as the slag is removed, reduced power should be turned on again and allowed to remain on during the tapping of the furnace. POURING DURING TAPPING SHOULD BE AS FAST AS POSSIBLE.

TEMPERATURE CONTROL

Properly calibrated optical pyrometers should be used to obtain temperatures of the molten bath during the melting operation. It is advisable to take numerous temperature readings during the superheating of the melt. The induction furnace produces a very rapid rate of energy input to the melt and this may result in overheating of the metal. Power should be reduced as the desired temperature is approached and an attempt made to reach the final temperature gradually.

DE-OXIDATION

The final de-oxidizers should be added in the ladle when the heat is tapped into one large ladle. When tapped into small ladles, aluminum (2 ounces per 100 pounds of melt) may be added in the furnace with a small extra addition in the small ladle. Power must be turned off and tapping must be rapid after the aluminum is added to the furnace.

POURING

Separate pouring cups or basins are necessary when pouring steel castings. A pouring cup cut into the top of the mold at the sprue does not have the necessary properties to resist the erosive action of the stream of molten steel. Pouring cups or basins for steel should be made from sand with extra binders (core oil or clay), baked, and coated with a refractory wash.

Steel should be poured as rapidly as possible without causing defects such as swells or shifts. These may be caused by excessively high pouring rates. When determining how fast a casting is to be poured, consideration must be given to the structure of the mold. A simple mold without cores can be poured much faster than a heavily cored mold requiring a more extensive gating system. For large steel castings, an average speed of pouring which results in metal rise in the mold of 1 inch per second is satisfactory.

CLEANING

Wire brushing and sand blasting are the best ways to remove adhering sand from steel castings. If sand adheres tightly to the casting, it can usually be removed with a chipping hammer. Many surface defects can be removed by chipping and grinding.

For Grade B and carbon-molybdenum steel, flame cutting with the oxyacetylene, oxyhydrogen, or oxypropane torch is the best method for removing gates and risers. Thorough cleaning is very important to facilitate starting the cut and to insure a uniform cut. The gates and risers should be cut about 3/16 or 3/8 inch from the casting. The remaining stub is then removed by grinding or by the use of power chipping hammers.

The gates and risers of stainless steel castings cannot be removed by flame cutting. They must be removed by mechanical means, such as sawing, chipping, shearing, or an abrasive cutoff wheel, or by melting off with the electric arc from a welding machine. In melting off, care must be taken to leave a stub of 1/4 to 1/2 inch on the casting.

If castings show a tendency to crack during cutting, risers should be removed while the castings are at a temperature over 400°F. For risers larger than 6 inches in diameter, it is well to preheat to 700°F. or higher. The heat may be that remaining in the casting from the mold, or it may be obtained by heating in a furnace.

CAUSES AND CURES FOR COMMON DEFECTS IN STEEL CASTINGS

The casting defects described in Chapter 11, "Causes and Cures for Casting Defects," are generally applicable to steel castings. Defects that pertain particularly to steel castings are described here.

Metal Composition. The control of sulfur and phosphorus content is important in steel castings because an excess of either element harms strength and toughness of the castings.

Sulfur should be kept below the upper limit stated in the specifications (0.06 percent). At this low level, manganese is able to combine with the sulfur and form manganese sulfide, which is not harmful to steel castings. An excess of sulfur will result in "hot shortness," which is brittleness at high temperatures. High sulfur, free machining, and screw stock should not be used in the charge. All sheared ship plate from GSSO will be accompanied by certified analyses. These analyses should be checked for sulfur content. An excess of phosphorus produces a similar effect, but the brittleness occurs at room temperature and is known as "cold shortness."

Pouring. Defects caused by pouring steel castings are generally the same as for other metals. Reference should be made to the section, "Pouring," Chapter 11, "Causes and Cures for Common Casting Defects."

Skimming is particularly important in the pouring of steel castings. Small amounts of slag that get into the mold will be found as isolated inclusions that form weak spots in the casting (nonmetallic inclusions), or as stringers of solidified slag that completely cross a section and make the casting useless. Therefore, extreme care should be taken to remove all slag from the metal surface before it is poured.

Melting practice. Excessive superheating of steel can be a major contributing factor to casting defects. Excessive superheating causes severe oxidation of the melt and the formation of iron oxide. Iron oxide entering the mold can react with carbon to produce gases that cause blows or porosity. Iron oxide can also combine with the molding sand and produce slag inclusions or cause a hard glassy surface of slag to adhere to the casting.

Miscellaneous. Steel castings are very susceptible to hot tearing (as has been mentioned in other parts of this chapter). The direct cause of hot tearing must be determined before corrective measures can be taken. Refer to the "Summary," Chapter 11, "Causes and Cures for Common Casting Defects," for the various causes.

It is important to have the proper degree of collapsibility in cores and molding sand to prevent hot tears. Large cores should be hollowed out before use to provide easier collapsibility.

The design of the casting should not be overlooked as a cause of hot tears. A small change in design can often prevent hot tearing and save a lot of extra work.

WELDING AND BRAZING

Repairs to steel casting by welding and brazing are not difficult but should be made only by qualified personnel. When repairs by welding are required, refer to the "General Specifications for Ships of the United States Navy," Section S9-1, "Welding," for general guidance.

SUMMARY

The production of good castings from steel is more difficult than from other cast metals because of the high temperatures involved. Melting control must be more rigid to prevent an oxidized heat or severe slagging of the furnace lining and its ruin. Sand properties must be controlled much closer than for other metals. In general, for repair ship work, the molder must be more alert to the possibility of defects and take immediate corrective measures while he is going through the various procedures for making a steel casting.

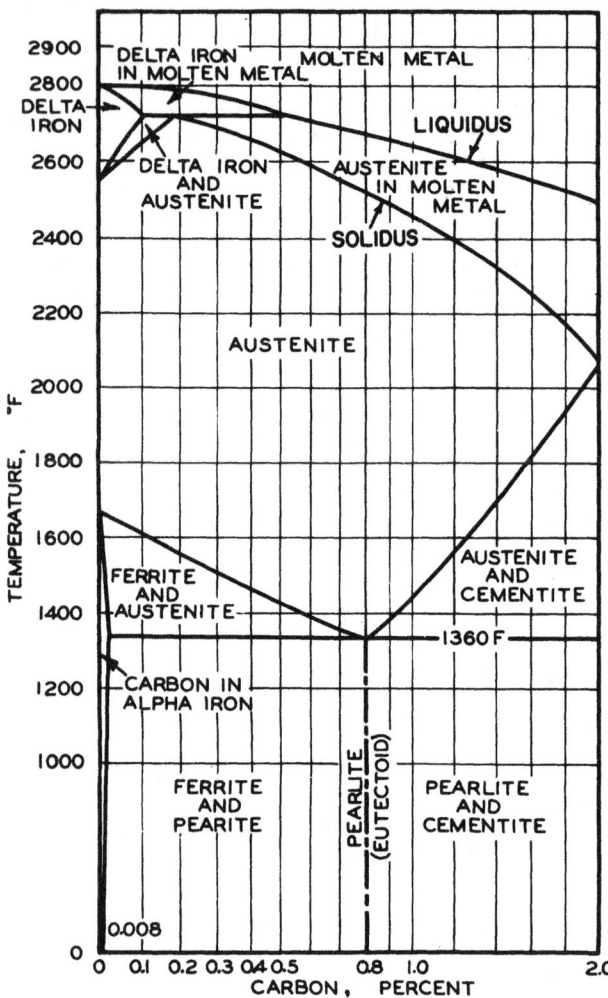

Figure 252. Iron-carbon diagram.

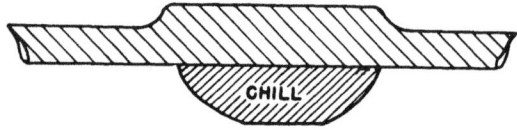

Figure 253. Tapered chill.

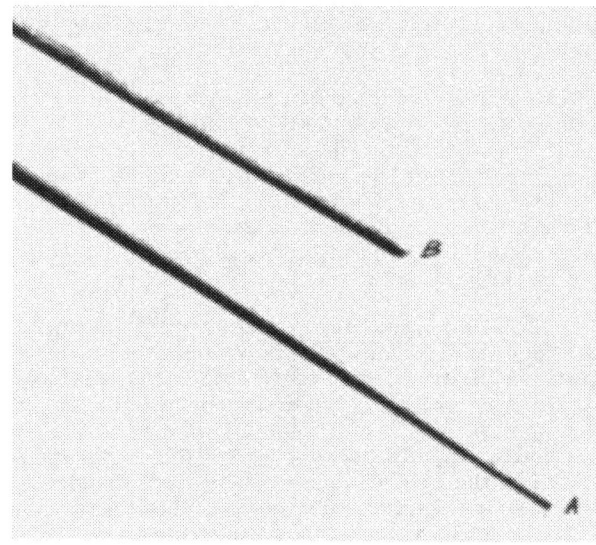

Figure 254. Steel rods used for determining the pouring temperature of steel.

(A) Is a sheared end of the rod before immersion in the molten steel.
(B) Is the rounded end of the rod, showing how the molten steel melts the end of the rod at the correct pouring temperature.

Chapter XIX
COPPER

Copper castings are not included in the Navy Specifications. Many times, however, the need arises for castings of this type. The following information is given as a guide for the production of these castings.

SELECTION OF METAL

Copper castings may be required where unusually high electrical or heat conductivity is needed. The addition of alloying elements to copper reduces its electrical conductivity. The effect of small amounts of various elements on the electrical conductivity of copper is illustrated in figure 255. The conductivities are compared to a specially prepared standard whose conductivity is taken as 100 percent. Notice the drastic reduction in conductivity with very small additions of phosphorus, silicon, or iron. Even brass and bronze castings have a lower electrical conductivity than pure copper.

Castings that must have an electrical conductivity greater than 80 percent must be made from electrolytic copper. Where strength is required and electrical conductivity may be as low as 50 percent, castings can be made from beryllium-nickel-copper (0.5 beryllium, 2.0 nickel, balance, copper) or chromium-silicon-copper alloys (0.08 silicon, 0.80 chromium, balance, copper). Heat treatment of these alloys is required to obtain their best properties.

HOW COPPER SOLIDIFIES

Copper solidifies by the nucleation and growth of crystals as described in Chapter 1, "How Metals Solidify." High-conductivity copper is a pure metal so far as foundry metals are concerned and solidifies at a single temperature instead of over a range of temperature as for most metals. Low-alloy copper has a very narrow solidification range. Because of their narrow solidification range, high-conductivity copper solidifies with strong piping characteristics and requires large risers.

PATTERNS

Pattern practice for copper castings is the same as for copper-base alloys. Refer to Chapter 14, "Copper-Base Alloys." Deep pockets should be avoided because it is hard to make such areas free of defects.

MOLDING AND COREMAKING

The molding and coremaking practices for copper castings are generally the same as for copper-base alloys.

SAND MIXES

Molding sands that are used for copper-base castings can be used for copper castings. However, the use of organic additives in molding sands for copper castings should be kept as low as possible. Organic additives (seacoal, wood, flour, etc.) are likely to generate large amounts of gas, which the copper absorbs readily.

PROCEDURES

Coremaking procedures for copper castings are in general the same as those described in Chapter 6, "Making Cores." Cores should have a high permeability and very good collapsibility.

Oil or cereal binders are usually used. Clay up to 15 percent may also be used as a binder. Regardless of which type of binder is used, it should be kept low. Use only enough binder to hold the core together. Physical strength of the cores is obtained by using core arbors and reinforcing wires. Strong, hard cores should not be used.

Molding procedures are also the same as for copper-base alloy castings. To reduce the possibility of porosity in copper castings, the molds should be skin dried or completely dried. Dry-sand molds are preferred for large castings.

Washes are not used on molds or cores for high-conductivity-copper castings. If a better casting surface is desired, the mold can be dusted with finely powdered graphite. Excess graphite should be carefully removed.

GATING

Top-pouring systems are preferred to establish proper temperature gradients and to obtain good directional solidification. If size or design do not permit top gating and pouring of castings, the best gating system obtainable should be used. In all cases, the gating system must establish proper temperature gradients for strong directional solidification. In this respect, the gating of copper castings is similar to that for steel.

RISERING

Because of the short solidification range of high-conductivity copper, it is often difficult to design risers that will feed the casting correctly. The risers are larger than those used for copper-base alloys. Each casting presents its own problem, but it is suggested that risers comparable in size to those for steel castings be used as a starting point.

CHILLS

Because strong directional solidification is necessary for producing good copper castings, chills should be used to help accomplish it. Chill practice, as described in Chapter 14, "Copper-Base Alloys," can be used as a guide. In this respect, too, copper castings are similar to steel in behavior.

VENTING

Copper in the molten stage absorbs gases readily. Molds should be thoroughly vented to provide easy escape for air or gases that may be generated in the mold.

MELTING

The melting procedure for copper is similar to that for copper-base alloys. The same routine can be followed but much closer control of the heat is necessary.

TYPE OF FURNACE

All of the melting units aboard repair ships can be used for melting copper. The oil-fired crucible furnace is the poorest melting unit because of the difficulty of maintaining close control over the furnace atmosphere and because it is slow.

PROCEDURE

The charge should consist of electrolytic copper, copper wire, bus bar stock, and gates and risers from other high-conductivity castings.

HIGH-CONDUCTIVITY

Gates and risers may be used up to 50 percent of the charge. SCRAP METAL SHOULD NEVER BE USED IN THE CHARGE FOR HIGH-CONDUCTIVITY-COPPER CASTINGS. If the alloyed compositions are to be melted, the alloy additions are made with high-purity master alloys. All melting should be done under slightly oxidizing conditions. Melting should be as fast as possible so that little gas is picked up.

Where the heat is to be poured from the melting crucible (such as those used in the oil-fired crucible furnace) 1/3 to 2/3 of the final weight is charged into the crucible. This may be electrolytic copper or returns. It is melted down and black copper oxide (1/2 pound per 100 pounds of melt) added and stirred into the melt. The balance of the charge is then melted, skimmed, deoxidized, and poured.

The procedure for electric melting furnaces is the same, except for the copper oxide addition. This addition is made in the bottom of the pouring ladle before filling it. The ladle is skimmed, the melt deoxidized in the ladle, and the molten metal poured.

TEMPERATURE CONTROL

Temperature control is important when melting high-conductivity copper because copper absorbs gas readily. The amount of gas dissolved increases very rapidly with increasing temperature and makes the problem of gas removal much more difficult. Control should be maintained with a properly calibrated immersion pyrometer. The metal should be melted fast with no overheating.

DEOXIDATION

Deoxidation is done in the melting crucible in the case of oil-fired crucible furnaces, or in the pouring ladle for the other types of furnaces. Deoxidation is usually done with 2 ounces of 15 percent phos-copper per 100 pounds of melt. For the alloyed compositions, deoxidation can be done by adding 1 percent of 2 percent lithium copper. If possible, this should be followed by bubbling dry nitrogen gas through the melt for approximately three minutes. This can be done with a stainless steel tube that is perforated in the immersion end of the tube.

As a check on a heat to determine whether it is properly deoxidized, the following test may be used. Before pouring a casting, a small sample approximately 2 inches high and 2 inches in diameter should be poured in a dry core-sand mold. If the heat is properly deoxidized, the sample will show a strong pipe as shown in figure 256. If it solidifies with a flat surface, or with a slight pipe as illustrated in figure 257, it requires further deoxidation. A puffed-up surface on the sample indicates a badly gassed heat, as shown in figure 258. Notice the gas holes in figure 257.

POURING

Pouring of copper castings should be done as quickly as possible to prevent cooling of the melt and occurrence of defects such as laps and cold shuts. Pouring temperatures range from 2,100°F. to 2,160°F. The castings should preferably be poured on the high side of this range. Pour fast and with the lip of the ladle close to the sprue. The most common defect that is encountered from pouring practice is the cold shut, caused by too low a pouring temperature. As mentioned previously, copper castings should be poured fairly hot, but the metal should not be overheated so that it absorbs gas.

CLEANING

Sand adherence is not a problem with copper castings. Sand can be easily removed by wire brushing and sandblasting. Gates and risers must be removed by cutting, usually by sawing. The castings should be permitted to cool in the mold to room temperature when high conductivity is required.

CAUSES AND CURES FOR COMMON CASTING DEFECTS IN COPPER

The defects described in Chapter 11, "Causes and Cures for Common Casting Defects," apply generally to copper castings. There are some that occur more frequently in high-conductivity-copper castings and will be discussed here.

METAL COMPOSITION

The principal defect caused by improper metal composition would not be called a casting defect, but can render a copper casting unfit for its intended service. This defect is the reduction of electrical conductivity. The use of phosphorus as a deoxidizer is particularly important. Phosphorus that dissolves in copper because of large additions for deoxidation causes a big reduction in electrical conductivity. Phosphorus is the worst offender in this respect (as shown in figure 255), but is still commonly used as a deoxidizer (as phos-copper).

MELTING PRACTICE

Gassy melts caused by poor melting practice produce badly gassed copper castings. Close control over the furnace atmosphere (by maintaining slightly oxidizing conditions) and proper deoxidation measures are the only way to avoid gassy melts and defective castings.

SUMMARY

The information given in this chapter is given as a starting point for personnel who may be required to make copper castings. The production of castings of this type is not in great demand. Copper castings are not strong and, because they are difficult to cast, should be used only when specifically needed for their high electrical conductivity.

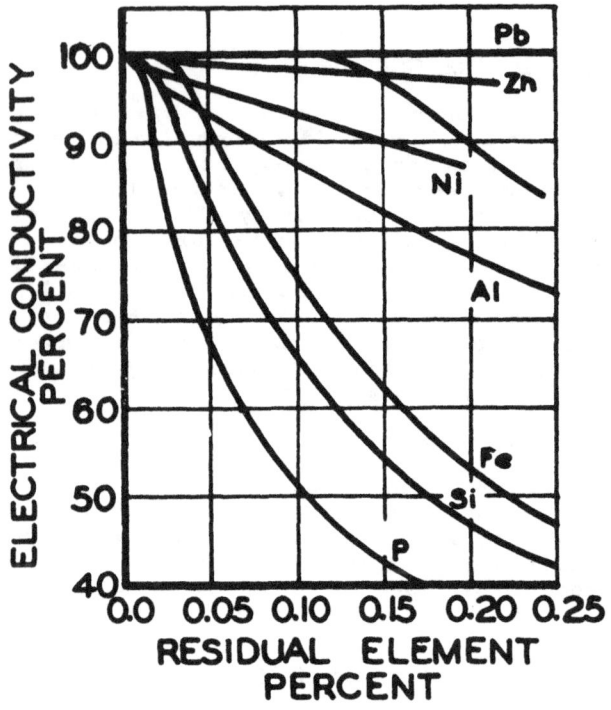

Figure 255. The effect of various elements on the electrical conductivity of copper.

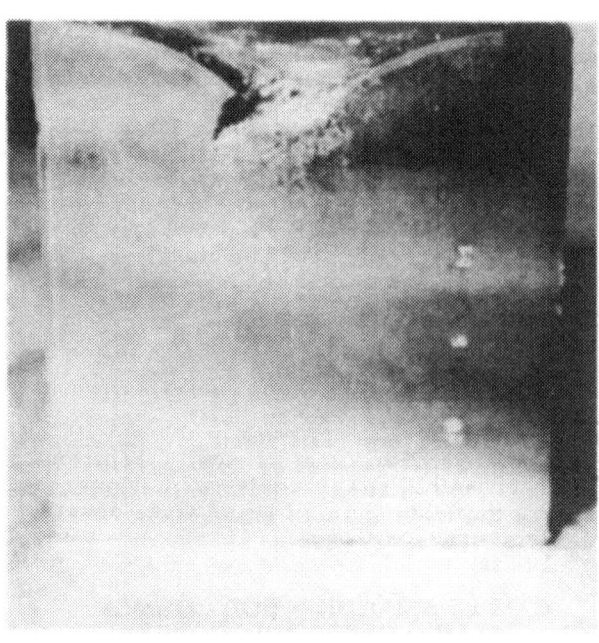

Figure 257. Partially deoxidized copper sample

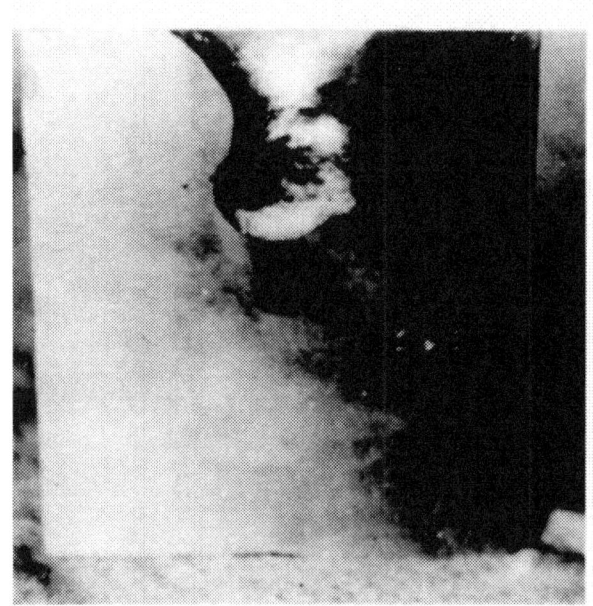

Figure 256. Properly deoxidized copper sample.

Figure 258. Gassy copper sample.

Chapter XX
BABBITTING WITH TIN-BASE BEARING METAL

Bearing metals must have certain special properties: (1) the ability to retain an oil film, (2) resistance to scoring and galling, (3) the ability to imbed foreign particles in themselves, and (4) the ability to deform within very slight limits.

The ability of a bearing material to retain an oil film on its surface is necessary for proper lubrication of the moving part in the assembly. Failure of the bearing material to retain an oil film or lack of its ability to re-establish the oil film when it is broken causes premature breakdown of the bearing.

Because shafts and bearings are not perfectly smooth, there are instances when there is metal-to-metal contact. Foreign particles in the lubricant cause a momentary breakdown of the oil film and permit metal-to-metal contact. Also, many bearings start under a load that causes a momentary metal-to-metal contact. During momentary periods of metal-to-metal contact, severe seizure or galling can occur between the surfaces in contact. It is during this critical period that resistance to seizure and galling is necessary.

Foreign particles (such as dirt or metal filings) are always present in a lubrication system. When these particles reach the bearing, it must have the ability to embed the particles in the bearing material. Lack of ability to do this can lead to serious damage or failure of the bearing assembly.

Perfect alignment of a bearing and shaft is impossible for normal equipment. The ability to deform is necessary in a bearing material to permit it to conform with slight misalignments in bearing and shaft assemblies. This property permits bearings to be "worked-in" to obtain good operation of the shaft and bearing.

SELECTION OF ALLOYS

There are several types of castable bearing materials that are in use for machinery applications. They are: (1) tin-base babbitts, (2) lead-base babbitts, (3) cadmium-base alloys, and (4) copper-lead alloys. The tin-base babbitts are the only ones covered in the Specifications and will be discussed here.

Tin-base babbitts contain 80 to 90 percent tin and are alloyed with copper and antimony. The alloying elements are added to increase the hardness of the babbitt. The bearing properties of tin-base babbitts are excellent. Their strength decreases as they get hotter.

PREPARATION OF BEARINGS

The proper cleaning of the shell cannot be stressed too strongly in bearing preparation. No matter how well the other operations may be conducted, an improperly cleaned shell will result in bearing failure.

REMOVAL OF OLD BABBITT

In the preparation of the bearing shell for rebabbitting, all old babbitt should be machined out of the bearing. Dirt and other foreign matter can be removed from the shell by warming it. Be careful to avoid overheating or the subsequent tinning will be difficult. Machining of the shell should be the last step in the mechanical cleaning of the shell. The finish cut should leave a fine machine-tool finish and the base metal of the shell should be completely exposed. No old babbit should be left on the shell.

CLEANING

Oil or grease should be removed by first washing in a degreasing solvent. The use of a vapor degreaser is recommended if available. An alternative method is to completely immerse the shell in a boiling alkaline-cleaner solution (4 to 6 ounces of Oakite in one gallon of water) or a lye solution (6 ounces of lye to one gallon water) for 20 minutes. The shell then should be washed in boiling water. Any oxide film deposited during the degreasing and washing operations should be removed by pickling the shell in a 25 percent hydrochloric acid solution (1 part HCl to 3 parts water). The shell should then be washed again in boiling water. During the cleaning operation, the shell should be handled with a pair of clean, grease-free tongs. Holes through the shell should be plugged with dry asbestos or magnesia. Surfaces which are not to be tinned should be coated with fire clay wash.

FLUXING

After cleaning thoroughly and pickling, the surface to be babbitted should be fluxed either by dipping or swabbing. A suitable flux can be prepared by mixing equal parts by weight of zinc chloride and water. A flux may also be made by mixing 11 parts of commercial grades of zinc chloride with 1 part of granulated ammonium chloride (sal ammoniac). Boiling water should be stirred into this mixture until a hydrometer reading of 52° Baumé is reached.

It is important that freshly prepared flux be used. Oil or weak flux will not properly clean the surface.

TINNING

Procedures. Bronze and steel shells should be coated with a thin layer of tin before babbitting. Tin will aid in producing a strong bond between the babbitt and shell. After the shell has been fluxed, it should be dipped in a bath of pure tin maintained at a temperature of 570° to 580°F. The bearing shell is dipped into the bath and held there for 55 seconds. The shell is then removed and inspected for flaws. Those areas which have not been adequately coated should be scraped, refluxed, and retinned. As soon as an adequate coat of solder is assured, the bearing shell should be set in the jig and mandrel for immediate pouring.

Jig and Mandrel Assembly. Figure 259 shows a simple jig for use in babbitting bearings. Before tinning, the two halves of the bearing are wired together with 1/8 to 1/4 inch-thick spacer pieces of metal wrapped with asbestos paper between them. A center core or mandrel of steel of the correct diameter is bolted to the base plate. Steel tubing or pipe thoroughly cleaned is satisfactory. The mandrel should be 3/8 to 3/4 inch smaller in diameter than the diameter of the finished bearing to allow for peening and subsequent machining. A cross-bar is bolted to the top of the mandrel after the bearing is set in place and wooden wedges are driven between the crossbar and the pouring lip to hold the bearing and lip in place during pouring. While the bearing shell is being cleaned, fluxed, and tinned, the mandrel must also be prepared for pouring.

PREHEATING

The mandrel should be preheated with a torch to at least 600°F. This reduces the chilling effect of the mandrel, permits solidification to start at the bearing shell and progress toward the mandrel, and prevents the babbitt from pulling away from the bearing shell during solidification.

Care must be taken also that no moisture remains on the bearing shell. Moisture will cause the hot babbitt to spatter dangerously and will cause porosity in the liner.

As soon as the mandrel has been thoroughly preheated and the bearing tinned, the bearing is set in the jig and centered. The pouring lip is set in place and the crossbar bolted in place and wedges driven in. The jig assembly should be carefully checked for leaks and made tight with calking clay or putty to prevent metal loss during pouring.

MELTING

Melting of any of the tin-base antifriction metals can be accomplished in a pressed steel, cast steel, or cast iron pot.

EQUIPMENT

Melting is done by means of an ordinary oil or gas torch or by a burner specially designed for the purpose. A muffle-type furnace may also be used for melting these metals.

PROCEDURE

A sufficient quantity of the proper grade of ingot should be charged in the pot and the heat applied slowly so that the melting does not take place too fast. As soon as the babbitt begins to melt, the surface may be covered with powdered charcoal. This will protect the metal from the air and retard the formation of oxides and accumulation of dross. Care must be taken to remove the charcoal before pouring to prevent it from becoming trapped in the babbitt during pouring.

When the bars or ingots are all melted, the bath should be stirred with a motion from bottom to top (not circular). No splashing should be permitted.

TEMPERATURE CONTROL

The quality of the babbitted bearing depends mainly upon the temperature at which the metal is poured. Too high a pouring temperature will increase the amount of shrinkage during solidification and this will create more severe shrinkage stresses. A high pouring temperature will heat up the bearing and mandrel and tend to keep metal in the "mushy" state for too long a time. The shrinkage stresses may produce cracks which will cause bearing failure during service. The temperature needed at the end of melting will depend on the distance from the melting position to the point of pouring. Generally, 675°F. to 690°F. is the best pouring temperature for Grade 2. An immersion-type pyrometer should be used in all cases to control the temperature of the bath and of pouring.

POURING

The babbitt should be poured as soon as possible after the assembly is ready. Good bonding of the babbitt to the shell will not be obtained if the tin bond has become solid. Pouring must be accomplished while the tinned surface on the shell is still hot.

LADLES

The size of the ladle and quantity of metal melted should be governed by the size of the bearing to be babbitted. More metal should always be melted than is actually required because the excess can be pigged and remelted. The ladle or ladles should always be large enough to hold more metal than required for the pour. Twice the amount of babbitt required should be melted to prevent shilling of the metal in the ladle. If an insufficient amount of metal is poured into the cavity, subsequent addition of metal will produce a defective bearing because the second metal will not bond with the originally poured metal that has solidified.

When a single ladle of sufficient size is not available, two ladles may be used if the contents of both are poured into the bearing at the same time. If there is insufficient room for both ladles over the pouring jig, the contents of one ladle may be poured directly into the bearing, while the metal from the second ladle is being poured into the first.

SKIMMING

A bottom-pour or self-skimming type of ladle is preferred because it prevents dross, dirt, oxides, or other impurities from entering the mold. If such a ladle is not available, a lip-pour ladle may be used. When using a lip-pour ladle, careful attention must be given to skimming of the surface of the molten metal to produce a bright surface free from oxides. Any scum which forms on the top of the molten metal must be pushed back from the lip with a wood or metal rod.

The metal should be stirred thoroughly and poured slowly and steadily in a thin stream. A fast heavy stream fills the opening too rapidly to allow the necessary escape of air and causes it to be trapped as bubbles or seams in the lining. A slow steady stream will prevent this.

Solidification shrinkage will take place in antifriction tin-base babbitts as in other casting metals. During pouring, the chilling action of the steel mandrel and the bearing shell will usually cause solidification to start before the mold is filled. Liquid and solidification shrinkage will take place in the first metal to enter the mold as soon as the freezing temperature is reached. Because the babbitt is generally poured slowly, automatic feeding will take place (the early shrinkage of the solidifying metal is compensated for by metal added during pouring). The amount of feeding is dependent upon the temperature of the metal, rate of pouring, and the thickness of sections involved.

PUDDLING

Pouring should be continued until the bearing cavity is filled to the top. The elimination of gas and air from the bearing is of vital importance. This is usually accomplished by thorough puddling of the metal during pouring. Generally, a flat steel rod 3/8 inch by 1/16 inch in cross section which has been previously preheated by immersing the rod in the molten babbitt is used. As soon as the first metal is poured, the puddling or churning should be started. This action, which permits efficient feeding, should be continued throughout pouring. Molten babbitt should be added as contraction occurs until the mold has been completely filled. Immediately after pouring is completed, cooling of the bearing should be started with a water spray. A circular spray jig made from a piece of pipe is ideal for this purpose.

POST POURING OPERATION

When the babbitt has solidified, it will be seen that the sides of the shell have been drawn together by the contraction of the lining. The shell may be returned to its original dimensions by peening the inner surface of the lining. The spacers are removed from between the shells and the halves separated by sawing through the babbitt on each side.

FINISHING OF BEARING

The bearings are then set up for machining to the correct dimensions. After machining, the bearing should be carefully fitted to the shaft whenever possible. This is done by coating the shaft with blue chalk and setting the bearing in place. High spots as indicated by the adherence of chalk to the babbitt are scraped off and the operation repeated until a close fit is obtained with maximum contact with the shaft.

BEARING FAILURES

Checked, cracked, or crumbling bearings were probably improperly bonded or poured at too high a temperature, thus causing high contraction stresses to occur during cooling. The proper relation between the temperatures of the metal, mandrel, and bearing shell is an important factor in producing sound babbitted bearings.

The presence of air pockets in the lining or between lining and shell can quickly produce bearing failure. Oil will collect in these pockets and prevent uniform transfer of heat to the shell. A hot spot can develop and progress until the bearing metal is hot enough to melt.

A properly cast babbitt liner with a homogeneous structure will give good service. During the running in of a bearing, the load and vibration will cause the metal to pack, closing any minute voids and making the metal more dense. This close structure will give the maximum bearing life.

An outline of the preceding procedures is also contained in paragraph 43-36, Chapter 43 of the Bureau of Ships Manual. The "United States Naval Engineering Experiment Station Report C-3230-B," dated 6 May 1949, recommends the use of a wood mandrel and wood riser. Heating of the mandrel and puddling are not required with this method.

SUMMARY

The production of good bonds in tin-base babbitts depends on: (1) proper cleaning of the shell, (2) correct tinning temperature, (3) proper time of immersion for the tinning operation, (4) correct pouring temperature, and (5) rapid cooling of the bearing when pouring is finished. Strict control of these steps is necessary for the production of good tin-base babbitt bearings.

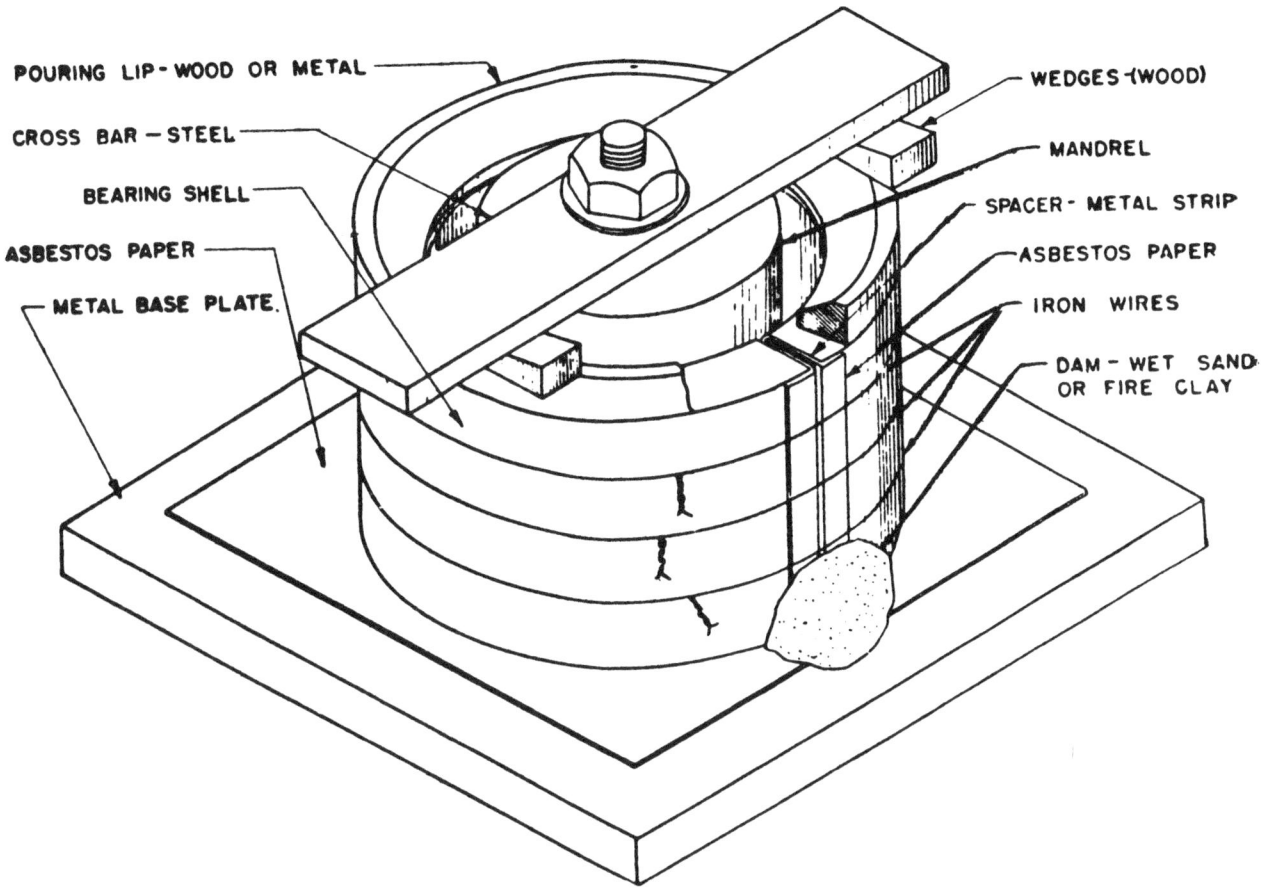

Figure 259. Jig for babbitting bearings.

Chapter XXI
PROCESS CONTROL

The ability to make good castings depends on experience. To make the most of experience, it is necessary to keep written records of what has been done. A manual such as this one could not have been written unless many people had made many castings and kept records of what they did. Good records contain information that enable a molder to produce a better casting without repeatedly going through the trial-and-error routine.

Some persons may believe that the keeping of records and their use in repair ship work is a waste of time and not worthwhile. Inspection of repair ship foundries has shown that the foundries with the best reputation for producing good castings are those that keep foundry records and refer to them continually when making new castings.

SAND

Sand records should include three types of information: (1) the properties of sand as it is received by the repair ship foundry, (2) day-to-day test results on the properties of sand mixtures prepared for particular alloys (gray iron, copper base, etc.), and (3) the properties of sand mixtures used successfully for making a particular casting. Records of the properties of sand received from a source of supply indicate the uniformity of the sand supplied. Many times, the sand-grain distribution or the clay content of a sand will change over a period of time because of some change in the supplier's source. A change in these properties can affect the castings made in that sand. The only way to check this as a possible source of trouble is to have records of the properties of each lot of sand received.

Day-to-day testing and recording of sand properties provides the molder with an up-to-date record of his molding sand. Any change in properties will be shown by routine tests. For example, a buildup of clay content or an increase in fines will result in a decrease in permeability and gas defects can be expected if corrective measures are not taken.

A record of sand properties applying to a particular type of casting provides information for repeated production of the same casting or similar castings. Such information can save a lot of time, which may be extremely valuable in an emergency.

MOLDING

Records should be kept of successful gating and risering arrangements for various castings. Photographs or simple snap shots are very helpful for showing gating and risering of loose patterns. Any information pertaining to the method of ramming the mold (whether it was rammed hard or soft), any additives used, venting, size of flask, and depth of flask are items of information that should be included in records of this kind. They are all helpful in determining the causes of various defects. It is a bad practice to depend on memory when correcting casting defects. Some defects may be very apparent, while others may be caused by factors that are not self-evident to the molder.

Gating and risering arrangements are particularly important for future work that may be done. By far the best method of recording a gating and risering system is to make a simple sketch or photograph with pertinent notes and dimensions. Very often, when a new casting is to be made, the gating and risering arrangement will be determined by a guess. If the guess is good, a good casting is the result; bad guess-scrap casting. The records might not show a casting identical in design or size to the one under consideration but good records will often give information that will take the guesswork out of the gating and risering of the new casting.

MELTING

The purpose of melting records is to supply information that permits day-to-day control when melting various metals. Records should contain such information as the type of alloy melted, size of the heat, how it was charged (all or in parts), when alloy additions were made, time to melt-down, holding time, and in the case of electrical melting equipment, the power input during various periods of the melting cycle.

Any control tests that are available for determining melt characteristics should be used as a routine test. Results of these tests should be incorporated as part of the melting record. If a melter is to make effective use of a control test, he must use it repeatedly to become familiar with it, and must have records to show him what the test means in terms of metal quality.

The chill test for gray cast iron is discussed in Chapter 17, "Cast Iron." The results of routine chill tests can be compared with casting defects that may be caused by poor melting practice, or with machining characteristics. After some experience is obtained in its use, a heat that would produce an iron hard to machine can be corrected before being poured into the mold.

Heats of many metals can also be checked by means of a fracture test. A trial heat can be melted under different conditions varying from strongly oxidizing to strongly reducing. A test sample can be poured when the melt has become stabilized under each condition. The test samples are then fractured and the fractured surface examined. This group of samples will then be a set of comparison standards that will be useful for showing the particular conditions under which other heats of the same alloy have been melted. If clear lacquer is applied to the fractured surface, the fractured sample can be used for a long time. Another procedure which is less satisfactory is to set up a record that describes the character of the various types of fractures. This should include the comparative grain size, color, and whether the fracture was light and solid or whether it had an open structure.

INSPECTION AND TEST

For inspection and test requirements refer to the "General Specifications for Ships of the United States Navy," Section S-1-O, "Castings." Inspection and tests required therein shall be performed within the capabilities of the inspection and test equipment available. Refer to the specification numbers in the material specifications. These cover detailed inspection requirements.

SUMMARY

Good records are not only a means of control for producing sand of uniform properties and melts of good quality, they are also the best source of information for inexperienced personnel. It is true that a molder can learn by experience, that is, by making molds, but he can learn still more in less time by profiting from someone else's work. This information can be obtained only from properly kept records. Records should also contain methods that resulted in poor results so that such errors may be avoided in future work.

FOUNDRY TERMINOLOGY

A.

Abrasion. Wearing away of material at its surface because of the cutting action of solids.

Abrasives. Materials for grinding, polishing, cleaning, etc., bonded to form wheels, bricks, and files, or applied to paper and cloth by means of glue. Natural abrasives include emery, corundum, garnet, sand, etc. Main manufactured abrasives are silicon carbide and aluminum oxide. Metallic shot and grit also are used as abrasives in cleaning castings.

Acetylene. A colorless, tasteless gas composed of carbon and hydrogen and used for welding and cutting operations in combination with oxygen. It is really generated by action of water on calcium carbide.

Acid. With reference to refractories, those materials high in silica or in minerals chemically similar to silica and low in bases such as lime or magnesia.

Acid lining. In a melting furnace, the lining composed of materials that have an acid reaction in the melting process - either sand, siliceous rock, or silica bricks.

Acid brittleness. Brittleness induced in steel when pickling (cleaning) in dilute acid to remove scale, or during electroplating; commonly attributed to absorption of hydrogen. Also commonly called hydrogen brittleness.

Acid refractories. Ceramic materials of high melting point consisting largely of silica.

Acid steel. Steel melted in a furnace that has an acid bottom and lining, under a slag that is mainly siliceous.

Aerator. A machine for fluffing or decreasing density of sand and for cooling sand by mixing with air.

Age hardening. An aging process that increases hardness and strength and ordinarily decreases ductility, usually following rapid cooling or cold working. See AGING.

Agglomeration (flocculation). Gathering together of small particles into larger particles in a liquid medium; usually used in connection with fineness test of clay or foundry sand.

Aging. In an alloy, a change in properties that generally occurs slowly at atmospheric temperature and more rapidly at higher temperatures. See AGE HARDENING, PRECIPITATION HARDENING.

Air compressor. Machine to build up air pressure to operate pneumatic tools and other equipment.

Air control equipment. Device for regulating volume, pressure, or weight of air.

Air drying. Surface drying of cores in open air before baking in an oven; also applies to molds which air dry when left open, thus causing crumbling or crushing when metal is poured; a core or mold dried in air without application of heat.

Air-dried strength. Compressive, shear, tensile, or transverse strength of a sand mixture after being air dried at room temperature.

Air furnace. Reverberatory-type furnace in which metal is melted by the flame from fuel burning at one end of the hearth, passing over the bath toward the stack at the other end of the hearth; heat is also reflected from the roof and side walls.

Air hammer. Chipping tool operated by compressed air.

Air hoist. Lifting device operated by compressed air.

Air hole. A cavity in a casting, caused by air or gas trapped in the metal during solidification. More commonly called "gas hole."

Alkali metals. Metals in Group IA of the periodic system, including lithium, sodium, potassium, rubidium, cesium, and francium.

Alkaline earth metals. Metals in Group IIA of the periodic system, including calcium, strontium, barium, and radium.

Allotropy. Occurrence of an element in two or more forms. For example, carbon occurs in nature as the hard crystalline diamond, soft flaky crystalline graphite, and amorphous coal (lamp black).

Alloy. A substance having metallic properties and composed of two or more chemical elements, of which at least one is a metal.

Alloying elements. Chemical elements constituting an alloy; in steels, usually limited to metallic elements added to modify the

A.—Continued

properties of the steel; chromium, nickel, molybdenum are examples.

Alpha iron. The form of iron that is stable below 1670°F. See TRANSFORMATION TEMPERATURE.

Alumel. A nickel-base alloy containing about 2.5 percent Mn, 2 percent Al, and 1 percent Si, used chiefly as a part of pyrometric thermocouples.

Anchor. Appliance for holding cores in place in molds.

Annealing. A process involving heating and cooling, usually applied to cause softening. The term also applies to treatments intended to alter mechanical or physical properties, produce a definite microstructure, or remove gases. Any process of annealing will usually reduce stresses, but if the treatment is applied for the sole purpose of such relief, it should be designated as stress relieving.

Annealing pots. Iron containers in which castings are packed for protection against the furnace atmosphere during annealing.

Arbors. Metal shapes embedded in and used to support cores.

Arc welding. Welding accomplished by using an electric arc that may be formed between a metal or carbon electrode and the metal being welded; between two separate electrodes, as in atomic hydrogen welding; or between the two separate pieces being welded as in flash welding.

Arrestor, dust. Equipment for removing dust from air.

Artificial aging. An aging treatment above room temperature. See PRECIPITATION HEAT TREATMENT.

Artificial sand. Product resulting from crushing a rock to the size of sand grains.

Asbestos. Hydrated magnesium silicate often used for insulation of risers, thus keeping metal molten for feeding purposes.

Austenite. Solid solution in which gamma iron is the solvent, the non-magnetic form of iron found in 18-8 stainless steels.

Austenitizing. Process of forming austenite by heating a ferrous alloy into the transformation range (partial austenitizing) or above the transformation range (complete austenitizing).

B.

Back draft. A reverse taper which prevents removal of a pattern from the mold.

Backing board. A second bottom board on which molds are opened.

Backing plate. See BACKING BOARD.

Backing sand. Reconditioned sand used for ramming main part of mold or loose molding sand used to support green cores while baking.

Bail. Hoop or connection between the crane hook and ladle.

Baked core. A sand core which has been heated.

Baked permeability. The property of a molded mass of sand which allows gases to pass through it after the sand mass is baked above 230°F. and cooled to room temperature.

Baked strength. The compressive, shear, tensile, or transverse strength of a sand mixture when baked at a temperature above 230°F. and then cooled to room temperature.

Balanced core. One with the core-print portion so shaped and dimensioned that it will overbalance that part of the core extending into the mold cavity.

Band, inside. A steel frame placed inside a removable flask to reinforce the sand.

Band, snap flask. See JACKET, MOLD.

Band saw. A saw in the form of an endless steel belt which runs over a pulley.

Bars. Ribs of metal or wood placed across the cope portion of a flask. Sometimes called "cleats."

Base permeability. That physical property which permits gas to pass through packed dry sand grains containing no clay or other bonding substance.

Basic. A chemical term which refers to a material which gives an alkaline reaction. In refractories, the chemical opposite of acid. See ACID.

Basic lining. In a melting furnace, the inner lining composed of materials that have a basic reaction in the melting process. Crushed burned dolomite, magnesite, magnesite bricks, and basic slag are examples.

- 272 -

B.—Continued

Basic pig iron. A special high-phosphorus (2.0 to 2.5 percent), low-sulfur (0.08 percent), low-silicon (0.80 percent) pig iron made for the basic open hearth process for steelmaking.

Basic refractory materials. Bodies containing basic oxides which react with acids to form salts. Magnesia and lime are examples.

Basic steel. Steel melted in a furnace that has a basic bottom and lining, and under a slag that is predominantly basic.

Bath. Molten metal contained in hearth of furnace during melting process.

Battens. Wooden bars or strips fastened to patterns for rigidity or to prevent distortion during ramming of the mold.

Baumé. A measure of specific gravity of liquids and solutions reduced to a simple scale of numbers.

Bauxite. An ore of aluminum consisting of moderately pure hydrated alumina, $Al_2O_3 \cdot 2H_2O$.

Bayberry wax. Wax made from bayberries, used for coating patterns.

Beam and sling. Tackle used in conjunction with a crane for turning over the cope or drag of a mold prior to assembly.

Bedding a core. Resting an irregular-shaped core on a bed of sand for drying.

Bed-in. Method whereby drag may be rammed in the pit or flask without necessity of rolling over; process used for production of heavy castings.

Bellows. A baglike hand-operated air blowing device for removing loose sand or dirt from molds or parting sand from patterns.

Bench molder. A craftsman who makes molds for smaller type castings, working at the molder's bench only.

Bentonite. A colloidal clay derived from volcanic ash and employed as a binder in connection with synthetic and silica sands, or added to ordinary natural (clay-bonded) sands where extra dry strength is required; found in South Dakota and Wyoming, also in Africa and India. (Western bentonites are sodium bentonites; southern bentonites are calcium bentonites.) Their properties are slightly different.)*

Binary alloy. An alloy containing two principal elements.

Binder. Material to hold sand grains together in molds or cores, such as cereal, pitch, resin, oil, sulfite by-product, etc.

Black lead. A form of graphite used for coating green sand molds and cores, applied as a water suspension to skin-dried molds (in some cases brushed on dry).

Black wash. Graphite applied as a water suspension to mold and core; it is customary to add a bonding agent such as bentonite, fireclay, dextrin, molasses, etc.

Blacking. A thin facing of carbonaceous materials, such as graphite or powdered charcoal, used to finish mold surfaces and protect the sand from the hot metal.

Blacking holes. Irregular-shaped surface cavities containing carbonaceous matter, sometimes found in defective castings.

Blacking scab. A casting defect formed by blacking flaking off because of sand expansion and being retained in or on the surface of the metal.

Blast cleaning. Removal of sand or oxide scale from castings by the impact of sand, metal shot, or grit projected by means of air, water, or centrifugal pressure. See BLASTING.

Blasting. A process for cleaning or finishing metal objects by use of air blast that blows abrasive particles against the surfaces of the work pieces. Small, irregular particles of steel are used as the abrasive in grit blasting, sand in sand blasting, and steel balls in shot blasting. See BLAST CLEANING.

Bleeder. A defect wherein a casting lacks completeness because of molten metal draining or leaking out of some part of the mold cavity after pouring has stopped.

Bleeding. See SLUSH CASTING.

Blended molding sands. Mixtures of molding sands made to produce desirable sand properties.

Blind riser. A riser which does not break through the top of the cope, and is entirely surrounded by sand; often combined with whirl gates, together forming an efficient method of gating and feeding a casting.

---

*Grim, Ralph E.; Clay Mineralogy; McGraw-Hill Book Company, Inc.; New York (1953).

B.—Continued

Blister. A shallow blow with a thin film of iron over it appearing on the surface of a casting.

Blow. Forcing of air into the molten metal due to insufficient venting.

Blower. Machine for supplying air under pressure to the melting unit.

Blow gun. Valve and nozzle attached to a compressed air line to blow loose sand or dirt from a mold or pattern; used also to apply wet blacking.

Blowhole. Irregular-shaped cavities with smooth walls produced in a casting when gas trapped while the mold is being filled or evolved during solidification of the metal fails to escape and is held in pockets.

Blows. Rounded cavities, that may be spherical, flattened, or elongated, in a casting caused by the generation or accumulation of gas or entrapped air.

Bond, bonding substance, or bonding agent. Any material other than water which, when added to foundry sands, imparts strength to them.

Bond strength. Property of a foundry sand by virtue of which it offers resistance to deformation and holds together.

Bonding clay. Any clay suitable for use as a bonding material.

Boss. A projection on a casting that can be used for various purposes, such as drilling and tapping holes for bolts.

Bot. A cone-shaped lump of clay attached to the end of an iron or wooden "bot stick" used to close the taphole of a furnace.

Bot stick. An iron rod, with a loop or long wooden handle at one end and a small round disk at the other, to receive the clay for botting off when the ladle is sufficiently full.

Bottom board. A flat base of wood or metal for holding the flask in making sand molds.

Bottom running or pouring. Filling of a mold from the bottom by means of gates from the runner.

Box. See FLASK.

Brass. Copper-base metal with zinc as the major alloying element.

Brazing. Joining metals by melting of nonferrous alloys that have melting points above 800°F. but lower than those of the metals being joined. This may be accomplished by means of a torch (torch brazing), in a furnace (furnace brazing), or by dipping in a molten flux bath (dip or flux brazing). The filler metal is ordinarily in rod form in torch brazing; whereas in furnace and dip brazing the work material is first assembled and the filler metal may then be applied as wire, washers, clips, bands, or may be integrally bonded.

Breeze. Coke or coal screenings.

Bridging. Local freezing across a mold before the metal solidifies. Also locking a part of the charge in the crucible above the molten metal while melting.

Brinell hardness. The hardness of a metal tested by measuring the diameter of the impression made by a ball of given diameter applied under a known load. Values are expressd in Brinell hardness numbers.

British thermal unit (B. t. u.). The quantity of heat required to raise the temperature of 1 lb of water 1°F.; a unit of heat measurement.

Bronze. Copper-base metals with tin as the major alloying element.

Buckle. An indentation in a casting resulting from expansion of the sand.

Built-up plate. A pattern plate with the cope pattern mounted on or attached to one side and the drag pattern on the other. See MATCHPLATE.

Bull ladle. A large ladle for carrying molten metal, has a shank and two handles, and may have a geared wheel for tilting.

Burned sand. Sand in which the binder has been destroyed by the heat of the metal.

Burning in. Rough surface of a casting due to metal penetrating into the sand.

Burnt. Term applied to a solid metal permanently damaged by having been heated to a temperature close to the melting point. Damage is caused by severe oxidiation.

Burnt-on sand. A phrase developed through common usage and indicating metal penetration into sand resulting in a mixture of sand and metal adhering to the surface of the casting. Properly called metal penetration.

Butt rammer. The flat end of the molder's rammer.

C.

Calcium carbide. A grayish-black, hard, crystalline substance made in the electric furnace by fusing lime and coke. Addition of water to calcium carbide forms acetylene and a residue of slaked lime.

Carbide. A compound of carbon with one metallic element. (If more than one metallic element is involved, the plural "carbides" is used.)

Carbonaceous. A material containing much carbon. Examples are coal, coke, charcoal, and graphite.

Carbon equivalent. A relationship of the total carbon, silicon, and phosphorus content in a gray iron expressed by the formula

$$C.E. = T.C.\% + \frac{(Si\% + P\%)}{3}$$

Carbonization. Coking or driving off the volatile matter from carbon-containing materials such as coal and wood. (Do not confuse this term with carburization.)

Carbon steel. Steel that owes its properties chiefly to the presence of carbon without substantial amounts of other alloying elements; also termed "ordinary steel," "straight carbon steel," and "plain carbon steel."

Casting (noun). Metal object cast to the required shape as distinct from one shaped by a mechanical process.

Casting (verb). Act of pouring molten metal into a mold.

Casting, open sand (noun). Casting poured into an uncovered mold.

Casting ladle. A crucible or iron vessel lined with refractory material for conveying molten metal from the furnace and pouring it into the mold.

Casting stress. Residual stresses resulting from the cooling of a casting.

Cast iron. Essentially an alloy of iron, carbon, and silicon in which the carbon is present in excess of the amount which can be retained in solid solution in austenite at the eutectic temperature. That is, some carbon is present as graphite flakes, as in gray cast iron, iron carbide, or in white cast iron. When cast iron contains a specially added element or elements in amounts sufficient to produce a measurable modification of the physical properties of the section under consideration, it is called alloy cast iron. Silicon, manganese, sulfur, and phosphorus, as normally obtained from raw materials, are not considered as alloy additions.

Cast steel. Any object shaped by pouring molten steel into molds.

Cast structure. The structure (on a macroscopic or microscopic scale) of a cast alloy that consists of cored dendrites and, in some alloys, a network of other constituents.

Cementite. A compound of iron and carbon known as "iron carbide" which has the approximate chemical formula Fe_3C.

Centerline. Well-defined gage line placed on the work to serve as a basis from which dimensions are to be measured.

Centrifugal casting. A casting technique in which a casting is produced by rotation of the mold during solidification of the casting; unusually sound castings may be produced by action of centrifugal force pressing toward the outside of the mold.

Chamfer. To bevel a sharp edge by machining or other methods.

Chamotte. Coarsely graded refractory molding material, prepared from calcined clay and ground firebrick mulled with raw clay.

Chaplets. Metallic supports or spacers used in molds to maintain cores in their proper position during the casting process; not used when a pattern has a core print which serves the same purpose.

Charcoal. Used in pulverized form as dry blacking, or in suspension with clay, as black wash.

Charge. A given weight of metal introduced into the furnace; the metal that is to be melted.

Cheeks. Intermediate sections of a flask that are inserted between cope and drag to decrease the difficulty of molding unusual shapes or to fill a need for more than one parting line.

Chill. A piece of metal or other material with high heat capacity and conductivity inserted in the mold to hasten solidification of heavy sections and introduce desired directional cooling.

Chill-cast pig. Pig iron cast into metal molds or chills.

Chilled iron. Cast iron in which sufficient combined carbon is retained to form a mottled or white structure. These conditions result

C.—Continued

from accelerated cooling that prevents normal graphitization in those areas.

Chipping. Removal of fins and other excess metal from castings by means of chisels and other suitable tools.

Chipping-out. The process of removing slag and refuse attached to the lining of a furnace after a heat has been run.

Clamp-off. An indentation in the casting surface caused by displacement of sand in the mold.

Clays. Hydrous silicates of alumina, more or less mixed with mineral impurities and colored by presence of metallic oxides and organic matter. Pure clay is an earth which possesses sufficient ductility and cohesion when kneaded with water to form a paste capable of being fashioned by hand and when suitably burned is capable of resisting intense heat.

Clay substance (AFS clay). Clay portion of foundry sand which when suspended in water fails to settle 1 in. per min and which consists of particles less than 20 microns (0.02 mm or 0.0008 in.) in diameter.

Clay wash. A mixture of clay and water for coating gaggers and the inside of flasks.

Cleaner. A tool of thin steel or brass, 16 to 18 in. long; one extremity has a bent spatula blade; the other, a short blade bent on the flat to a right angle. It is used for smoothing the molded surfaces and removing loose sand; also called a slick.

Coal dust. Used largely in green sand molding compositions for cast iron. A bituminous type of coal is selected with a high percentage of volatile matter, crushed to various grades (coarse to super fine) and additions of 5 to 10 percent added to the facing sand. Also called sea coal.

Cold short. A characteristic of metals that causes them to be brittle at ordinary or low temperatures.

Cold shot. Small globule of metal embedded in but not entirely fused with the casting; a casting defect. (Sometimes confused with cold shut.)

Cold shut. A casting defect caused by imperfect fusing of molten metal coming together from opposite directions in a mold.

Collapsibility. The tendency of a sand mixture to break down under the conditions of casting. A necessary property to prevent hot tears in castings.

Colloids, colloidal material. Finely divided material less than 0.5 micron (0.00002 in.) in size, gelatinous, highly absorbent and sticky when moistened. Clays are colloids.

Columnar structure. A coarse structure of parallel columns of grains caused by highly directional solidification resulting from sharp thermal gradients.

Combination core box. A core box that may be altered to form a core of another shape.

Combined carbon. Carbon in iron and steel which is joined chemically with other elements; not in the free state as graphitic carbon.

Combined water. Water in mineral matter which is chemically combined and driven off only at temperatures above 230°F.

Combustion. Chemical change as a result of the combination of the combustible constituents of the fuel with oxygen to produce heat.

Compressive strength (sand). Maximum stress in compression which a sand mixture is capable of withstanding.

Compressive strength. Maximum stress that a material subjected to compression can withstand.

Continuous phase. In alloys containing more than one phase, the phase that forms the matrix or background in which the other phase or phases are present as isolated units. For example, in a concrete aggregate, the cement paste is the continuous phase and the gravel is a discontinuous phase.

Contraction cracks. A crack formed by the metal being restricted while it is contracting in the mold; may occur during solidification (called a hot tear).

Contraction rule. A scale divided in excess of standard measurements, used by patternmakers to avoid calculations for shrinkage. (Usually called a shrink rule.)

Controlled cooling. A process of cooling from an elevated temperature in a predetermined manner to avoid hardening, cracking, or internal damage or to produce a desired microstructure.

C.—Continued

Cooling stresses. Stresses developed by uneven contraction or external constraint of metal during cooling after pouring.

Cope. Upper or topmost section of a flask, mold, or pattern.

Core. A preformed baked sand or green sand aggregate inserted in a mold to shape the interior, or that part of a casting which cannot be shaped by the pattern.

Core-blowing machine. A machine for making cores by blowing sand into the core box.

Core box. Wood or metal structure, the cavity of which has the shape of the core to be made.

Core cavity. The interior form of a core box that gives shape to the core.

Core compound. A commercial mixture which when mixed with sand supplies the binding material needed in making cores.

Core driers. Sand or metal supports used to keep cores in shape while being baked.

Core frame. Frame of skeleton construction used in forming cores.

Core irons. Bars of iron embedded in a core to strengthen it (core rods or core wires).

Coremaker. A craftsman skilled in the production of cores for foundry use.

Core marker. A core print shaped or arranged so that the core will register correctly in the mold.

Core oil. Linseed or other oil used as a core binder.

Core ovens. Low-temperature ovens used for baking cores.

Core paste. A prepared adhesive for joining sections of cores.

Core plates. Heat resistant plates used to support cores while being baked.

Core print. A projection on a pattern which forms an impression in the sand of the mold into which the core is laid.

Core raise. A casting defect caused by a core moving toward the cope surface of a mold, causing a variation in wall thickness.

Core sand. Sand free from clay; nearly pure silica.

Core shift. A variation from specified dimensions of a cored section due to a change in position of the core or misalignment of cores in assembling.

Coring. Variable composition in solid-solution dendrites; the center of the dendrite is richer in one element.

Cover core. A core set in place during the ramming of a mold to cover and complete a cavity partly formed by the withdrawal of a loose part of the pattern.

Critical points. Temperatures at which changes in the phase of a metal take place. They are determined by liberation of heat when the metal is cooled and by absorption of heat when the metal is heated, thus resulting in halts or arrests on the cooling or heating curves.

Crossbar. Wood or metal bar placed in a cope to give greater anchorage to the sand.

Cross section. A view of the interior of an object that is represented as being cut in two, the cut surface presenting the cross section of the object.

Crucible. A ceramic pot or receptacle made of graphite and clay, or clay and other refractory material, used in melting of metal; sometimes applied to pots made of cast iron, cast steel, wrought steel, or silicon carbide.

Crush. An indentation in the casting surface due to displacement of sand in the mold.

Cupping. The tendency of tangential sawed boards to curl away from the heart of the tree.

Cuts. Defects in a casting resulting from erosion of the sand by metal flowing over the mold or cored surface.

Cutting over. Turning over sand by shovel or otherwise to obtain a uniform mixture.

D.

Damping capacity. Ability of a metal to absorb vibrations.

Daubing. Filling cracks in cores, or plastering linings to fill cracks.

Decarburization. Loss of carbon from the surface of a ferrous alloy as a result of heating in a medium that reacts with the carbon.

D.—Continued

Deformation (sand). Change in a linear dimension of a sand mixture as a result of stress.

Dendrite. A crystal formed usually by solidification and characterized by a treelike pattern composed of many branches; also termed "pine tree" and "fir tree" crystal.

Dezincification. Corrosion of an alloy containing zinc (usually brass) involving loss of zinc and a surface residue or deposit of one or more active components (usually copper).

Diffusion. Movement of atoms within a solution. The net movement is usually in the direction from regions of high concentration towards regions of low concentration in order to achieve homogeneity of the solution. The solution may be a liquid, solid, or gas.

Dispersion (deflocculation). Separation or scattering of fine particles in a liquid; usually used in connection with the fineness test of clay.

Distribution (sand). Variation in particle sizes of foundry sand as indicated by a screen analysis.

Dowel. A pin used on the joint between sections of parted patterns or core boxes to assure correct alignment.

Draft, pattern. Taper on vertical surfaces in a pattern to allow easy withdrawal of pattern from a sand mold.

Drag. Lower or bottom section of a mold or pattern.

Drawback. Section of a mold lifted away on a plate or arbor to facilitate removal of the pattern.

Draw bar. A bar used for lifting the pattern from the sand of the mold.

Draw peg. A wooden peg used for drawing patterns.

Drawing. Removing pattern from mold.

Draw plate. A plate attached to a pattern to make drawing of pattern from sand easier.

Drop. A defect in a casting when a portion of the sand drops from the cope.

Drop core. A type of core used in forming comparatively small openings above or below the parting; the print is so shaped that the core is easily dropped into place.

Dross. Metal oxides in or on molten metal.

Dry fineness. The fineness of a sample of foundry sand from which the clay has not been removed and which has been dried at 221° to 230°F.

Dry permeability. The property of a molded mass of sand that permits passage of gases resulting during pouring of molten metal into a mold after the sand is dried at 221° to 230°F. (105° to 110°C.) and cooled to room temperature.

Dry sand mold. A mold which has been baked before being filled with liquid metal.

Dry strength, dry bond strength. Compressive, shear tensile, or transverse strength of a sand mixture which has been dried at 221° to 230°F. and cooled to room temperature.

Ductility. The property permitting permanent deformation in a material by stress without rupture. A ductile metal is one that bends easily. Ductile is the opposite of brittle.

Durability (sand). Rate of deterioration of a sand in use, due to dehydration of its contained clay.

E.

Elastic limit. The greatest stress which a material can withstand without permanent deformation.

Elongation. Amount of permanent extension in the vicinity of the fracture in the tensile test; usually expressed as a percentage of original gage length, as 25 percent in 2 in. Simply, the amount that a metal will stretch before it breaks.

Erosion scab. A casting defect occurring where the metal has been agitated, boiled, or has partially eroded away the sand, leaving a solid mass of sand and metal at that particular spot.

Etching. In metallography, the process of revealing structural details by preferential attack of reagents on a metal surface.

Eutectic or eutectic alloy. An alloy that melts at a lower temperature than neighboring compositions. For example, ordinary solders, made up of a tin and lead combination, are eutectic alloys that melt at a lower temperature than either lead or tin alone.

Eutectoid reaction. A reaction of a solid that forms two new solid phases (in a binary alloy

E.—Continued

system) during cooling. In ferrous alloys, the product of the eutectoid reaction is pearlite.

Eutectoid steel. A steel of eutectoid composition. This composition in pure iron-carbon alloys is 0.80 percent C., but variations from this composition are found in commercial steels and particularly in alloy steels, in which the carbon content of the eutectoid usually is lower.

Expansion (sand). The increase in volume which a sand undergoes when heated.

Expansion scab. A casting defect; rough thin layers of metal partially separated from the body of the casting by a thin layer of sand and held in place by a thin vein of metal.

F.

Facing, facing material. Coating material applied to the surface of a mold to protect the sand from the heat of the molten metal. See MOLD WASH.

Facing sand. Specially prepared molding sand used in the mold next to the pattern to produce a smooth casting surface.

False cheek. A cheek used in making a three-part mold in a two-part flask.

False cope. Temporary cope used only in forming the parting; a part of the finished mold.

Fatigue. Tendency for a metal to break under repeated stressing considerably below the ultimate tensile strength.

Fatigue crack. A fracture starting from a nucleus where there is a concentration of stress. The surface is smooth and frequently shows concentric (sea shell) markings with a nucleus as a center.

Fatigue limit. Maximum stress that a metal will withstand without failure for a specified large number of cycles of stress.

Feeder, feeder head. A reservoir of molten metal to make up for the contraction of metal as it solidifies. Molten metal flowing from the feed head, also known as a riser, minimizes voids in the casting.

Feeding. The passage of liquid metal from the riser to the casting.

Ferrite. Solid solution in which alpha iron is the solvent.

Ferroalloys. Alloys of iron and some other element or elements.

Ferroboron. An alloy of iron and boron containing about 10 percent boron.

Ferrochromium. An alloy of iron and chromium available in several grades containing from 66 to 72 percent chromium and from 0.06 to 7 percent carbon.

Ferromanganese. Iron-manganese alloys containing more than 30 percent manganese.

Ferromolybdenum. An alloy of iron and molybdenum containing 58 to 64 percent molybdenum.

Ferrophosphorus. Alloy of iron and phosphorus for the addition of phosphorus to steel.

Ferrosilicon. Alloy of iron and silicon for adding silicon to iron and steel.

Ferrotitanium. An alloy of iron and titanium available in several grades containing from 17 to 38 percent titanium.

Ferrous. Alloys in which the main metal is iron.

Ferrovanadium. Alloy of iron and vanadium containing 35 to 40 percent vanadium.

Fillet. Concave corner at the intersection of surfaces.

Fin. A thin projection of metal attached to the casting.

Fineness (sand). An indication of the grain-size distribution of a sand. See GRAIN FINENESS NUMBER.

Fines. In a molding sand, those sand grains smaller than the predominating grain size. Material remaining on 200 and 270-mesh sieves and on the pan in testing for grain size and distribution.

Fine silt. Sand particles less than 20 microns in diameter (0.02 mm or 0.0008 in.). This is included in AFS clay and by itself has very little plasticity or stickiness when wet.

Finish allowance. Amount of stock left on the surface of a casting for machine finish.

Finish mark. A symbol (f) appearing on the line of a drawing that represents a surface of the casting to be machine finished.

F.—Continued

Fire brick. Brick made from highly refractory clays and used in lining furnaces.

Fire clay. A clay with a fusion temperature of not less than 2,770°F.

Flask. A metal or wood box without top or bottom; used to hold the sand in which a mold is formed; usually consists of two parts, cope and drag. Remains on mold during pouring.

Flask pins. Pins used to assure proper alignment of the cope and drag of the mold during ramming and after the pattern is withdrawn.

Flowability. The property of a foundry sand mixture which enables it to fill pattern recesses and to move in any direction against pattern surfaces under pressure.

Fluidity. Ability of molten metal to flow and reproduce detail of the mold.

Fluorspar. Commercial grade of calcium fluoride (CaF_2).

Flux. Material added to metal charges during melting to form a slag.

Fluxing. Applying a solid or gaseous material to molten metal in order to remove oxide dross and other foreign materials, or a cover slag to protect the melt from oxidizing.

Follow board. A board which conforms to the pattern and locates the parting surface of the drag.

Fusion. Melting. Also used to designate a casting defect caused when molding sand softens and sticks to the casting to give a rough glossy appearance.

Fusion point. Temperature at which a material melts.

G.

Gaggers. Metal pieces used to reinforce and support sand in deep pockets of molds.

Gamma iron. The form of iron stable between 1,670° and 2,550°F.

Ganister. A siliceous material used as a refractory, particularly for furnace linings.

Gas holes. Rounded or elongated cavities in a casting, caused by the generation or accumulation of gas or trapped air.

Gate. End of the runner where metal enters the casting; sometimes applied to the entire assembly of connected channels, to the pattern parts which form them or to the metal which fills them; sometimes restricted to mean the first or main channel.

Gated patterns. Patterns with gates or channels attached.

Grain fineness number (AFS). An arbitrary number used to designate the grain fineness of sand. It is calculated from the screen analysis and is essentially the "average" grain size. The higher the number, the finer the sand.

Grain (sand). The granular material of sand left after removing the clay substance in accordance with the AFS fineness test.

Grains. Individual crystals in metals.

Grain refiner. Any material added to a liquid metal for producing a finer grain size in the casting.

Graphite. A soft form of carbon existing as flat hexagonal crystals and black with a metallic luster. It is used for crucibles, foundry facings, lubricants, etc. Graphite occurs naturally and is also made artificially by passing alternating current through a mixture of petroleum and coal tar pitch.

Graphitizing, graphitization. A heating and cooling process by which the combined carbon in cast iron or steel is transformed to graphite or free carbon. See TEMPER CARBON.

Gravity segregation. Variable composition caused by the settling of the heavier constituents where the constituents are insoluble or only partially soluble.

Gray cast iron. Cast iron which contains a relatively large percentage of its carbon in the form of graphite and substantially all of the remainder of the carbon in the form of eutectoid carbide. Such material has a gray fracture.

Green permeability. The ability of a molded mass of sand in its tempered condition to permit gases to pass through it.

Green sand. Molding sand tempered with water.

Green sand mold. A mold of prepared molding sand in the moist tempered condition.

Green strength. Tempered compressive, shear, tensile, or transverse strength of a tempered sand mixture.

G.—Continued

Grinding. Removing excess materials, such as gates and fins, from castings by means of an abrasive grinding wheel.

H.

Hardness (sand). Resistance of a sand mixture to deformation in a small area.

Head. Riser. Also refers to the pressure exerted by the molten metal.

Heap sand. Green sand usually prepared on the foundry floor.

Heat. Metal obtained from one period of melting in a furnace.

Holding furnace. A furnace for maintaining molten metal from a melting furnace at the right casting temperature, or provide a mixing reservoir for metals from a number of heats.

Homogenization. Prolonged heating in the solid solution region to correct the microsegregation of constituents by diffusion.

Horn gate. A semicircular gate to convey molten metal over or under certain parts of a casting so that it will enter the mold at or near its center.

Hot deformation (sand). The change in length or shape of a mass of sand when pressure is applied while the sand is hot.

Hot permeability (sand). The ability of a hot molded mixture of sand to pass gas or air through it.

Hot shortness. Brittleness is hot metal.

Hot spruing. Removing gates from castings before the metal has completely cooled.

Hot strength (sand). Compressive, shear, tensile, or transverse strength of a sand mixture determined by any temperature above room temperature.

Hot tear. Fracture caused by stresses acting on a casting during solidification after pouring and while still hot.

Hypereutectoid steel. A steel containing more than the eutectoid percentage of carbon. See EUTECTOID STEEL.

Hypoeutectoid steel. A steel containing less than the eutectoid percentage of carbon. See EUTECTOID STEEL.

I.

Impact strength. The ability to withstand a sudden load or shock.

Inclusions. Particles of impurities (usually microscopic particles of oxides, sulfides, silicates, and such) that are held in a casting mechanically or are formed during solidification.

Induction heating. Process of heating by electrical induction.

Ingate. The part of the gating system connecting the mold cavity to the runner.

Inhibitor. (1) A material such as fluoride, boric acid or sulfur used to prevent burning of molten magnesium alloys or for restraining an undesirable chemical reaction. (2) An agent added to pickling solutions to minimize corrosion.

Inoculant. Material which is added to molten cast iron to modify the structure and change the physical and mechanical properties to a degree not explained on the basis of the change in composition.

Inoculation. Addition to molten metal of substances designed to form nuclei for crystallization.

Internal stresses. A system of forces existing within a part.

Inverse chill. This, also known as reverse chill, internal chill, and inverted chill in gray cast iron, is the condition in a casting where the interior is mottled or white, while the outer sections are gray.

Iron, malleable. See MALLEABLE CAST IRON.

J.

Jacket. A wooden or metal form or box which is slipped onto a mold to support the sides of the mold during pouring.

Jig. A device arranged so that it will expedite and improve the accuracy of a hand or machine operation.

K.

Killed steel. Steel deoxidized with a strong deoxidizing agent such as silicon or aluminum to reduce the oxygen content to a minimum so that no reaction occurs between carbon and oxygen during solidification.

K.—Continued

Kiln-drying. Artificial drying of lumber by placing it in a specially designed furnace called a kiln.

Kish. Free graphite which has separated from molten iron.

Knockout. Operation of removing sand cores from castings or castings from a mold.

L.

Ladles. Metal receptacles lined with refractories and used for transporting and pouring molten metal.

Layout board. A board on which a pattern layout is made.

Ledeburite. Cementite-austenite eutectic structure.

Life (of sand). See DURABILITY.

Light metals. Metals and alloys that have a low specific gravity, such as beryllium, magnesium, and aluminum.

Liquid contraction. Shrinkage occurring in metal in the liquid state as it cools.

Liquidus. The temperature at which freezing begins during cooling or melting is completed during heating. The lowest temperature at which a metal is completely molten.

Loam. A mixture of sand, silt, and clay that is about 50 percent sand and 50 percent silt and clay.

Loam molds. Molds of bricks, plates, and other sections covered with loam to give the form of the castings desired.

Loose piece. Part of a pattern that is removed from the mold after the body of the main pattern is drawn.

Low-heat-duty clay. A refractory clay which fuses between 2768° and 2894°F.

Lute. (1) Fire clay used to seal cracks. (2) To seal with clay or other plastic material.

M.

Machinability. The relative ease with which a metal can be sawed, drilled, turned, or otherwise cut.

Machine finish. Turning or cutting metal to produce a finished surface.

Macroscopic. Visible either with the naked eye or under low magnification up to about ten diameters.

Macrostructure. Structure of metals as revealed by macroscopic examination.

Malleable cast iron. The product obtained by heat treatment of white cast iron which converts substantially all of the combined carbon into nodules of graphite. Differs from gray cast iron because it contains nodules of graphite instead of flakes.

Malleability. The property of being permanently deformed by compression without rupture.

Malleabilizing. Process of annealing white cast iron in such a way that the combined carbon is wholly or partly transformed to graphitic or free carbon or, in some instances, part of the carbon is removed completely. See TEMPER CARBON.

Marking a core. Shaping the core and its print so that the core cannot be misplaced in the mold.

Martensite. An unstable constituent often formed in quenched steel. Martensite is the hardest structure formed when steels are quenched.

Mass hardness. A condition in which the entire casting is hard and unmachinable.

Master pattern. A pattern with a special contraction allowance in its construction; used for making castings that are to be employed as patterns in production work.

Matched parting. A projection on the parting surface of the cope half of a pattern and a corresponding depression in the surface of the drag.

Match plate. A plate on which patterns split along the parting line are mounted back to back with the gating system to form a complete pattern unit.

Mechanical properties. The properties of a metal which determine its behavior under load. For example, strength.

Melting loss. Loss of metal in the charge during melting.

Melting rate. The tonnage of metal melted per hour.

M. —Continued

Metal penetration. A casting defect resulting from metal filling the voids between the sand grains.

Metallography. The science of the constitution and structure of metals and alloys as revealed by the microscope.

Microporosity. Extremely fine porosity caused in castings by shrinkage or gas evolution.

Microshrinkage. Very finely divided shrinkage cavities seen only by use of the microscope.

Microstructure. The structure of polished and etched metal as revealed by the microscope.

Misch metal. An alloy of rare earth metals containing about 50 percent cerium with 50 percent lanthanum, neodymium, and similar elements.

Misrun. Cast metal that was poured so cold that it solidified before filling the mold completely.

Modification. Treatment of aluminum-silicon alloys in the molten state with a small percentage of an alkaline metal or salt such as sodium fluoride to develop maximum mechanical properties in the metal.

Modulus of elasticity. The ratio of stress to the corresponding strain within the limit of elasticity of a material, a measure of the stiffness of a material. A high modulus of elasticity indicates a stiff metal that deforms little under load.

Moisture. Water which can be driven off of sand by heating at 221° to 230°F.

Mold. The form (usually made of sand) which contains the cavity into which molten metal is poured.

Mold board. Board on which the pattern is placed to make the mold.

Mold cavity. Impression left in the sand by a pattern.

Mold clamps. Devices used to hold or lock cope and drag together.

Mold shift. A casting defect which results when a casting does not match at parting lines.

Mold wash. Usually an emulsion of water and various compounds, such as graphite or silica flour; used to coat the face of the cavity in the mold. See FACING.

Mold weights. Weights placed on top of molds to withstand internal pressure during pouring.

Molder's rule. A scale used in making patterns for casting; the graduations are expanded to allow for contraction of the metal being cast. See PATTERNMAKER'S SHRINKAGE.

Molding, bench. Making sand molds from loose or production patterns at a bench not equipped with air or hydraulic action.

Molding, floor. Making sand molds from loose or production patterns of such size that they cannot be satisfactorily handled on a bench or molding machine; the equipment is located on the floor during the entire operation of making the mold.

Molding machine. May refer to squeezer or jolt squeezer machines on which one operator makes the entire mold, or to similar or larger machines including jolt-squeeze-strippers, and jolt and jolt-rollover-pattern-draw machines on which cope and drag halves of molds are made.

Molding material. Substance that is suitable for making molds into which molten metal can be cast.

Molding sand. Sand containing sufficient refractory clay substance to bond strongly when rammed to the degree required.

Mottled cast iron. Cast iron which consists of a mixture of variable proportions of gray cast iron and white cast iron; such a material has a mottled fracture.

Muller. A machine for mixing foundry sand, in which driven rolls knead the sand mixture against suitable plates.

Multiple mold. A composite mold made up of stacked sections. Each section produces a complete casting and is poured from a single downgate.

N.

Naturally bonded molding sand. A sand containing sufficient bonding material for molding purposes in the "as-mined" condition.

Neutral refractory. A refractory material which is neither definitely acid nor definitely basic. The term is merely relative in most cases, since at high temperature, such a material will usually react chemically with a strong base functioning as a weak acid, or with a strong acid functioning as a weak base.

N.—Continued

Chrome refractories are the most nearly neutral of all commonly used refractory materials. Alumina (Al_2O_3) is also nearly neutral and often serves as a neutral refractory.

Nodular iron. Cast iron which has all or part of its graphitic carbon content in the nodular or spherulitic form as cast. In nodular iron, the graphite is in tighter balls than in malleable iron. Nodular iron does not require heat treatment to produce graphite but malleable iron always requires heat treatment.

Nonferrous. Alloys in which the predominant metal is not iron.

Normalizing. A process in which a ferrous alloy is heated to a suitable temperature above the transformation range and cooled in still air at room temperature.

Normal segregation. Concentration of alloying constituents that have low melting points in those portions of a casting that solidify last.

Nucleus. The first particle of a new phase. The first solid material to form during the solidification of molten metal, occurring in a very microscopic size.

O.

Off-Grade metal. Metal whose composition does not meet the specification.

Open hearth furnace. A furnace for melting metal. The bath is heated by hot gases over the surface of the metal and by radiation from the roof. The preheating of air in a separate checker work distinguishes this furnace from an air furnace or reverberatory furnace.

Operating stress. The stress or load placed on a metal or structure during service.

Optical pyrometer. A temperature measuring device through which the observer sights the heated object and compares its brightness with that of an electrically heated filament whose brightness can be regulated.

Optimum moisture. The moisture content which results in the maximum development of any property of a sand mixture.

Overaging. Aging under conditions of time and temperature greater than that required to obtain maximum strength.

Overhang. The extension of the end surface of the cope half of a core print beyond that of the drag in order to provide clearance for the closing of the mold.

Overheated. A term applied when, after exposure to an excessively high temperature, a metal develops an undesirably coarse grain structure but is not permanently damaged. Unlike burnt structure, the structure produced by overheating can be corrected by suitable heat treatment, by mechanical work, or by a combination of the two.

Oxidation. Any reaction where an element reacts with oxygen.

P.

Packing or packing material. Sand, gravel, mill scale, or other similar material used to support castings to prevent warpage during annealing.

Pad. Shallow projection on a casting, usually added to provide a taper that will benefit directional solidification.

Parted pattern. Pattern made in two or more parts.

Parting. The separation between the cope and drag portions of mold or flask in sand casting.

Parting compound. A material dusted or sprayed on patterns to prevent adherence of sand. It is also used on the drag surface of a mold at the parting line.

Parting line. A line on a pattern or casting corresponding to the separation between the cope and drag portions of a sand mold.

Parting sand. Finely ground sand for dusting on surfaces that are to be separated when making a sand mold.

Pattern. A form of wood, metal, or other materials around which sand is packed to make a mold for casting metals.

Pattern board. Board having a true surface upon which the pattern is laid for ramming of the drag.

Pattern checking. Verifying dimensions of a pattern with those of the drawing.

Pattern layout. Full-sized drawing of a pattern showing its arrangement and structural features.

P.—Continued

Pattern coating. Coating material applied to wood patterns to protect them against moisture and abrasion of molding sand.

Pattern letters and figures. Identifying symbols fastened to a pattern as a means for keeping a record of the pattern and for the identification of the casting.

Pattern members. Components parts of a pattern.

Pattern plates. Straight flat plates of metal or other material on which patterns are mounted.

Patternmaker's shrinkage. Shrinkage allowance made on patterns to allow for the change in dimensions as the solidified casting cools in the mold from freezing temperature of the metal to room temperature. Pattern is made larger by the amount of shrinkage characteristic of the particular metal in the casting. Rules or scales are available for use. See MOLDER'S RULE.

Plaster pattern. A pattern made from plaster of Paris.

Pattern record card. A filing card giving description, location in storage, and movement of a pattern.

Pearlite. Laminated mixture of ferrite and iron carbide in the microstructure of iron and steel.

Pearlitic malleable cast iron. The product obtained by a heat treatment of white cast iron which converts some of the combined carbon into graphite nodules but which leaves a significant amount of combined carbon in the product.

Peen. Rounded or wedge-shaped end of a tool used to ram sand into a mold. It is also the act of ramming sand or surface working metals.

Peg gate. A round gate leading from a pouring basin in the cope to a basin in the drag.

Penetration, metal. A casting defect in which it appears as if the metal has filled the voids between the sand grains without displacing them.

Permanent mold. A metal mold, usually refractory coated, used for production of many castings of the same form; liquid metal is poured in by gravity; not an ingot mold.

Permeability (sand). A property which permits gas to pass through a molded mass of sand.

Phase diagram. A graph showing the equilibrium temperatures and composition of phases and reactions in an alloy system.

Phosphorus. A chemical element; symbol is P.

Photomicrograph. A photographic reproduction of any object magnified more than 10 times. The term micrograph may be used.

Pickle. Chemical or electrochemical removal of surface oxides from metal surfaces to clean them; usually done with acids.

Pig. An ingot of virgin or secondary metal to be remelted.

Pig bed. Small excavation or open mold in the floor of the foundry to hold excess metal.

Pig iron. Iron produced by the reduction of iron ore in a blast furnace; contains silicon, manganese, sulfur, and phosphorus.

Pinhole porosity. Small holes scattered through a casting.

Pipe. A cavity formed by contraction of metal during solidification of the last portion of liquid metal in a riser. It is usually carrot shaped.

Plumbago. Graphite in powdered form. Plumbago crucibles are made from plumbago plus clay.

Porosity. Unsoundness in cast metals. The term is used generally and applies to all types of cavities (shrinkage, gas, etc.).

Porosity (sand). Volume of the pore spaces or voids in a sand.

Pot. A vessel for holding molten metal.

Poured short. Casting which lacks completeness because the mold was not filled.

Pouring. Transfer of molten metal from ladle to molds.

Pouring basin. A cup may be cut in the cope or a preformed receptacle placed on top of the cope to hold molten metal prior to its entrance into the gate.

Precipitation hardening. A process for hardening an alloy; a constituent precipitates from a supersaturated solid solution. See AGE HARDENING and AGING.

P.—Continued

Precipitation heat treatment. In nonferrous metallurgy, any of the various aging treatments conducted at elevated temperatures to improve certain mechanical properties through precipitation from solid solution. See AGING.

Pressure tight. A term that describes a casting free from porosity of the type that would permit leaking.

Primary crystals. The first crystals that form in a molten alloy during cooling.

Proportional limit. The greatest stress (load) which a material can stand and still return to its original dimensions when the load is removed.

Pull cracks. Cracks in a casting caused by contraction, generally associated with stresses due to the irregular shape of the casting. See HOT TEAR.

Pull down. A buckle in the cope; sometimes severe enough to cause a scab.

Push-up. An indentation in the casting surface because of displacement of sand in the mold.

Pyrometer. An instrument used for measuring temperatures.

Pyrometer tube. A metal, ceramic or carbon tube sealed at one end and containing the thermocouple for measurement of temperatures, used as protection for the thermocouple.

Q.

Quench hardening. Process of hardening a ferrous alloy of suitable composition by heating within or above the transformation range and cooling at a rate sufficient to increase the hardness.

Quenching. A process of rapid cooling from an elevated temperature by contact with liquids, gases, or solids.

Quenching crack. A fracture resulting from thermal stresses induced during rapid cooling or quenching. Frequently encountered in alloys that have been overheated and are "hot short."

R.

Radioactive metals. A group of metals with high atomic weights and with atomic nuclei that decompose slowly giving off continual radiations of positively charged alpha particles (which are relatively slow), negatively charged beta particles (which are faster and lighter), and gamma rays. The gamma rays are similar to X-rays but are more penetrating and are used for radiography of very thick sections. Bombardment by neutrons can make any metal radioactive. Small concentrations of such metals are used as "tracers" in the study of diffusion and other phenomena.

Radiography. A nondestructive method of examination in which metal objects are exposed to a beam of X-ray or gamma radiation. Differences in metal thickness, caused by internal defects or inclusions, are seen in the image either on a fluorescent screen or on photographic film placed behind the object.

Ram. To pack the sand in a mold.

Rammer, air. A pneumatic tool used for packing sand around a pattern.

Rammer, hand. A wooden tool with a round mallet-shaped head at one end and a wedge-shaped head at the other, used to pack sand around the pattern when making a mold.

Ramming. The operation of packing sand around a pattern in a flask to form a mold.

Ram-off. A casting defect resulting when a section of the mold is forced away from the pattern by ramming after the sand has conformed to the pattern.

Rapping. Knocking or jarring the pattern to loosen it from the sand in the mold before withdrawing the pattern.

Rapping plate. A metal plate attached to a pattern to prevent injury to the pattern and to assist in loosening it from the sand.

Rat. A lump on the surface of a casting caused by a portion of the mold face sticking to the pattern.

Rat-tail. A casting defect caused by a minor buckle in the sand mold, occurs as a small irregular line or lines.

Rebonded sands. Used or reclaimed molding sand restored to usable condition by the addition of a new bonding material.

Reclamation, sand. See SAND RECLAMATION.

Recuperator. Equipment for recovering heat from hot gases and using it for the preheating of incoming fuel or air.

R.—Continued

Red shortness. Brittleness in hot metal.

Refractory (noun). A heat-resistant material (usually nonmetallic) used for furnace linings and such.

Refractory (adj.). The quality of resisting heat.

Refractory clay. A clay which fuses at 2890°F. or higher.

Residual stress. See CASTING STRESS and STRESS.

Retained strength. Compressive, shear, tensile, or transverse strength of a sand mixture after being subjected to a cycle of heating and cooling which approximates foundry practice.

Reverberatory furnace. A furnace with a vaulted ceiling that deflects the flame and heat toward the hearth or to the surface of the charge to be melted.

Riddle. Screen or sieve operated manually or by power for removing large particles of sand or foreign material from foundry sand.

Riddled sand. Sand that has been passed through a riddle or screen.

Riser. A reservoir designed to supply molten metal to compensate for shrinkage of a casting during solidification.

Rockwell hardness. The hardness value of a metal as determined by measuring the depth of penetration of a 1/16 inch steel ball ("B" scale) or a diamond point ("C" scale) using a specified load.

Rolling over. The operation of reversing the position of a flask in which the drag part of the pattern has been rammed with the parting surface downward.

Rollover board. A wood or metal plate on which the pattern is laid top face downward for ramming the drag half mold; the plate and half mold are turned over together before the cope is rammed.

Runner. A channel through which molten metal is passed from one receptacle to another; in a mold, the portion of the gate assembly that connects the downgate or sprue with the ingate.

Runner box. Device for distributing molten metal around a mold by dividing it into several streams.

Runner riser. A channel which permits flow of molten metal to the ingates and also acts as a reservoir to feed the casting.

Runout. A casting defect where a casting lacks completeness due to molten metal draining or leaking out of some part of the mold cavity during pouring; escape of molten metal from a furnace, mold, or melting crucible.

S.

Sag. A decrease in metal thickness in a casting caused by settling of the cope or core.

Sand. A loose material consisting of small but easily distinguishable grains, usually of quartz from the disintegration of rock. When used as a molding material, the grains should pass a No. 6 and be retained on a No. 270 sieve. Sometimes used to designate a sand-clay mixture appearing naturally in proper proportions for molding.

Sand, bank. Sand from a bank or pit usually low in clay content.

Sandblast. Sand driven by a blast of compressed air and used to clean castings.

Sand burning. Formation of a hard glassy surface on a sand casting by reactions between the sand of the mold and the hot metal or metallic oxides.

Sand castings. Metal castings poured in sand molds.

Sand control. Procedure where various properties of foundry sand (such as fineness, permeability, green strength, and moisture content) are adjusted to obtain castings free from blows, scabs, veins, and similar defects.

Sand cut. Erosion of sand from the mold surfaces by running metal.

Sand cutting. Preparing sand for molding, either by hand or by a machine.

Sand holes. Cavities of irregular shape and size; the inner surfaces plainly show the imprint of a granular material.

Sand reclamation. Processing of used foundry sand, normally wasted, by thermal or hydraulic methods so that it may be used in place of new sand.

Sand tempering. Adding sufficient moisture to sand to make it satisfactory for molding purposes.

S.—Continued

Scab. A defect on the surface of a casting; usually appears as a rough mass of metal attached to the normal surface of the casting and often contains sand; caused by generation of gas in the mold, poor ramming, or high expansion of the sand when heated.

Scarfing. Cutting off surface areas, such as gates and risers, from castings by using a gas torch.

Scrap. Material unsuitable for direct use but usable for reprocessing; metal to be remelted. Includes sprues, gates, risers, defective castings, and scrapped machinery.

Screen. Perforated metal placed between the gate and the runner of a mold to minimize the possibility of oxides passing through into the casting.

Sea coal. Finely ground soft coal often mixed with molding sand. See COAL DUST.

Seam. A surface defect on a casting; related to, but of lesser degree than, a coldshut.

Season cracking. Stress-corrosion cracking of copper-base alloys; involves residual stresses and specific corrosive agents (usually ammonia or compounds of ammonia).

Sectional core. A core made in two or more parts and pasted or wired together.

Segregation. In a casting, concentration of alloying elements at specific regions, usually as a result of the primary crystallization of one phase with the subsequent concentration of other elements in the remaining liquid. Microsegregation refers to normal segregation on a microscopic scale where material rich in the alloying element freezes in successive layers on the dendrites (coring). Macrosegregation refers to gross differences in concentration (for example, from one area of a casting to another).

Selective heating. A process by which only certain portions of a casting are heated in a way that will produce desired properties after cooling.

Selective quenching. A process by which only certain portions of a casting are quenched.

Semi steel. Incorrect name sometimes mistakenly used for high-strength gray iron made from a charge containing considerable steel scrap.

Semikilled steel. Steel incompletely deoxidized to permit evolution of sufficient carbon monoxide to offset solidification shrinkage.

Shakeout. The operation of removing castings from the mold.

Shank. The handle of a ladle. (The metal form that holds the ladle.)

Sharp sand. A sand that is substantially free of bond. The term has no reference to the grain shape.

Shear strength. Maximum shear stress which a sand mixture is capable of developing.

Shift. A casting defect caused by mismatch of cope and drag.

Shrink rule. See MOLDER'S RULE.

Shrinkage cavity. A void left in cast metals as a result of solidification shrinkage and the progressive freezing of metal toward the center.

Shrinkage, patternmaker's. Shrinkage allowance on patterns to compensate contraction when the solidified casting cools in the mold from the freezing temperature to room temperature.

Shrinkage cracks. Cracks that form in metal as a result of the pulling apart of grains by thermal contraction before complete solidification.

Silica. The hard mineral part of natural sand. Chemical formula SiO_2.

Silicious clay. A clay containing a high percentage of silica.

Silt. Very fine particles that pass a No. 270 sieve, but which are not plastic nor sticky when wet.

Sintering point. The temperature at which the molding material begins to adhere to the casting.

Sizing. Primary coating of thin glue applied to end-grain wood to seal the pores.

Skeleton pattern. A framework representing the interior and exterior form and the metal thickness of the required casting; not a solid pattern.

Skimgate. See SCREEN.

Skimmer. A tool for removing scum and dross from molten metal.

S.—Continued

Skimming. Holding back or removing the dirt, slag, or scum in the molten metal before or during pouring to prevent it from entering the mold.

Skin. A thin surface layer that is different from the main mass of an object in composition, structure, or other characteristics.

Skin-drying. Drying of the surface of the mold by direct application of heat.

Skull. A film of metal or dross remaining in a pouring vessel after the metal has been poured.

Slab core. Flat plain core.

Slag. A nonmetallic covering which forms on the molten metal as a result of the impurities contained in the original charge, ash from the fuel, and silica and clay eroded from the refractory lining. It is skimmed off prior to tapping of the heat.

Slip jacket. A frame to place around a snap-flask mold after the flask is removed.

Slick, slicker, smoother. A tool used for mending and smoothing the surfaces of a mold.

Slurry. A thin flowable mixture of clay or bentonite in water; used to fill cracks in linings or to fill joints in cores.

Slush casting. A casting made from an alloy that has a low melting point and freezes over a wide range of temperature. The metal is poured into the mold and brought into contact with all surfaces so as to form an inner shell of frozen metal. Then the excess metal is poured out. Castings that consist of completely enclosed shells may be made by using a definite quantity of metal and a closed mold.

Snap flask. A flask used for small work; differs from the ordinary flask in that it has hinges and latches or some other device so that it can be opened or held together as desired.

Soaking. Prolonged heating of metal at a selected temperature.

Soda ash. Commercial sodium carbonate.

Soldiers. Thin pieces of wood used to strengthen a body of sand or to hold it in place.

Solid contraction. Shrinkage occurring in metal in the solid state as it cools from the solidus temperature to room temperature.

Strain. The change in dimension of a structure. Strain may be caused by stress, or strain may cause stress. Usually expressed in inches per inch.

Stress. The load applied to or existing in a structure; usually expressed in pounds per square inch (p. s. i.).

Superheating. Heating a molten metal to a temperature above its melting point.

Superimposed core. A core placed on a pattern and rammed up with it.

Swab. A sponge or piece of waste, hemp, or other material used in dampening sand around a pattern before withdrawing it. Sometimes used in blacking molds which might be broken by a brush.

Sweep. A board having the profile of the desired mold; when revolved around a stake or spindle, it produces the outline of the mold.

Sweep-work. Molds made of pieces of patterns and sweeps instead of patterns.

Swell. A casting defect consisting of an increase in metal thickness because of the displacement of sand by metal pressure.

Synthetic molding sand. Any sand compounded from selected individual materials which when mixed together produce a mixture having the proper physical properties from which to make foundry molds.

T.

Tally mark. A mark or combination of marks indicating the correct location of a loose piece of a pattern or core box.

Tapping. Opening the hole at the spout to permit molten metal to run from the furnace. It also applies to tilting a furnace to pour molten metal.

Temper (verb). (1) Mixing sand with sufficient water to develop desired molding properties. (2) Reheating of a quenched casting to reduce internal stresses and reduce hardness.

Temper (noun). The moisture content of a sand at which any desired property is obtained; for example, temper with respect to green compressive strength, permeability, retained compressive strength, etc.

Temper carbon. The free or graphitic carbon that precipitates during the graphitizing or malleableizing of white cast iron.

T.—Continued

Temper water. Water added to molding sand to give proper molding consistency.

Tempering sand. Dampening and cutting over or otherwise mixing sand to produce uniform distribution of moisture.

Template. Thin piece of material with the edge contour made in reverse to the surface to be formed or checked.

Temporary pattern. A pattern used to produce one or two castings and made economically as the case will permit.

Tensile. Pertains to pulling of a structure as compared with pushing it (compression) or twisting it (torsion).

Tensile strength. Maximum pulling stress (load) which a material is capable of withstanding. (Also known as ultimate strength.)

Ternary alloy. An alloy that contains three principal elements.

Test bar. A specimen having standard dimensions designed to permit determination of mechanical properties of the metal from which it was poured.

Thermal contraction. The decrease in length accompanying a change of temperature.

Thermal expansion. The increase in length accompanying a change of temperature.

Thermal stresses. Stresses resulting from nonuniform distribution of temperature.

Thermit reactions. Heat-producing processes in which finely divided aluminum powder is used to reduce metal oxides to free metals.

Thermocouple. A device for measuring temperatures by the use of two dissimilar metals in contact; the junction of these metals gives a measurable voltage change with changes in temperature that is recorded and read on a meter.

Tie piece (bar). Bar or piece built into a pattern and made a part of the casting to prevent distortion caused by uneven contraction between separated members.

Tilting furnace. A melting furnace that can be tilted to pour out the molten metal.

Tin sweat. Beads of tin-rich low-melting phase that are found on the surface of bronze castings when the molten metal contains hydrogen.

Top board. A wooden board which is used on the cope half of the mold to permit squeezing of the mold.

Transfer ladle. A ladle that may be supported on a monorail or carried in a shank and is used to transfer metal from melting furnace to holding furnace.

Transformation range or transformation temperature range. The temperature or range of temperatures at which metals undergo phase changes while still solid. The existence of these ranges and different phases make it possible to harden or soften iron, steel, and some nonferrous alloys almost at will by proper selection of heat treatment.

Transverse strength. The load required to break a test specimen that is supported at both ends and loaded in a direction perpendicular to the longitudinal axis.

Trimming. Removing fins and gates from castings.

Trowel. A tool for slicking, patching, and finishing a mold.

Tucking. Pressing sand with the fingers under flask bars, around gaggers, and other places where the rammer does not give the desired density.

Tumbling. Cleaning castings by rotation in a cylinder in the presence of cleaning materials.

Tumbling barrels. Rotating barrels in which castings are cleaned; also called rolling barrels and rattlers.

U.

Ultimate strength. See TENSILE STRENGTH.

Undercut. Part of a mold requiring a drawback. See DRAWBACK.

Upset. Frame to increase the depth of a flask.

V.

Vent. A small opening in a mold to permit escape of gases when pouring metal.

Vent wire. A wire used to make vents or small holes in the mold to allow gas to escape.

Vibrator. A device operated by compressed air or electricity for loosening and withdrawing patterns from a mold.

V.—Continued

Virgin metal. Metal obtained directly from the ore rather than by remelting.

Vitrification point. The temperature at which clays reach the condition of maximum density and shrinkage when heated.

W.

Warm strength (of a core). Strength of a core at temperatures of 150° to 300°F.

Warpage. Deformation other than contraction that develops in a casting between solidification and room temperature; also, distortion of a board through the absorption or expulsion of moisture.

Wash. Defect in a casting resulting from erosion of the sand by metal flowing over the mold or core surface.

Wax. Class of substances of plant, animal, or mineral origin, insoluble in water, partly soluble in alcohol, and miscible in all proportions with oils and fats. Common waxes are beeswax, paraffin wax, ozokerite, ceresin, and carnauba. Mixtures are formed into rods and sheets and used for forming vents in cores and molds.

Weak sand. Sand lacking the proper amount of clay or bond.

Web. A plate or thin member lying between heavier members.

Welding. A process used to join metals by the application of heat. Welding requires that the parent metals be melted. This distinguishes welding from brazing.

Welding stress. Stress resulting from localized heating and cooling of metal during welding.

Whirl gate. A gate or sprue arranged to introduce metal into a mold tangentially, thereby imparting a swirling motion.

White cast iron. Cast iron in which substantially all the carbon is present in the form of iron carbide. Such a material has a white fracture.

Wood's metal. A low-melting alloy containing 25 percent lead, 12.5 percent tin, 50 percent bismuth, and 12.5 percent cadmium, melting temperature is 154.4°F.

Workable moisture. The range of moisture content within which the sand fills, rams, draws, and dries to produce a satisfactory mold; also the range in which the sand does not dry out too fast to mold and patch.

Working face. Surface of a piece of material that has been planed true and that is to be used as a basis for the dressing of other surfaces.

X.

X-ray. Form of radiant energy with extremely short wavelength which has the ability to penetrate materials that absorb or reflect ordinary light.

Y.

Yield strength. See PROPORTIONAL LIMIT.